AiGX6
RESEARCH LAB

Transformation Blueprint Series

Published by EDGE Publish, www.EdgePublish.com

Workforce Transformation for the AI Era, First Edition, 2026
ISBN-13: 978-1-971148-01-4, Printed in the United States of America.

AI6X6 RESEARCH LAB

Transformation Blueprint Series

WORKFORCE TRANSFORMATION FOR THE AI ERA

Elevating Human Capability in a Tech-Driven Era.
The 6x6 Blueprint for Talent Architecture and Cultural Agility.

SUBRATA MUKHERJEE

EDGE PUBLISH
www.EdgePublish.com

"For the grace of God, who lights the path.
And, in loving memory of my mother Rupali and my father Dilip, who taught me the importance of kindness and perseverance. This work is born from the love you gave, the sacrifices you made, and the quiet wisdom that still echoes in my life. This one is for you."

Personal message from our Founder & Executive Director

SUBRATA MUKHERJEE

For over twenty-five years, I have sat at the heart of global digital transformations—from shifting a 180-year-old institution like *The Economist* into a digital powerhouse, to architecting hyper-automation for Fortune 500 leaders. I have seen the rise of the internet, the mobile revolution, and the cloud. But none of these shifts compare to the era we have just entered.

The Ai6x6 Research Lab was born out of a singular, urgent observation: Most organizations are trying to solve tomorrow's problems with yesterday's mental models. Traditional transformation is often treated as a "project"—a linear path with a start and an end. But in an era of Agentic AI and self-directing systems, transformation is a living organism. It is a system of behaviors, decisions, and capabilities that must evolve in unison. I founded this Lab to move beyond the "hype" and the "tools." My motivation is to provide leaders with the Cohesion, Schema, and Custody required to navigate a world where the boundary between human intuition and machine intelligence is disappearing. People often ask why "6×6." Throughout history, the number six has been the sacred geometry of our existence—from the sexagesimal systems of ancient Sumerian astronomy to the hexagonal patterns found in the farthest reaches of our cosmos. Mathematically, six is the first "perfect number" because its divisors (1, 2, and 3) add up to itself, making it uniquely balanced. In my journey across the U.S., Europe, and Asia, I realized that this ancient symmetry holds the key to modern enterprise. By distilling organizational complexity into six fundamental pillars and viewing them through six transformative lenses, we activate a collective "sixth sense" for the organization—the ability to perceive and act on what others miss. This architecture forms the foundation of our three flagship methodologies—the "*guided navigation system every transformation needs*".

My journey to this blueprint was forged in the engine rooms of the world's most complex organizations. I have led wholesale business transformations for *The Economist*, reinvented Bloomberg's digital presence following the *Businessweek* acquisition, revitalized the iconic *Dummies* brand within Wiley, and optimized the global ratings processes within S&P Global and McGraw-Hill. This strategic work was balanced by deep technical execution: implementing RPA, AI, IoT, and Digital Twins within Raytheon, and architecting CRM, BPM, and digital workplace automations for the global clients of Cognizant and Unisys.

This perspective was further refined through rigorous academic pursuit, drawing on my MBA in Marketing from Jadavpur University, my Master's in Technology Management from Columbia University, and executive certifications from Harvard University and the Indian Institute of Management (IIM-C) to bridge the gap between high-level theory and industrial-scale execution.

Today, through the Ai6x6 Research Lab, I lead the development of the frameworks that are now being adopted by executive teams globally. At the center of this work is three methodologies: 6X6 Blueprint, IGNITE Methodology and REVEAL Methodology our multidimensional navigation system that unifies architecture, execution, and evaluation across the enterprise.

However, my background as a Tech Journalist for *The Telegraph* for over a decade and a columnist for *CIO Magazine* has taught me that true insight requires total independence. I have built this Lab as a space where my team and I can express our opinions freely, without the interference of corporate sponsors. Our goal is not just to report the news, but to "fold" it into the Ai6x6 Blueprints—giving you the critical "so-what" and the "what-next" to make global shifts actionable for your organization.

We are democratizing this knowledge because I believe that the ability to thrive in the AI era should not be reserved only for those who can afford expensive global consulting firms. Our mission is to empower businesses of all sizes with the targeted, practical, and battle-tested information that I often wish I had at my disposal during my own leadership assignments. Through the Ai6x6 Transformation Blueprints and our Executive Mentor Network, we are scaling AI literacy to ensure that as systems become more autonomous, leaders become more intentional. We are not just adopting new tools; we are reimagining the nature of the enterprise itself.

Ai6X6 RESEARCH LAB

NEW YORK | LONDON | KOLKATA

To our esteemed Industry Thought Leaders: Thank you!

Your generous time and profound insights have been the bedrock of this book. Each interview has been converted into sections and one-page cards that have been sprinkled across six books. You have provided a vital perspective, transforming our work from theory into a deeply practical guide rooted in real-world leadership in the AI-Era. Your willingness to share your hard-won wisdom, navigating complex challenges and revealing the future of your industries, enriches every page. You all have successfully navigated the AI transition in your careers, your companies and your industries. This book is a testament to your expertise. We are immensely grateful for your unwavering support and belief in this project. Thanks a lot, from me and my team at Ai6x6 and Executive Mentor Network!

TABLE OF CONTENTS

9	SECTION A	INTRODUCTION	"GUIDED NAVIGATION"	BLUEPRINT
17	Chapter A1	The Prologue to a Purpose:	*Ai6x6 Research Lab & products*	*Lab, Series & Book*
25	Chapter A2	Understand Pillars & Lenses	*Introduce Elements Of The Blueprint*	*6 Pillars x 6 Lens*
45	Chapter A3	Why Transform Workforce?	*Critical Need For Change*	*12 business situations*
51	Chapter A4	Power of Diversified Insight	*Thought leadership Interview excerpts*	*15 Renowned Leaders*
69	**SECTION B**	**HUMAN-AI SYNERGY**	**THE "SPARK"**	**SYNERGY LENS**
73	Chapter B1	Evolution of Humanity	*Purpose Of Life & Learnings*	*4 Periods Of Civilization*
87	Chapter B3	Human-AI Personas	*6 Types of AI, Steps and Use cases*	*6 Human-AI pairs*
115	**SECTION C**	**BUSINESS MATURITY**	**THE "STRATEGY"**	**BUSINESS LENS**
121	Chapter C1	Foundation | Workforce Planning	*Future talent demand plan*	*Business maturity lifecycle Total 36 topics - 6 Topics per chapter; What, why, how and what leaders need to be aware of?*
129	Chapter C2	Initiation | Employee Onboarding	*Welcome & new hire setup*	
137	Chapter C3	Cohesion | Employee Experience	*Holistic day to day journey*	
145	Chapter C4	Momentum| Performance Review	*Optimize Career Growth*	
153	Chapter C5	Optimization | Offboarding	*Transition and Exit process*	
161	Chapter C6	Authority | Employee Loyalty	*Succession planning & Retain staff*	
169	**SECTION D**	**DIGITAL EVOLUTION**	**THE "ENGINE"**	**DIGITAL LENS**
175	Chapter D1	Remote-work Infrastructure	*Remote readiness & access*	*Digital evolution lifecycle Total 36 technologies - 6 technologies per chapter; With brief implementation steps, challenges, products to consider, and cost guidance.*
183	Chapter D2	People Intelligence	*Talent & Skill Gap Analysis*	
191	Chapter D3	Teamwork Integration	*Collaboration & IT Connectivity*	
199	Chapter D4	Agile-led Innovation	*Incubation & Upskilling*	
207	Chapter D5	Workplace Interface	*Employee experience platforms*	
215	Chapter D6	Workplace Integrity	*Ethics and Team values*	
223	**SECTION E**	**EXECUTION MECHANICS**	**THE "SPRINT"**	**EXECUTION LENS**
233	Chapter E1	Change Management frameworks	*Reduce resistance & friction*	*1 Common + 6 methodologies*
247	Chapter E2	The Unlearning Rate (UR)	*How fast can a worker unlearn?*	*Unlearn past behavior*
251	**SECTION F**	**CULTURAL COMPASS**	**THE "SOUL"**	**CULTURAL LENS**
257	Chapter F1	Workforce Stagnation Beliefs	*Unconscious assumptions*	*Cultural maturity lifecycle Total 36 topics – 6 topics per chapter; 216 culture situations with suggested remedial actions, 216 Best practice advices (Do's & don'ts)*
265	Chapter F2	Workforce Systemic Biases	*Irrational Preconceived decisions*	
273	Chapter F3	Workforce Deficiency Barriers	*Isolated, low trust workplace*	
281	Chapter F4	Workforce Resilient Behaviors	*Strong connections, meet potential*	
289	Chapter F5	Workforce Evolution Bridges	*Diverse, inclusive space*	
297	Chapter F6	Workforce Harmony Bliss	*Profound Satisfaction*	
305	Chapter F7	*Workforce Diversity (Bonus)*	*6 types of workforce diversity*	*Steps to take, challenges faced*
313	**SECTION G**	**MORAL COMPASS**	**THE "AUDIT"**	**TRUST LENS**
319	Chapter G1	Workplace Ethics	*Ethical Behavior Norms*	*Moral evolution lifecycle Total 36 topics – 6 topics per chapter; 216 different moral dilemmas and practical remedial actions.*
327	Chapter G2	Worker Safety	*Hazard Prevention Protocols*	
335	Chapter G3	Opportunity Equity	*Bias-Free Opportunities*	
343	Chapter G4	Talent Risk	*Capability Gap Exposure*	
351	Chapter G5	Role Control	*Responsibility Boundary Clarity*	
359	Chapter G6	Legal Protections	*Employee Rights Safeguards*	
367	Chapter G7	*The Humanity Paradox (Bonus)*	*NetZero Sustainability, Green policy*	*36 Project Examples*

guided navigation for transformation

BLUEPRINT SERIES

SECTION A

Introduction

Ai6x6™ Transformation Blueprint

Ai6x6 Research Lab is a global intellectual institution dedicated to advancing the science and practice of organizational transformation. We develop original frameworks, diagnostic systems, capability playbooks, and research models that help leaders navigate complexity with clarity, precision, and confidence. Our work spans **Leadership, Workforce, Data, Customer, Product, and Operations**—the six pillars that define modern enterprise maturity. We are based in New York, London & Kolkata.

OUR MISSION FOR ORGANIZATIONAL TRANSFORMATION

To build a globally recognized body of knowledge that helps organizations think clearly, act decisively, and transform responsibly in an age of AI acceleration. We essentially play **six roles** for the industry:

THE CATALYST BRING ALIGNMENT TO LEADERSHIP	**THE STRATEGIST** BRING STRUCTURE TO AMBIGUITY	**THE ARCHITECT** BRING CLARITY TO COMPLEXITY
THE ENGINEER BRING DISCIPLINE TO TRANSFORMATION	**THE ETHICIST** BRING INTEGRITY TO OUTCOMES	**THE GUARDIAN** BRING OVERSIGHT TO AUTONOMY

WHAT WE DO?

1. Build Transformation Frameworks: We architect the Ai6x6 Transformation Blueprints— a structured, multi-lens transformation model that helps orgs decode patterns, risks, behaviors, and decision pathways.
2. Develop Diagnostic Systems: Our maturity diagnostics and radar models translate organizational signals into actionable insights, identifying Growth, Risk, Kill, and Shift zones.
3. Publish Research & Methodologies: We codify best practices in enterprise design through multi-volume series and research playbooks, available in bookstores or on our website ai6x6.com.
4. Create Executive Tools: From RACI matrices to transformation roadmaps, we build high-utility models for immediate executive application. We have many accelerated off-the-shelf offerings on our website.
5. Support Global Organizations: Our frameworks empower enterprises, consulting teams, and innovation labs. Our advisory team worldwide including our Executive Mentor Network is here to help.

ACCOMPLISHMENTS

Subrata Mukherjee has delivered enterprise-scale transformation across the Americas, Europe, and Asia, shaping modernization programs for some of the world's most influential organizations. His work spans AI/ML, automation, digital product strategy, enterprise architecture, and large-scale organizational transformation. He has led digital reinvention in global publishing, driven automation and AI adoption in aerospace and defense, modernized ratings and compliance processes in financial services, and guided product and data transformation across multiple industries. In 2016, he was recognized as one of Folio's Top 100 People in Publishing for his impact on digital innovation.

EDUCATION

Subrata holds a Master's in Technology Management from Columbia University, an MBA in Marketing from Jadavpur University, and a Business Communications certification from Harvard University. He also holds a Management by Values certification from the Indian Institute of Management, Calcutta, deepening his grounding in leadership behavior, organizational culture, and values-driven transformation. He is a Certified Product Owner. His academic foundation spans technology, strategy, communication, and organizational leadership, informing his multidisciplinary approach to AI-era transformation.

INDUSTRIES

His career spans aerospace and defense, financial services, publishing, healthcare, consulting, education, manufacturing, and technology services. He has delivered transformation in highly regulated environments, mission-critical engineering contexts, global media organizations, and fast-moving digital businesses. This cross-industry exposure gives him a multidimensional perspective on how AI reshapes strategy, operations, and organizational identity.

COMPANIES

Across his career, Subrata has delivered transformation programs for organizations including The Economist, Raytheon Technologies, S&P Global, Bloomberg, Wiley, Cognizant, Unisys, and Gujarat Ambuja Cement. His work has ranged from leading The Economist's digital transformation and modernizing Bloomberg's subscription and data infrastructure, to launching Raytheon's enterprise-wide Citizen Development program for Automation and AI, transforming the Dummies brand at Wiley, modernizing S&P's Ratings Process after the sub-prime mortgage crisis, scaling CRM and Automation/AI consulting portfolios at Cognizant and Unisys, and establishing the technology backbone and smart-building systems for the Gujarat Ambuja conglomerate.

EXPERTISE

Subrata's expertise spans the full arc of enterprise transformation, beginning with early work in CAD/CAM for steel plants and smart-city infrastructure, which shaped his systems-thinking foundation. He then became a specialist in enterprise CRM and BPM transformation, leading large-scale modernization programs that re-architected customer operations, workflow orchestration, and enterprise process design. Over time, he transitioned into a strategist, product thinker, and transformation architect, guiding organizations on designing, adopting, and scaling next-generation digital capabilities. Today, he is recognized as a leading voice in AI and Emerging Technologies, with deep expertise across AI/ML, Generative AI, LLMs, RAG architectures, intelligent automation, hyper-automation, conversational AI, chatbots, agentic systems, and decision-automation frameworks. His work is rooted in strategy, product design, operating-model transformation, and enterprise-scale adoption—not coding or engineering execution—and focuses on translating emerging AI capabilities into practical, scalable, enterprise-grade systems that reshape how organizations operate and compete.

THOUGHT LEADERSHIP

Subrata is a prolific speaker who has presented at more than 100 global conferences and events. His writing has appeared in CIO Magazine, and he served as a weekly technology columnist for The Telegraph. Through the Ai6x6 Research Lab, he publishes frameworks, diagnostics, and transformation blueprints that help organizations navigate the AI economy. In 2026–27, he will launch the Executive Mentor Network to develop a global community of AI Strategy–certified coaches and expand AI literacy among senior leaders. Beyond his professional work, he is an avid traveler who has explored more than 40 countries and an accomplished portrait artist.

www.linkedin.com/in/sm360

1. GIVE **VALUE**, REPEATEDLY, BECOME A GOTO BOOK

   The 18-book series brings in a lot of content from a practitioner perspective. The advice we are providing have worked for us in our past projects across industries; so we are preaching what we practiced before

2. BE **BRIEF**, PROVIDE CONTENT IN SMALL NUGGETS

   I have never read a book cover to cover. I have short attention span, and more and more people are suffering from the same syndrome. Hence each micro-topic we have introduced is covered in two to three pages

3. BE **COMPREHENSIVE** FOR REAL WORLD IMPACT

   Transformation is not easy. In the real world there are a lot of inter-connected challenges. We have proposed detailed frameworks that will connect the dots and provide comprehensive solutions to your team/client

4. BE **PRACTICAL**, PROVIDE EXAMPLES TO DEMYSTIFY

   Most companies and leaders know they and their companies need to transform, but don't know where to start. I wanted to provide a step-by-step guide or a MOSCOW priority list on getting started

5. PUSH **BOUNDARIES** AND DARE LEADERS TO DREAM

   Real transformation is not tinkering around the edges and playing safe; its often about self disrupting and pushing boundaries and reinventing oneself. Applies to individuals, teams and organizations.

6. PUSH LEADERS TO **INTROSPECT** & ASK QUESTIONS

   Since the world around us is moving so fast, leaders often have a fear of been exposed about their lack of knowledge and ask "stupid" questions. These book educates you enough, so you can ask intelligent questions

7. BE **EXCITING**, DON'T BE BORING

   I have tried to keep the book interesting, small text, lot of structured tables, quite a few images, bold formatting, case studies, Tips and Tricks, Interviews, RACI matrix, buying guides, assignment questions etc.

8. BRING **OTHERS**, HEAR FROM MY COMMUNITY

   Tech sector worldwide is very collaborative. We learn from each other every day and we use the knowledge to impact lives positively. This book attempts to bring a few voices from my network and hear their perspective

9. PROMOTE **AI ECONOMY**, USE AI TOOLS YOURSELF

   Generative AI has been used extensively for formatting content in a repeatable format & page size control. Without GenAI such an extensive book series would not have been possible in a short time.

10. BE **DIFFERENT**, CHARTER YOUR OWN COURSE

    Most management self-help books are pages and pages of long text with guidelines provided by publishers that are a bin antiquated. Modern learning happens byte sized and is multi-modal and multi-channel..

The **Guided Navigation for Transformation** framework is a comprehensive, end-to-end methodology designed to de-risk and accelerate AI adoption within an organization. It moves away from "random acts of digital" by providing a disciplined, three-stage approach that balances human capability with technical execution. The core of the system is the **6x6 Lens Sequence**, which ensures that every strategic move is evaluated through six critical dimensions: **Leadership, Workforce, Data, Customer, Product, and Operations**.

By applying these lenses consistently across the Blueprint, Reveal, and Ignite phases, the framework creates a "Golden Thread" that links high-level strategy to on-the-ground deliverables. This is not just a consulting model, but an educational ecosystem with an 18-book series that ensures that AI transformation is both measurable and sustainable.

"Guided Navigation for Transformation" in the AI Era

We provide guided navigation through three distinct phases: First, we establish your **Transformation Blueprint** to align skills. Second, we run the **Reveal Diagnostic** to assess your current maturity. Finally, we execute the **Consulting Playbook** to turn the 'Art of the Possible' into a 'Science of Deliverables'.

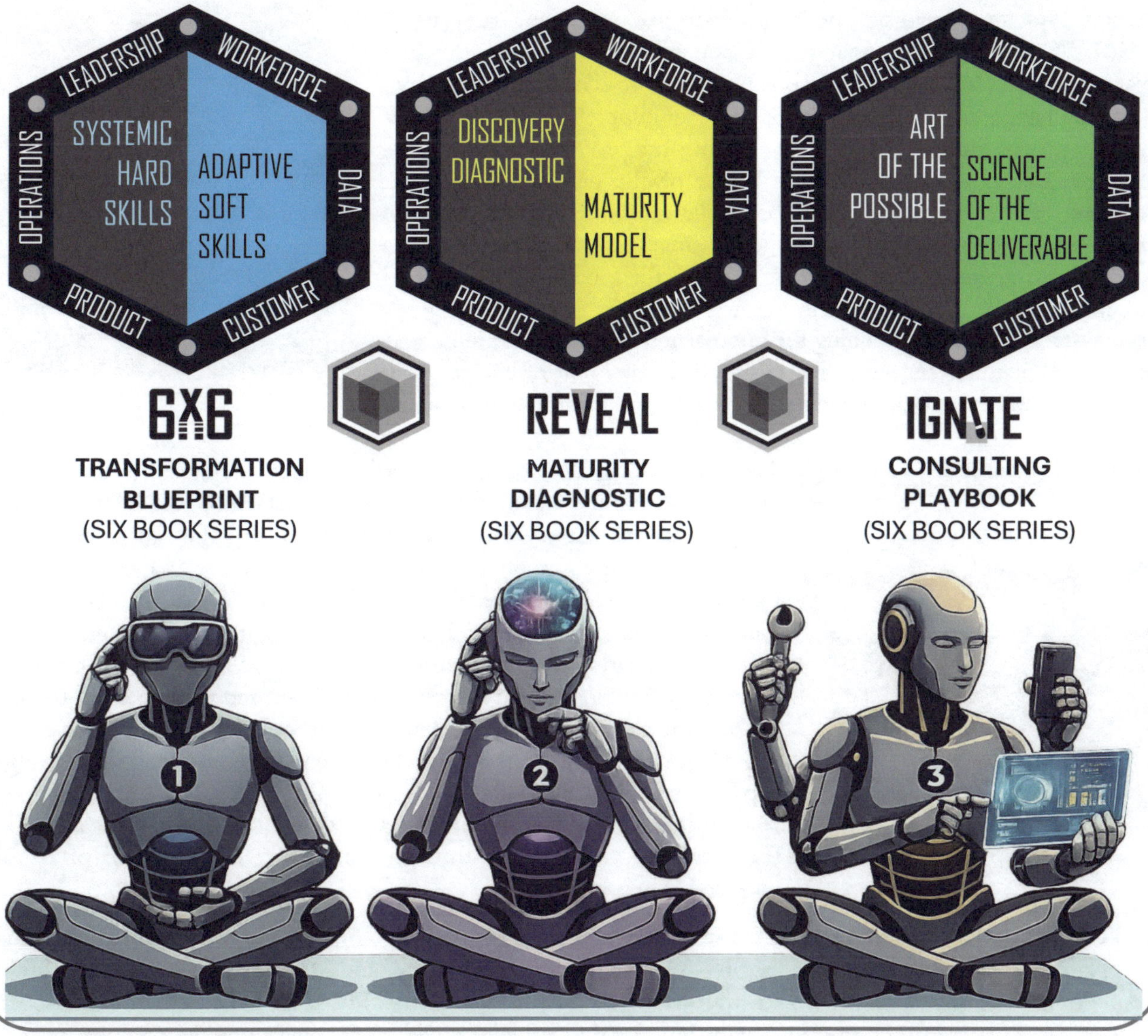

The **AI6x6 Transformation Engine™** is the flagship construct of the AI6x6 Research Lab—a precision-engineered ecosystem that serves as the definitive "Operating System" for the modern enterprise. Built on a proprietary 6×6 geometry, it unifies architecture, execution, and calibration into a single, high-performance unit. The Engine provides leaders with the high-torque machinery and **Guided Navigation** required to convert organizational inertia into scalable momentum in the era of autonomous systems.

The **Transformation Blueprint Series** defines the engine's architecture and establishes the structural logic of the entire system. By introducing the **Six Pillars** and **Six Lenses**, it forms the foundational geometry of enterprise modernization. Written for CEOs, Boards, and Strategy Leaders, this series acts as the **Navigation Deck**, providing the "Architectural North Star." It ensures that every move is strategically calibrated and ethically anchored before the gears are engaged, offering a shared mental model for aligning leadership teams. This intelligence layer removes the guesswork from transformation, providing a rigid yet adaptive framework that allows organizations to visualize their future state with absolute clarity. It is the vital starting point where complex, high-stakes ambition is codified into a disciplined, institutional reality that can withstand the pressures of rapid technological shift.

The **Maturity Diagnostic Series** provides the engine's accountability and calibration system, ensuring peak performance through proprietary **REVEAL ™** methodology. . Acting as the **Sensor-Driven Intake**, it utilizes 36 diagnostic cylinders to quantify organizational readiness and prioritize investments based on evidence rather than intuition. This series allows executive teams to identify hidden gaps and measure progress with scientific clarity, functioning as the **Measuring Layer** of the entire ecosystem. It ensures the engine remains tuned for resilience through a data-driven maturity model that tracks evolution over time. By providing rigorous assessment instruments and scoring logic, it offers the accountability required to sustain a long-term transformation journey. This layer acts as the engine's nervous system, constantly scanning the internal and external environment to calibrate the Blueprint's trajectory and ensure that every unit of energy spent drives the organization toward sustained excellence and institutional stability.

The **Consulting Playbook Series** translates architecture into action by operationalizing the engine through the proprietary **IGNITE™** methodology. It serves as the **Heavy-Duty Gearbox**, providing the repeatable consulting patterns, execution models, and directional force required to activate transformation across diverse domains. Designed for PMOs and delivery leads, these playbooks convert high-level strategy into tangible enterprise-scale outcomes, ensuring that the "torque" of the engine is effectively applied to the road. They function as the essential doing layer, reducing internal friction and shifting the organization from abstract concept to deep-seated conviction. By providing step-by-step pathways and intervention models, the series bridges the gap between executive intent and operational reality. This transmission layer ensures that momentum is not just created, but sustained and scaled, allowing the enterprise to accelerate through complex transitions with precision and measurable impact.

The **Transformation Engine** is built for scale across the global economy. Future **Industry Evolution** volumes will apply this 6×6 engine to specific sectors—such as Healthcare, Banking, and Manufacturing—revealing how the model drives specialized, high-velocity transformation. By adapting the Pillars and Lenses to sector-specific demands, these extensions ensure the Engine remains the definitive standard for navigating the complexities of the age of intelligent systems.

SERIES 1 6 BOOKS

SERIES 2 6 BOOKS

SERIES 3 6 BOOKS

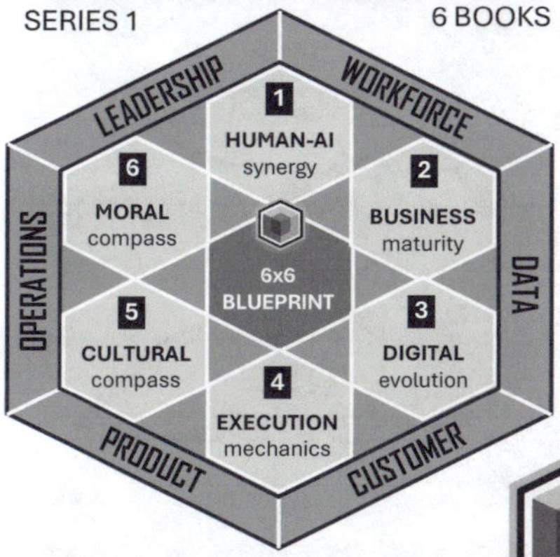

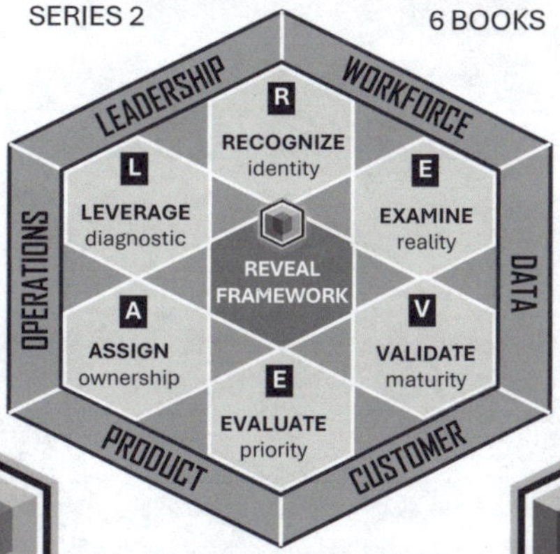

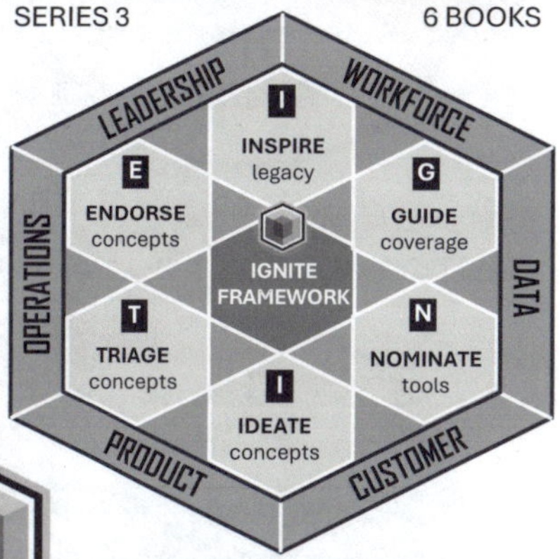

6X6 TRANSFORMATION BLUEPRINT

REVEAL MATURITY DIAGNOSTIC

IGNITE CONSULTING PLAYBOOK

see

think

act

The Ai6x6 palette is a semantic system, not a decorative choice. Each color encodes the emotional and operational signature of its lens: **pink** captures the human spark that ignites potential; **green** reflects the strategic discipline required for maturity; **purple** signals the depth and computational power of digital evolution; **orange** embodies the kinetic urgency of execution; **blue** anchors the cultural soul in trust and cohesion; and **yellow** illuminates the ethical clarity of the moral audit. Together, these colors form a coherent visual language that guides the reader through the transformation cycle with intuitive, emotionally aligned cues.

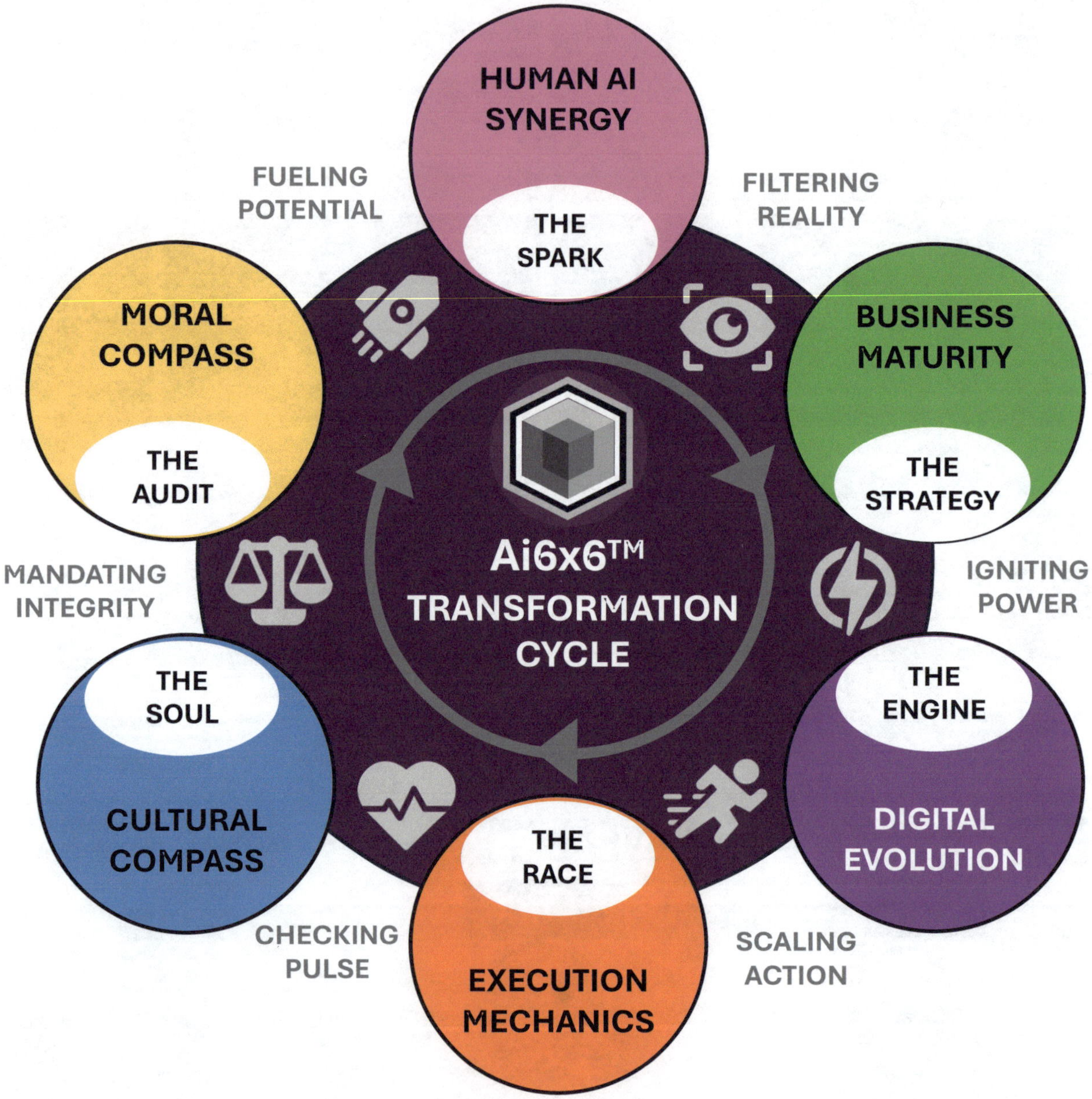

The **COLOR TABS ALONG THE PAGE EDGES IN THE MIDDLE** serve as the physical navigation system of the Ai6x6 series. Each tab mirrors the chromatic identity of its lens, allowing readers to locate themselves instantly within the transformation cycle. This creates a spatial reading experience where orientation is continuous and intuitive. The tabs turn our book into a modular interface—one where Human–AI Synergy, Business Maturity, Digital Evolution, Execution Mechanics, Cultural Compass, and Moral Compass become distinct territories with their own visual signatures. The result is a playbook that reads like a designed system, not a linear text.

6x6™ Transformation Blueprint

The Prologue to a Purpose

"The only way to go forward is to let go of the past."
The Myth of Orpheus and Eurydice

The tragic lesson of this myth is a cautionary tale for business. Orpheus's failure to trust his journey and looking back results in losing everything, underscoring the importance of fully committing to a forward-looking transformation.

The mission of this book series is simple: to transform leaders from passive observers of the AI Era into its active pioneers. We will ignite your thinking with exciting, highly practical content that pushes boundaries and provides the essential tools and frameworks you need to build real-world impact. We won't just tell you what to do; we'll show you how, using a diverse community's experiences and real-world examples that bring the abstract to life. This isn't a passive read; it's an active partnership and a dynamic dialogue designed to turn your natural curiosity into a definitive competitive advantage. This book is the first of a six-part series : Leadership Transformation; which governs your journey on how you should lead from the top and how you should inspire others to take on the AI Era and excel in it.

The **AI6x6 Transformation Engine** is a precision-engineered ecosystem designed to deliver **Guided Navigation** for enterprises facing high-stakes digital evolution. By integrating three proprietary systems—the **Transformation Blueprint**, the **Consulting Playbook**, and the **Maturity Diagnostic**—the Engine provides the robust "Operating System" required to dominate the AI era.

The architecture begins with the **Transformation Blueprint**, the Engine's intelligence layer. Here, the system utilizes **Six Lenses**—including the Cultural and Moral Compasses—to calibrate your trajectory and codify a master architectural plan. This strategic map then engages the **Consulting Playbook**, where the gears of the **IGNITE** methodology translate executive intent into immediate directional force through rapid-response execution cycles. Finally, the **Maturity Diagnostic** acts as a sensor-driven intake, utilizing six diagnostic cylinders to compress raw organizational data into high-pressure strategic insights that validate the transformation's impact. Bolted to a rigid **Chassis of Six Pillars**, the Engine ensures the enterprise remains resilient, balanced, and unshakeable even at peak acceleration. It is more than a strategy; it is a proprietary architecture for sustained institutional excellence.

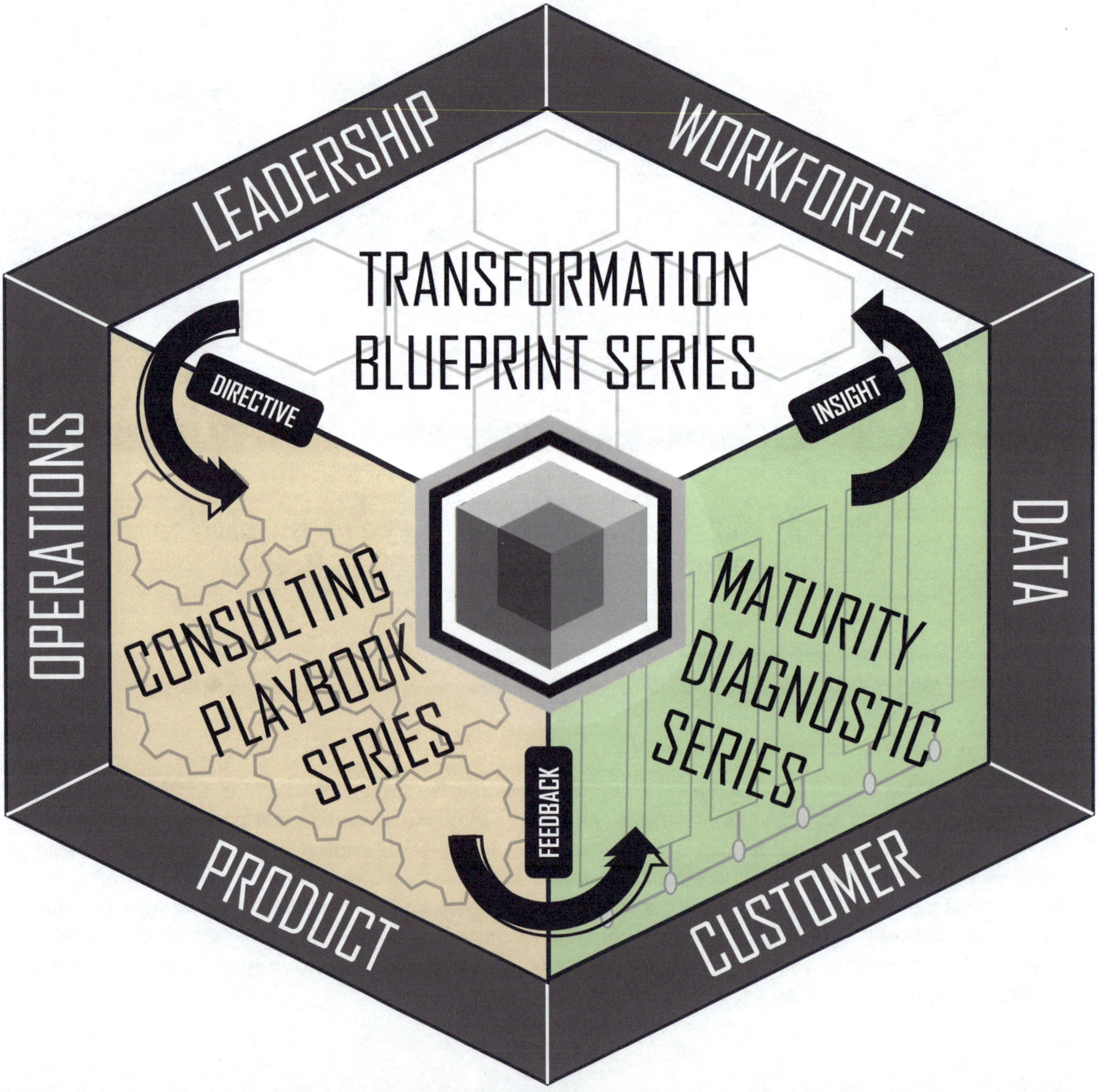

The AI6x6 Transformation Engine is a precision-engineered ecosystem that replaces guesswork with **Guided Navigation**. It is designed to convert organizational inertia into scalable momentum through a disciplined, mechanical approach to digital evolution.

CHASSIS (SIX PILLARS)
The entire system is anchored to a rigid structural frame. This chassis ensures that even as you reach peak acceleration, your enterprise remains resilient and unshakeable.

NAVIGATION DECK (SIX LENSES):
This is the intelligence layer. By applying our **Six Lenses**, we provide the **Guided Navigation** required to calibrate your trajectory, preventing costly drifts in strategy.

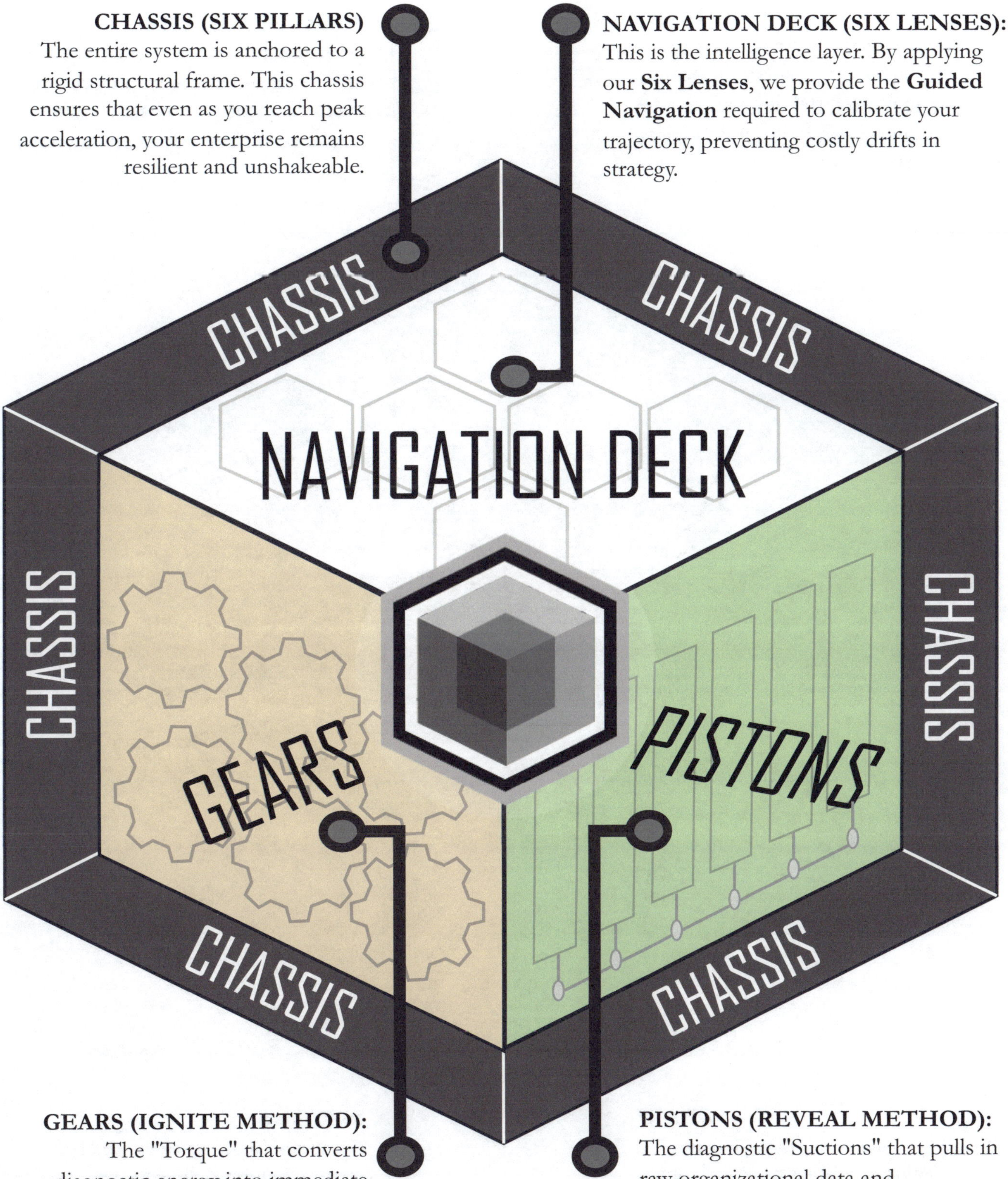

GEARS (IGNITE METHOD):
The "Torque" that converts diagnostic energy into immediate directional force through pilot programs and playbooks.

PISTONS (REVEAL METHOD):
The diagnostic "Suctions" that pulls in raw organizational data and compresses it into high-pressure actionable insights.

A car works best when its actions are coordinated through systems that generate power, transmit motion, and guide direction. Each component plays a precise mechanical role that keeps the vehicle stable, responsive, and moving with purpose.

CHASSIS
Supports vehicle structure, absorbs road stress, anchors mechanical systems, maintains stability, and ensures safe, balanced handling across varying speeds, terrains, and driving conditions.

NAVIGATION DECK
Displays real-time driving information, maps, routes, and alerts, helping the driver maintain direction, anticipate conditions, and make confident, informed decisions.

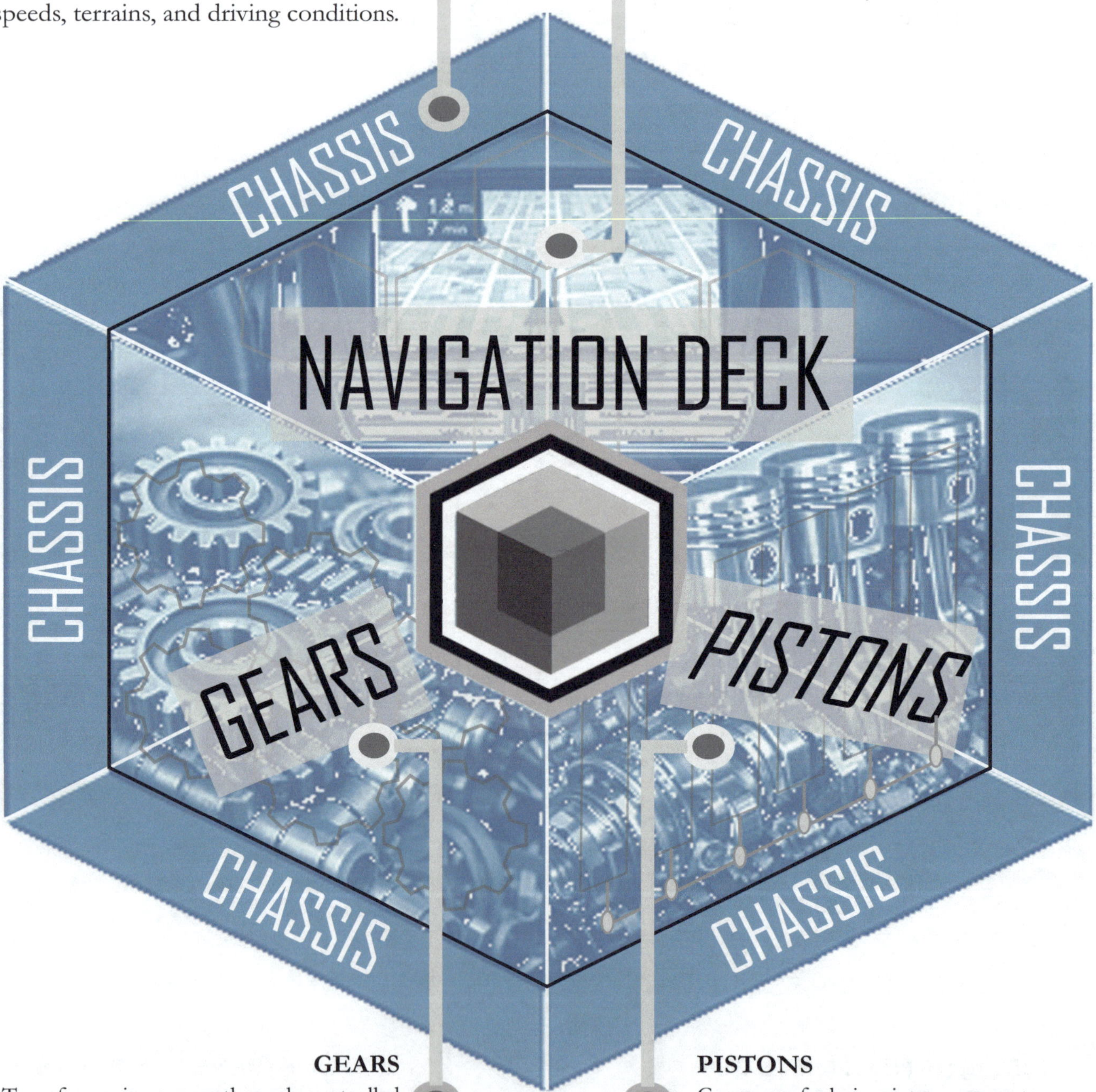

GEARS
Transfer engine power through controlled ratios, regulating speed and torque, ensuring smooth acceleration, efficient performance, and seamless coordination between mechanical components.

PISTONS
Compress fuel-air mixture, generate combustion force, drive the crankshaft, & convert explosive energy into continuous motion powering the vehicle's entire drivetrain system.

The **AI6x6 Transformation Engine** replaces the ambiguity of traditional consulting with mechanical certainty, converting organizational inertia into scalable momentum. By integrating three proprietary systems—the **REVEAL** diagnostic intake, the **IGNITE** consulting transmission, and the **6x6 Blueprint** navigation deck—the Engine provides the robust "Operating System" required to dominate the AI era. The **REVEAL** methodology acts as a sensor-driven intake, utilizing 6x6 diagnostic cylinders to compress raw organizational data into high-pressure strategic insights. This "data fuel" then powers the **IGNITE** phase, where consulting playbooks serve as a heavy-duty gearbox, translating executive intent into immediate directional force through rapid-response execution cycles. This entire power plant is governed by the **6x6 Blueprint**, the intelligence layer that uses **Six Lenses**—including the Cultural and Moral Compasses—to calibrate your trajectory and ensure ethical, operational alignment. Bolted to a rigid **Chassis** of **Six Pillars**, the Engine ensures the enterprise remains resilient, balanced, and unshakeable even at peak acceleration. It is more than a strategy; it is a proprietary architecture for sustained institutional excellence.

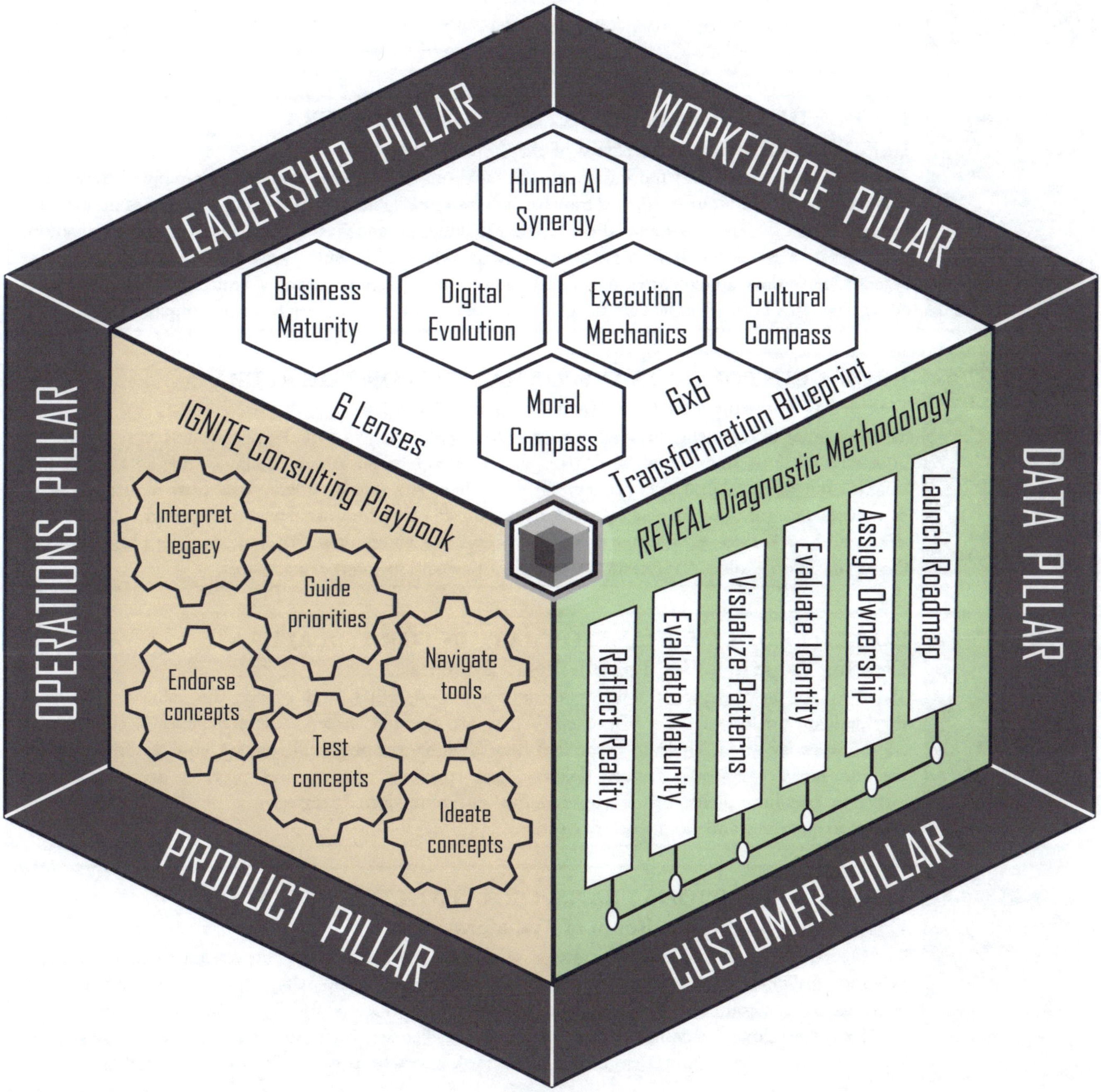

6 BOOKS OF THE Ai6x6 TRANSFORMATION BLUEPRINT SERIES

BOOK 1: LEADERSHIP TRANSFORMATION FOR THE AI ERA

Designing the Vision, Not Just Directing It

The Leadership book reframes what it means to lead in an era defined by intelligence, acceleration, and ambiguity. It explores how leaders think, decide, communicate, and shape meaning across the enterprise, emphasizing that leadership is not a role but an architecture. The narrative examines how leaders make decisions systemically, exercise judgment in an AI-augmented world, create alignment through narrative clarity, and uphold ethical stewardship as the foundation of trust. This book sets the tone for the entire series by positioning leadership as the integrator of all transformation forces.

BOOK 2: WORKFORCE TRANSFORMATION FOR THE AI ERA

Elevating Human Capability in a Technology-Driven Era

The Workforce book examines the human system of the enterprise — the skills, roles, mindsets, structures, and experiences that determine how people show up to the work of transformation. It explores how organizations support people as AI reshapes every function, and how leaders build environments where individuals feel empowered rather than displaced. The narrative dives into the evolution of adaptive talent ecosystems, the emergence of AI-augmented roles, the importance of team cohesion and psychological safety, and the recognition of human capability as a strategic asset.

BOOK 3: DATA TRANSFORMATION FOR THE AI ERA

Building the Intelligence Fabric of the Enterprise

The Data book establishes the informational backbone of the organization. It explores how data is collected, governed, interpreted, and transformed into intelligence, emphasizing that data is no longer a technical asset but the raw material of decision-making, personalization, automation, and innovation. The narrative examines how trustworthy data governance is built, how intelligence flows across systems, how data architectures are designed, and how data becomes a source of enterprise-level advantage. This book gives leaders the clarity to build an intelligent, insight-driven organization.

BOOK 4: CUSTOMER TRANSFORMATION FOR THE AI ERA

Designing Experiences That Earn Trust and Loyalty

The Customer book anchors the enterprise in the lived experience of the people it serves. It explores how organizations understand, anticipate, and respond to customer needs — not as a marketing exercise but as a cultural and operational discipline. The narrative examines how emotion shapes experience design, how AI enables personalization at scale, how transparency strengthens customer trust, and how end-to-end journey coherence becomes a competitive differentiator. This book ensures the organization evolves around the customer, not around its internal structures.

BOOK 5: PRODUCT TRANSFORMATION FOR THE AI ERA

Creating Value in a World of Intelligent Systems

The Product book explores how organizations design, build, and evolve the value they deliver. Products are no longer static offerings — they are dynamic systems of intelligence, behavior, and continuous adaptation. This book examines how AI enhances product behavior, how continuous value creation replaces episodic feature delivery, how product strategy and product operations must integrate, and how digital ecosystems redefine differentiation. It transforms product thinking from output to outcome, and from features to value.

BOOK 6: OPERATIONS TRANSFORMATION FOR THE AI ERA

Engineering the Engine Room of Transformation

The Operations book addresses the systems, workflows, and processes that determine whether strategy becomes reality. It explores how organizations orchestrate complexity, manage risk, and maintain resilience in environments defined by acceleration and uncertainty. The narrative examines how intelligent workflows are designed, how operational resilience is built, how AI optimizes processes, and how execution scales with precision. This is the book where transformation becomes tangible.

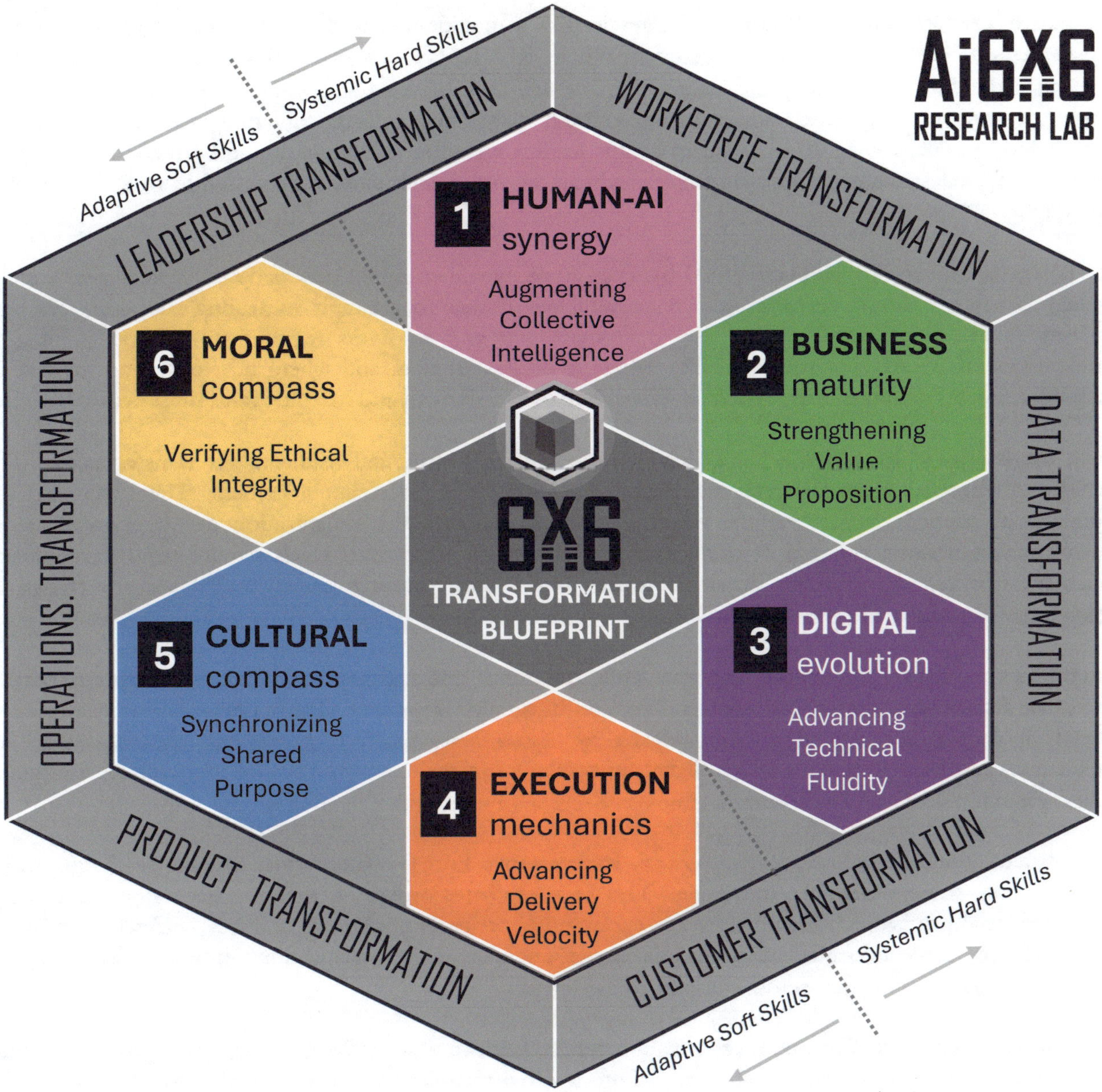

The **Ai6x6 TRANFORMATION BLUEPRINT™** provides the multidimensional structural integrity required for sustainable, high-stakes evolution. Serving as the **Navigation Deck** and **Chassis** of the Transformation Engine, the BLUEPRINT™ moves beyond episodic change to establish a permanent, scalable "Operating System" for the enterprise. The architecture is defined by the intersection of **Six Pillars** and **Six Lenses**. The Pillars—Leadership, Workforce, Data, Operations, Product, and Customer—provide the horizontal breadth and load-bearing strength of the organization. These are bisected vertically by the Six Lenses to reveal the true depth of the enterprise. Leaders apply **Human–AI Synergy** to augment collective intelligence and **Business Maturity** to instill systemic discipline. The **Digital Evolution** lens ensures technical fluidity, while **Execution Mechanics** masters the art of precision delivery. This is anchored by the **Cultural Compass**, which synchronizes shared purpose, and the **Moral Compass**, which verifies ethical integrity.

By aligning these twelve dimensions, the BLUEPRINT™ eliminates strategic drift and structural fragmentation. It ensures that every initiative is not just an isolated project, but a calibrated component of a unified, resilient, and unshakeable corporate architecture. Also remember to follow the Tab colors on the edges to navigate quickly.

The **Ai6x6 Transformation Blueprint™** gives leaders a multidimensional, architecture-grade view of how transformation behaves inside their enterprise. Where REVEAL™ diagnoses the present and IGNITE™ accelerates the future, the Blueprint defines the *structural truth* of the organization. Built on six proprietary Lenses, it converts fragmented observations, cultural noise, and operational ambiguity into a coherent, systems-level understanding. The Blueprint equips leaders with the clarity, depth, and structural intelligence required before designing any transformation roadmap, ensuring decisions are grounded in reality rather than aspiration.

The Blueprint begins with the **Human–AI Synergy Lens**, which examines how people and intelligent systems interact across workflows, decisions, and responsibilities. Rather than simply measuring adoption, this Lens analyzes augmentation, displacement risk, trust dynamics, and readiness for human-machine teaming. It distinguishes where AI enhances capability, where it creates friction, and where new operating models are required. Leaders begin with a grounded understanding of how human and machine roles must evolve together.

Next, the **Business Maturity Lens** evaluates the enterprise's structural readiness for transformation. This includes governance, financial discipline, decision velocity, and leadership alignment. The Lens separates foundational strength from structural fragility, revealing whether the organization can support sustained transformation without collapse. It clarifies which capabilities are stable, which require reinforcement, and which must be redesigned before acceleration. Business Maturity ensures leaders understand the chassis on which transformation will operate.

The third Lens, **Digital Evolution**, maps the depth, coherence, and scalability of the digital ecosystem. Leaders often face fragmented platforms, technical debt, and integration gaps that silently undermine transformation. Digital Evolution cuts through this complexity by assessing architecture, interoperability, data flow, and platform extensibility. It clarifies which digital foundations can scale, which must be modernized, and which create systemic risk. This Lens ensures technology decisions are grounded in architectural truth.

With structural and digital realities established, leaders move into the **Execution Mechanics Lens**, which examines how strategy becomes movement. This Lens evaluates processes, rhythms, accountability systems, and operational precision. It identifies where execution is consistent, where it is brittle, and where it breaks under pressure. Execution Mechanics ensures leaders understand the organization's true capacity to deliver, adapt, and sustain momentum across complex transformation cycles.

The fifth Lens, **Cultural Compass**, explores the beliefs, behaviors, incentives, and informal power structures that shape how people respond to change. Culture is often the silent accelerator or the invisible brake. This Lens reveals alignment, resistance patterns, psychological safety, and leadership influence networks. It distinguishes cultural strengths from cultural liabilities, giving leaders a clear view of the human forces that determine whether transformation will be embraced or rejected.

The final Lens, **Moral Compass**, evaluates ethics, trust, transparency, and responsible AI principles. As organizations adopt intelligent systems, the stakes of ethical decision-making rise dramatically. This Lens ensures transformation is not only effective but also equitable, accountable, and aligned with long-term societal and organizational values. It clarifies where safeguards are strong, where risks exist, and where governance must evolve.

Together, these Six Lenses form a complete, repeatable, consulting-grade architectural framework. The Ai6x6 Transformation Blueprint™ is not a checklist or maturity model; it is a structural intelligence system that moves leaders from assumption to understanding, from understanding to design, and from design to confident transformation planning. It gives leaders the conviction to move forward — not because the future is predictable, but because their decisions are grounded in multidimensional truth, structural clarity, and a deep understanding of how transformation behaves across the enterprise.

Ai6x6™ Transformation Blueprint

Understand Pillars & Lenses

"We don't see things as they are, we see them as we are." — Anaïs Nin

The **Ai6x6 TRANSFORMATION BLUEPRINT** is fundamentally structured around two core concepts: **Pillars and Lenses**. Together, they create a dual-axis architecture that brings both breadth and depth to enterprise transformation. The Pillars provide the WIDTH of the framework, defining the major domains where transformation must take root for the organization to evolve coherently. These domains—Customer, Product, Operations, Data, Workforce, and Leadership—represent the full span of enterprise behavior. Each Pillar receives its own dedicated book, allowing leaders to explore the complexities, tensions, and opportunities of that domain without dilution or compromise.

This ensures that transformation is not treated as a single initiative but as a system of interconnected disciplines that must advance in parallel. Complementing this horizontal structure is the DEPTH created by the Six Lenses, the universal imperatives that cut vertically through every Pillar. These Lenses represent the non-negotiable dimensions of modern transformation—forces that must be activated in every domain, regardless of industry, maturity, or strategic ambition. The six Lenses are: Human–AI Synergy (L1), Business Maturity (L2), Digital Evolution (L3), Execution Mechanics (L4), Cultural Compass (L5), and Moral Compass (L6). Each Lens exposes a different layer of organizational capability, ensuring that no domain is approached superficially, and no critical dimension is overlooked. The goal of the blueprint is to be holistic and very comprehensive so that executives are not blind-sighted by a key aspects being overlooked. Since a lot of ground has to be covered, the books may look very dense, but I have tried my best to keep it brief & engaging.

Transformation has become one of the most OVERUSED words in modern leadership, yet one of the least understood. Organizations declare it, teams attempt it, consultants package it—but few enterprises ever truly achieve it. Not because they lack ambition or intelligence, but because they approach transformation as a sequence of disconnected initiatives rather than a shift in how the organization thinks, behaves, and evolves. I created the **Ai6x6 Transformation Blueprint** because I have seen the same pattern repeat across industries, functions, and eras of technology: transformation fails not at the point of strategy, but at the point of structure. It fails when organizations modernize technology but ignore culture. It fails when they redesign processes but overlook trust. It fails when they invest in AI but fail to prepare people. It fails when they build strategy but cannot execute it. It fails when leaders underestimate the complexity of aligning human capability with machine intelligence.

The Ai6x6 Transformation Blueprint exists to solve this fragmentation. It brings together **six foundational lenses** and **six transformation pillars** into a single, coherent architecture. Each lens represents a dimension of capability that every organization must master. Each pillar represents a domain— Leadership, Workforce, Data, Customer, Product, Operations —where transformation must take root. Together, they form a 6×6 matrix that reveals not only what must change, but how change becomes sustainable, repeatable, and scalable.

This blueprint is intentionally multidimensional. Transformation is not linear. It is not sequential. It is not a checklist. It is a living system—human, technical, cultural, operational, ethical—moving together or not at all. The Ai6x6 model acknowledges this complexity and gives leaders a structure strong enough to hold it.

- It is called a blueprint because it is both **architectural** and **actionable**.
- This blueprint is not a methodology per se. It is not a maturity model in the traditional sense. It is not a checklist as you commonly see it. It is a **HOLISTIC** way of seeing things that are often overlooked.
- It is called Ai6x6 because it integrates **AI-driven intelligence** with **six human-centered disciplines**, applied across **six transformation pillars**.
- It is a system designed for organizations that want to evolve with intention rather than react with urgency.

The Six Lenses emerged from a simple but unavoidable truth: "**no single discipline can carry transformation on its own"**. Not AI. Not culture. Not governance. Not execution. Not digital modernization. Not ethics. Each is necessary, but none is sufficient. Real transformation requires all six lenses working together—reinforcing one another, correcting one another, and elevating one another.

It reveals the forces that shape organizational behavior, the tensions that undermine progress, and the capabilities that determine whether transformation becomes sustainable or collapses under its own ambition. It provides leaders with a structural map—one that makes complexity navigable and makes evolution repeatable.

Each lens represents a dimension of organizational life:

1. Human–AI Synergy: how intelligence is created, shared, and amplified.
2. Business Maturity: how discipline, readiness, and structure turn ambition into capability.
3. Digital Evolution: how technology becomes fluid, integrated, and trustworthy.
4. Execution Mechanics: how strategy becomes action and action becomes advantage.
5. Cultural Compass: how beliefs, behaviors, and meaning shape the human system.
6. Moral Compass: how ethics, transparency, and responsibility anchor everything else.

Together, they form a complete operating philosophy for modern transformation—one that is as human as it is technical, as strategic as it is operational, as aspirational as it is grounded. The Ai6x6 Transformation Blueprint is not about predicting the future. It is about preparing organizations to shape it. It is about giving leaders a language for what they instinctively know but rarely articulate. It is about building enterprises that can evolve continuously, coherently, and confidently. Most of all, it is about honoring the truth that **transformation is not something organizations *do*—it is something they *become*.**

The Pillars of the **Ai6x6 Transformation Blueprint** define the WIDTH of enterprise evolution. They represent the six domains where transformation must become real, measurable, and operationally grounded. While the Lenses provide the depth of capability, the Pillars provide the breadth of application—ensuring that transformation is not isolated to a single function but becomes a structural shift across the entire enterprise. Each Pillar receives its own dedicated book, allowing leaders to confront the full complexity of that domain with precision and without compromise.

The Leadership Pillar sits at the top of the architecture because leadership is the force that shapes every other domain. It defines how decisions are made, how priorities are set, how accountability is distributed, and how meaning is communicated. This Pillar examines the evolution of leadership in an era defined by AI, acceleration, and ambiguity. It challenges leaders to move from managing activities to designing systems, from directing work to enabling capability, and from protecting the past to shaping the future. Leadership is not a separate function—it is the integrator of the entire Blueprint.

The Workforce Pillar focuses on the human system that powers the enterprise. It explores how roles evolve, how teams collaborate, how skills are developed, and how people experience their work. In a world where AI augments every function, the Workforce Pillar examines how organizations support human capability, build adaptive talent ecosystems, and create environments where people feel empowered rather than displaced. This Pillar forces leaders to confront the truth that transformation succeeds or fails based on whether people feel equipped, supported, and connected to the mission.

The Data Pillar establishes the informational foundation of the enterprise. It examines how data is collected, governed, interpreted, and transformed into intelligence. In an AI-driven world, data is no longer a byproduct of operations—it is the raw material of decision-making, personalization, automation, and innovation. This Pillar explores how organizations build trustworthy data ecosystems, eliminate silos, and ensure that intelligence flows across the enterprise. It challenges leaders to treat data not as a technical asset but as a strategic capability that shapes every other Pillar.

The Customer Pillar anchors the organization in the lived experience of the people it serves. It forces leaders to confront the reality that customer expectations evolve faster than internal processes, and that loyalty is earned through coherence, trust, and emotional resonance. This Pillar examines how organizations understand, anticipate, and respond to customer needs—not as a marketing exercise, but as a cultural and operational discipline. It challenges leaders to build enterprises where customer experience is not a department, but a shared responsibility embedded in every decision.

The Product Pillar focuses on how organizations design, build, and evolve the value they deliver. Products are no longer static offerings—they are dynamic systems of intelligence, behavior, and continuous adaptation. This Pillar explores how product strategy, product design, and product operations must work together to create experiences that feel alive and responsive. It examines how AI reshapes product behavior, how digital ecosystems redefine differentiation, and how organizations must shift from feature delivery to continuous value creation.

The Operations Pillar addresses the engine room of the enterprise—the systems, workflows, and processes that determine whether strategy becomes reality. It examines how organizations orchestrate complexity, manage risk, and maintain resilience in environments defined by acceleration and uncertainty. This Pillar forces leaders to recognize that operational excellence is not about efficiency alone; it is about adaptability, intelligence, and the ability to respond to change without losing coherence. Operations is where transformation becomes tangible.

Together, these six Pillars form the structural WIDTH of the Ai6x6 Transformation Blueprint. They define the terrain on which the Six Lenses operate, ensuring that transformation is comprehensive, coherent, and enterprise-wide.

SIX PILLARS OF ENDURING TRANSFORMATION

4
Discerning Diners
CUSTOMER
"The Patrons"
5
Food, Drinks, Ambiance
PRODUCT
"The Feast"
6
Ingredients, Purchase
OPERATIONS
"The Logistics"
2
Cooks, Stewards, Servers
WORKFORCE
"The other staff"
3
Menu, Recipe & Reviews
DATA
"The Menu"
1
Lead Culinary aspect
LEADERSHIP
"The Exec Chef"

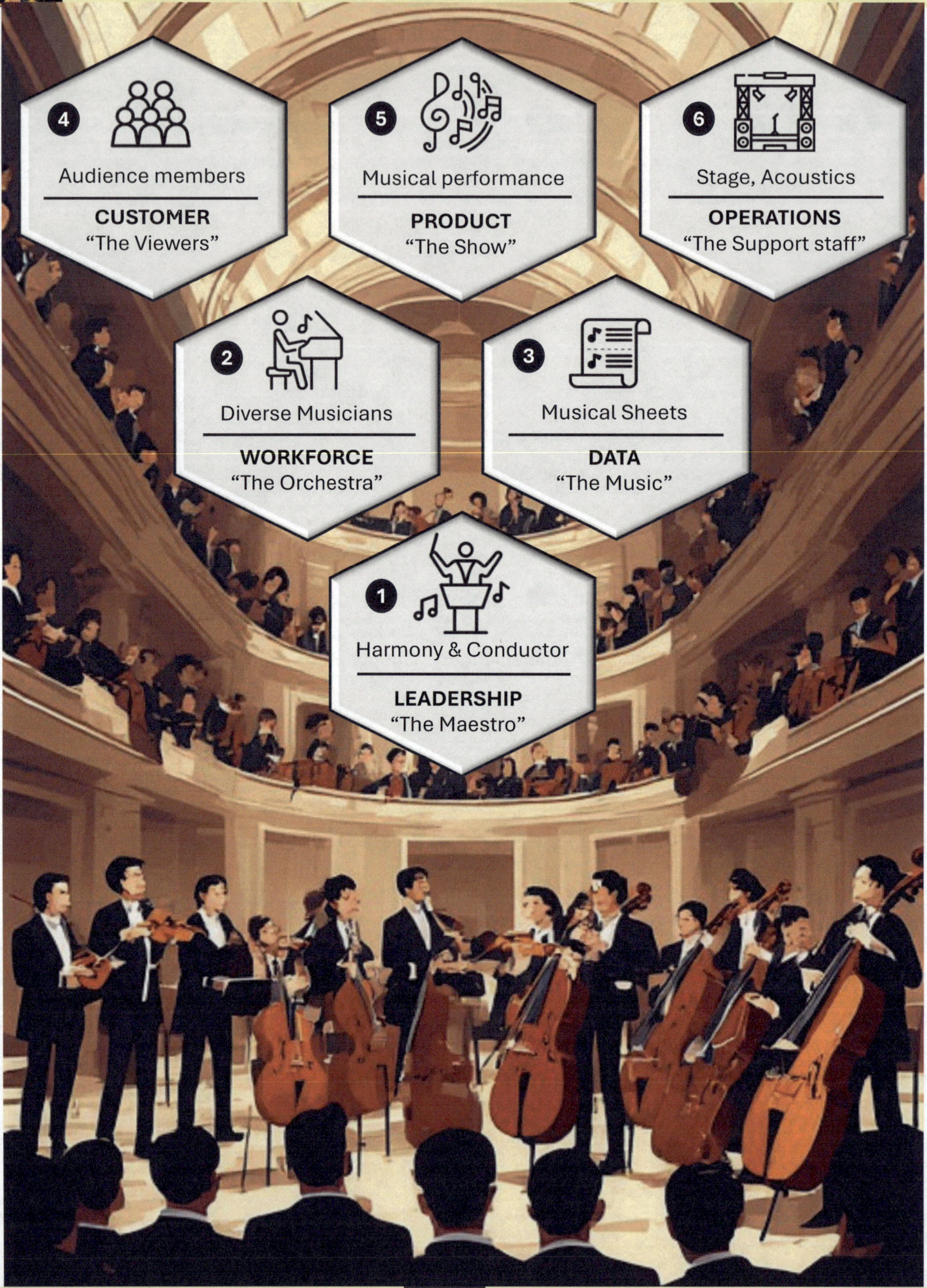
4
Audience members
CUSTOMER
"The Viewers"
5
Musical performance
PRODUCT
"The Show"
6
Stage, Acoustics
OPERATIONS
"The Support staff"
2
Diverse Musicians
WORKFORCE
"The Orchestra"
3
Musical Sheets
DATA
"The Music"
1
Harmony & Conductor
LEADERSHIP
"The Maestro"

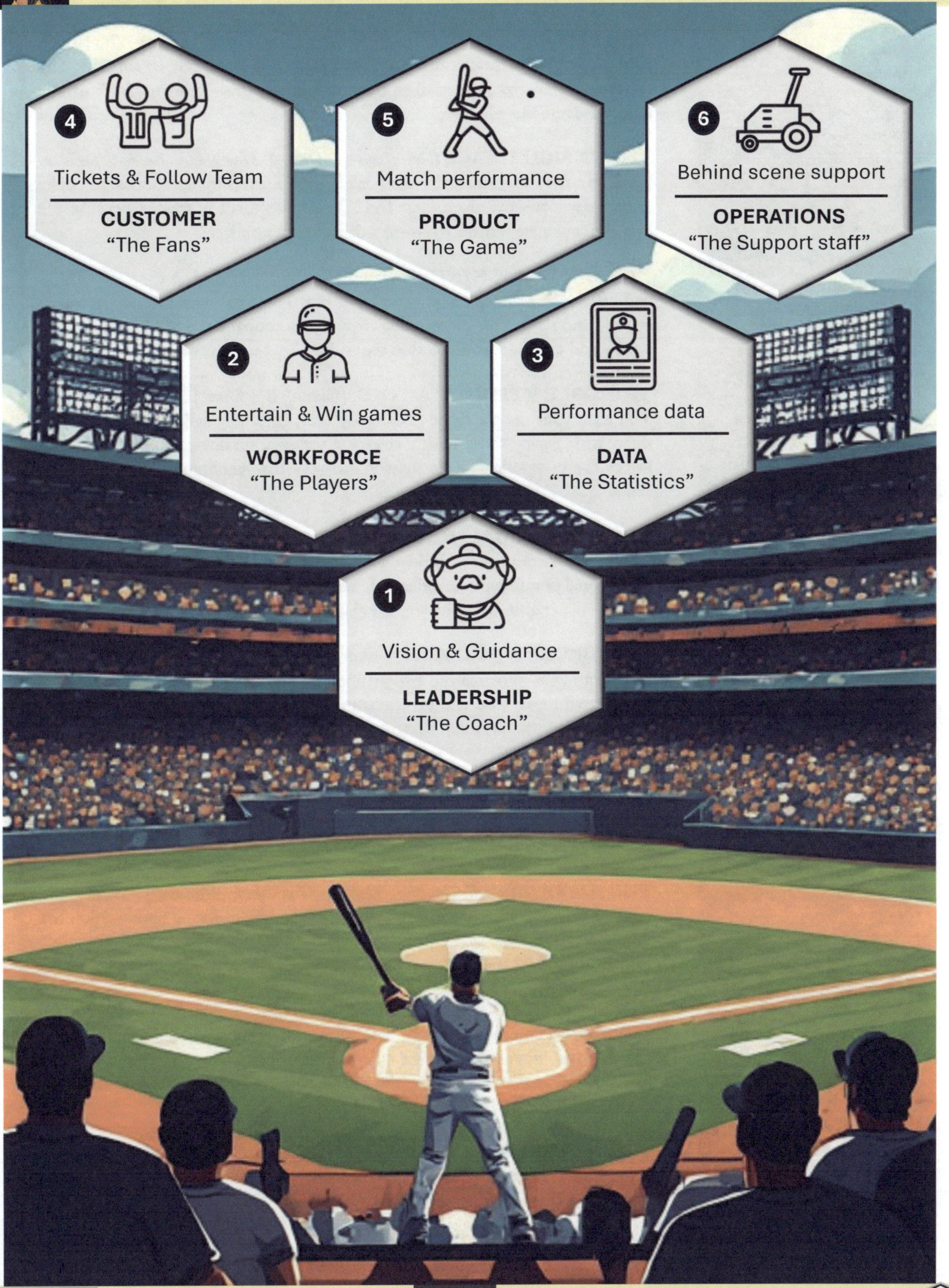
4
Tickets & Follow Team
CUSTOMER
"The Fans"
5
Match performance
PRODUCT
"The Game"
6
Behind scene support
OPERATIONS
"The Support staff"
2
Entertain & Win games
WORKFORCE
"The Players"
3
Performance data
DATA
"The Statistics"
1
Vision & Guidance
LEADERSHIP
"The Coach"

THE LITERACY GAP (*Invested: Product | Overlooked: Workforce*): You launch genius AI-driven features that your staff cannot explain, sell, or support. The technology fails to gain market traction because the "human layer" of the enterprise remains obsolete.

THE INSIGHT VACUUM (*Invested: Data | Overlooked: Leadership*): You build massive, high-integrity data lakes, but leadership continues to make critical decisions based on "gut feel." The investment is wasted because the upper-tier culture refuses to adopt a data-driven mindset.

THE COLD MACHINE (*Invested: Operations | Overlooked: Customer*): Back-end efficiency reaches an all-time high, but the actual user journey becomes robotic and rigid. You successfully reduce operational costs but lose the brand loyalty and emotional connection that drives long-term revenue.

THE SKILL WITHOUT A TOOL (*Invested: Workforce | Overlooked: Product*): You train your people to be "AI-Ready" and culturally agile, but you provide them with legacy tools and a stagnant product roadmap. Your best talent leaves for competitors who actually possess the technology to match their skills.

THE HALLUCINATION (*Invested: Leadership | Overlooked: Operations*): Executives announce a bold "Digital First" vision, but the underlying operational plumbing is decades old. The strategy never leaves the boardroom because the engine lacks the structural integrity to turn the wheels.

THE BROKEN PROMISE (*Invested: Customer | Overlooked: Data*): You market a hyper-personalized experience, but your internal data silos remain a fragmented mess. You end up alienating customers with irrelevant offers and inaccurate information that destroys their trust.

THE PROTOTYPE TRAP (*Invested: Product | Overlooked: Operations*): A brilliant AI-driven tool is developed in a vacuum, but the legacy supply chain and logistics cannot fulfill the demand it creates. The product succeeds in the lab but collapses the moment it hits the real-world operational floor.

THE DARK ASSET (*Invested: Data | Overlooked: Workforce*): You spend millions cleaning and structuring your data, but no one in the organization has the literacy to query or interpret it. The data sits idle and "dark," providing zero ROI because the people aren't equipped to find the value.

THE ROGUE ENGINE (*Invested: Operations | Overlooked: Leadership*): Middle management automates workflows to save time, but leadership has no visibility or governance over these new "black box" processes. The organization moves faster, but the steering wheel is no longer connected to the captain's chair.

THE INTERNAL ECHO (*Invested: Workforce | Overlooked: Customer*): Employees are empowered and "AI-literate," but they become obsessed with internal process improvements rather than user needs. The company becomes a high-performing bubble that is perfectly optimized for a customer that no longer exists.

COLLISION PATHS TO AVOID IN YOUR TRANSFORMATION

THE BLIND CAPTAIN (*Invested: Leadership | Overlooked: Data*): A visionary leader attempts to steer the company through a massive pivot based on bold assumptions. Because the data pillar was ignored, they navigate straight into a market reef that the "missing data" would have clearly identified.

THE FANCY WRAPPER (*Invested: Customer | Overlooked: Product*): Your marketing and customer service are world-class, but the actual product has not evolved to keep up with competitors. You win the initial sale through great branding, but the churn rate sky-rockets once the customer realizes the product is hollow.

THE ANALYTICS ISLAND (*Invested: Data | Overlooked: Operations*): The data team produces perfect, real-time reports on every aspect of the business. However, the operational workflows are so slow and manual that by the time the business acts on the data, the opportunity has already vanished.

THE ORPHAN PROJECT (*Invested: Product | Overlooked: Leadership*): Innovation teams build cutting-edge AI tools that solve real problems, but leadership never provides a permanent budget or strategic home for them. These "pilot projects" eventually wither and die because they lack the top-down air cover to survive.

THE FRUSTRATED EXPERT (*Invested: Workforce | Overlooked: Operations*): You hire the best AI researchers and data scientists, but you force them to work within manual, "broken" internal workflows. Your experts spend 90% of their time fighting the system rather than building the future.

THE IVORY TOWER (*Invested: Leadership | Overlooked: Customer*): The board develops a strategy based purely on shareholder value and financial engineering. By ignoring the Customer pillar, they fail to see a massive shift in user sentiment that makes their entire three-year plan irrelevant.

THE HUMAN BOTTLENECK (*Invested: Operations | Overlooked: Workforce*): You automate 80% of your workflows, but the remaining 20% require human intervention that your staff isn't trained for. The entire system grinds to a halt because you optimized the machine but forgot to upskill the operator.

THE REACTIVE PANIC (*Invested: Customer | Overlooked: Leadership*): The company chases every new customer trend and social media "vibe" to stay relevant. Without a strategic anchor from Leadership, the organization becomes a fragmented mess of contradictory ideas that stand for nothing.

THE WASTEFUL HARVEST (*Invested: Data | Overlooked: Product*): You collect trillions of data points and pay massive cloud storage fees, but none of that data is ever used to iterate or improve the actual product. You are "Data Rich" but "Value Poor," hoarding information with no path to monetization.

THE GRASSROOTS REBELLION (*Invested: Workforce | Overlooked: Leadership*): Staff begins using "Shadow AI" and unsanctioned tools because leadership is too slow or fearful to provide guidance. You have an AI-ready workforce that is operating outside of the company's governance, creating a massive security and legal risk.

If the Pillars define the WIDTH of enterprise transformation, the **Six Lenses** define its DEPTH. They are the universal imperatives that cut vertically through every domain—Leadership, Workforce, Data, Customer, Product, and Operations. Each Lens exposes a different dimension of capability, ensuring that transformation is not superficial, fragmented, or episodic, but deeply rooted in the systems that shape how organizations think, behave, and evolve. These Lenses are not optional. They are not enhancements. They are the structural forces that determine whether transformation becomes sustainable or collapses under its own ambition. Each Lens reveals a layer of truth about the enterprise—its intelligence, its discipline, its technology, its execution, its culture, and its integrity.

The Human–AI Synergy Lens (L1) examines how intelligence is created, shared, and amplified across the enterprise. It explores the evolving relationship between human judgment and machine capability, ensuring that AI augments rather than replaces human potential. This Lens forces leaders to confront how decisions are made, how insights flow, and how people interact with intelligent systems. It challenges organizations to build environments where humans and AI operate as partners—each elevating the other. Human–AI Synergy is the starting point because it defines the cognitive foundation of modern transformation.

The Business Maturity Lens (L2) establishes the structural discipline required for transformation to take hold. It examines governance, alignment, prioritization, funding, risk management, and organizational readiness. This Lens reveals whether the enterprise has the scaffolding to support change or whether ambition is outpacing capability. Business Maturity is where leaders confront the operational truth: transformation is not powered by inspiration alone—it is powered by systems, discipline, and repeatable mechanisms that ensure consistency across the enterprise.

The Digital Evolution Lens (L3) focuses on the technological backbone of the organization. It examines how systems integrate, how data flows, how platforms scale, and how digital environments adapt to new demands. This Lens ensures that technology becomes an enabler rather than a bottleneck. It challenges organizations to build digital ecosystems that are fluid, intelligent, secure, and trustworthy. Digital Evolution is not about tools—it is about creating a technological environment capable of supporting continuous change.

The Execution Mechanics Lens (L4) converts strategy into action. It examines the methodologies, operating models, workflows, and performance systems that determine whether the organization can deliver consistently and at scale. This Lens exposes the gap between what leaders intend and what teams experience. It ensures that execution is not left to improvisation but is engineered with precision. Execution Mechanics is where transformation becomes tangible—where ideas become outcomes and outcomes become advantage.

The Cultural Compass Lens (L5) explores the beliefs, behaviors, and shared meaning that shape how people think and collaborate. It examines psychological safety, communication norms, leadership behaviors, and the emotional climate of the enterprise. This Lens reveals whether the culture accelerates transformation or resists it. Cultural Compass ensures that people feel connected, supported, and aligned—because no transformation can succeed in a culture that does not want it.

The **Moral Compass Lens (L6)** anchors the Blueprint in ethics and transparency. Moving beyond a static matrix, it defines the **True North** for AI governance, data privacy, and leadership accountability. This Lens ensures transformation is principled, not just efficient. In the AI era, a Moral Compass is the boundary condition for legitimacy—without it, innovation loses its direction; with it, trust becomes an unbreakable competitive advantage.

Together, these Six Lenses form the DEPTH of the Ai6x6 Transformation Blueprint. They ensure that every Pillar—Leadership, Workforce, Data, Customer, Product, and Operations—is approached with rigor, coherence, and integrity. They transform transformation from a collection of projects into a complete operating philosophy.

SIX LENSES OF TRANSFORMATION : Human Body Analogy

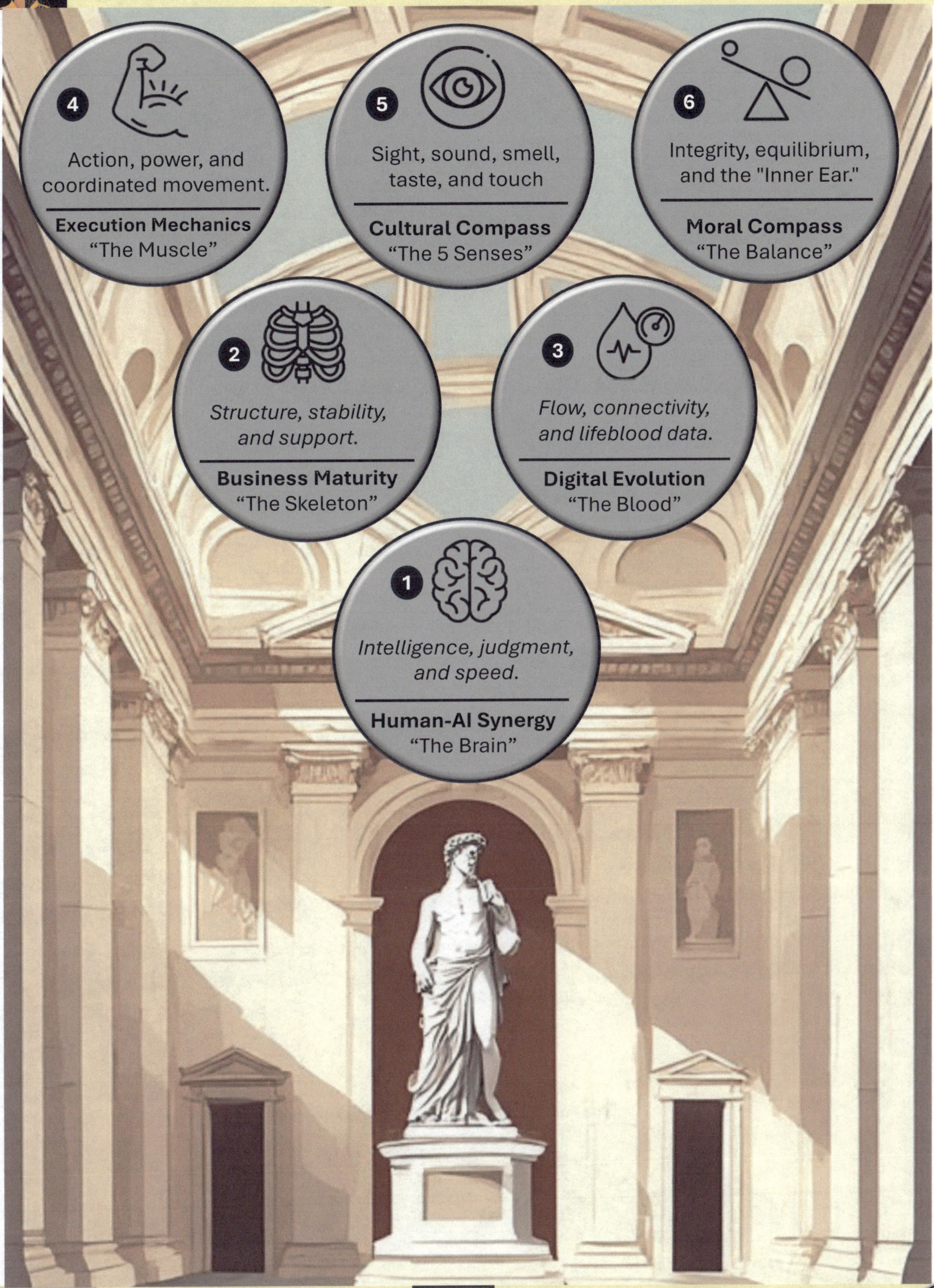

4
Gears that Convert energy power to motion
Execution Mechanics
"The Engine"
5
Monitor car health, temperature, audio
Cultural Compass
"The Dashboard"
6
Alignment, stability and stay in 'right lane'
Moral Compass
"The Steering"
2
Structural Frame and body of the car
Business Maturity
"The Chassis"
3
Gasoline, Electric energy & battery
Digital Evolution
"The Fuel"
1
Judgement, GPS, decision-making
Human-AI Synergy
"The Driver"

THE ETHICAL IMPLOSION (*Invested: Digital evolution | Overlooked: Moral compass*): You build a high-speed AI infrastructure with zero ethical guardrails or governance. One biased algorithm creates a public relations nightmare and a massive legal liability that can bankrupt the firm.

THE BUREAUCRATIC FREEZE (*Invested: Business maturity | Overlooked: Cultural compass*): You have established perfect governance and checklists, but you are operating within a terrified culture. People follow the rigid "process" of transformation but stop taking risks, effectively killing all real innovation.

THE THEORETICAL DREAM (*Invested: Human–AI synergy | Overlooked: Execution mechanics*): The organizational "brain" develops brilliant ideas for AI collaboration, but there is no "muscle" to act on them. Innovation stays trapped as a pilot project and never scales to produce enterprise value.

THE AUTOMATED ERROR (*Invested: Execution mechanics | Overlooked: Human–AI synergy*): You build a powerful execution engine but ignore the necessity of human judgment and intervention. The system begins executing "bad" decisions or flawed logic at 100x speed without anyone knowing how to stop it.

THE FUNDED HOBBY (*Invested: Cultural compass | Overlooked: Business maturity*): The internal culture is happy, safe, and collaborative, but there is no structural discipline or funding prioritization. Massive amounts of energy and capital are wasted on uncoordinated "passion projects" that never deliver ROI.

THE PRINCIPLED STANDSTILL (*Invested: Moral compass | Overlooked: Digital evolution*): You have world-class ethics and deep integrity, but you lack the "blood flow" of modern data and systems. You remain a "good" organization, but your technology is too outdated to actually compete with faster rivals.

THE FERRARI IN THE GARAGE (*Invested: Digital evolution | Overlooked: Execution mechanics*): You invest millions in an elite, high-performance tech stack that stays idle because no one has built the workflows to use it. You have purchased the "Engine" but never bothered to build the "Transmission" to turn the wheels.

THE LEGALISTIC LOOPHOLE (*Invested: Business maturity | Overlooked: Moral compass*): You strictly follow every legal rule and regulation while completely violating the spirit of ethics and trust. Your "Governance" is perfect on paper, but your brand is dying because your customers feel exploited by your "technically legal" actions.

THE INTELLIGENT OUTCAST (*Invested: Human–AI synergy | Overlooked: Cultural compass*): You develop AI tools that are technically brilliant and offer high synergy, but the workforce finds them creepy or intrusive. Because you ignored the "Senses" of the culture, the tools are rejected by the very people they were meant to help.

THE EFFICIENT COLLAPSE (*Invested: Execution mechanics | Overlooked: Business maturity*): The organization is moving with incredible power and speed, but it is moving toward a goal that isn't funded or strategically prioritized. You are winning the race, but you've realized too late that you're on the wrong track entirely.

COLLISION PATHS TO AVOID IN YOUR TRANSFORMATION

THE NICE LIAR (*Invested: Cultural compass | Overlooked: Moral compass*): The culture is polite, collaborative, and highly "engaged," but it is built on a foundation of hidden ethical skeletons. People are "happy" because they are protecting each other from the truth, leading to an eventual systemic implosion.

THE ANALOG SAINT (*Invested: Moral compass | Overlooked: Human–AI synergy*): The leadership has high integrity but lacks the "AI-IQ" to understand the risks of the modern digital landscape. They are making ethical decisions based on a world that no longer exists, leaving them vulnerable to new forms of disruption.

THE EXPENSIVE CHAOS (*Invested: Digital evolution | Overlooked: Business maturity*): You layer high-tech AI tools on top of disorganized, immature governance and disorganized funding. The tech doesn't solve your problems; it simply makes your existing chaos move much faster and cost much more.

THE STAGNANT FORTRESS (*Invested: Business maturity | Overlooked: Execution mechanics*): The organization has perfect defenses, massive funding reserves, and rigid discipline, but zero kinetic movement. You are a perfectly "safe" business that is being slowly surrounded and suffocated by smaller, faster competitors.

THE DR. FRANKENSTEIN (*Invested: Human–AI synergy | Overlooked: Moral compass*): You achieve brilliant "Brain-to-Machine" innovation that pushes the boundaries of what is possible. However, because you ignored the "Inner Ear" of balance and ethics, you create a system that causes more social harm than organizational good.

THE BURNOUT FACTORY (*Invested: Execution mechanics | Overlooked: Cultural compass*): You optimize your "Muscle" for maximum output but completely ignore your "Senses" and emotional climate. You fail to detect internal fatigue and resentment until your highest-performing talent quits en masse.

THE LUDDITE OASIS (*Invested: Cultural compass | Overlooked: Digital evolution*): You have created the best place in the world to work, with high safety and great vibes. However, because you ignored the "Blood Flow" of digital evolution, your company is being disrupted out of existence by competitors you can't even see.

THE TOOTHLESS POLICY (*Invested: Moral compass | Overlooked: Execution mechanics*): You have beautiful "Values" posters and world-class ethics policies, but no actual workflows or mechanics to enforce them. Your integrity is a theory that falls apart the moment a team is pressured to hit a quarterly target.

THE GHOST SHIP (*Invested: Digital evolution | Overlooked: Human–AI synergy*): You have successfully implemented full automation and "lights out" data systems across the enterprise. However, because there is no human judgment in the loop, the system continues to run on outdated logic that is driving the company into a ditch.

THE PRECISE DINOSAUR (*Invested: Business maturity | Overlooked: Human–AI synergy*): You have spent years optimizing and documenting every single process to a level of "Six Sigma" perfection. Unfortunately, you ignored the "Brain" of AI synergy, and those processes are now perfectly optimized for a market that has already moved on.

Human judgment amplified by AI

Human–AI Synergy defines how intelligence is created, shared, and amplified across the enterprise. It examines the evolving partnership between human judgment and machine capability, ensuring that AI elevates rather than replaces human potential. This Lens forces leaders to confront how decisions are made, how insights flow, and how people interact with intelligent systems. It exposes whether the organization is building a truly augmented workforce or simply layering tools on top of old thinking. Ignoring this Lens leads to fragmented automation, poor adoption, and inconsistent decision-making. Human–AI Synergy is essential because it establishes the cognitive foundation of transformation—how the enterprise thinks, learns, and adapts—and without it, AI becomes a disconnected tool rather than a strategic multiplier.

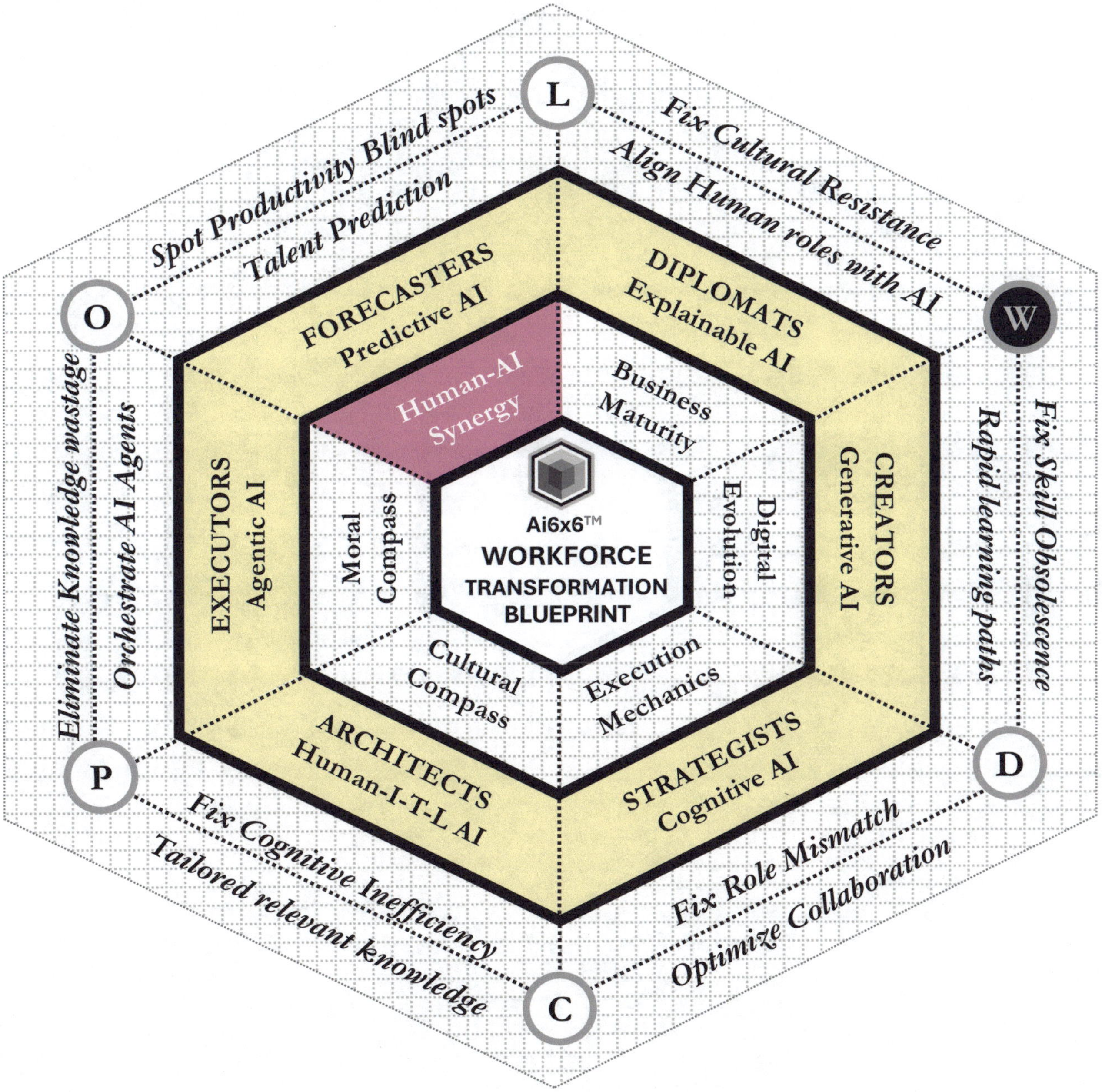

Ai6x6 HUMAN-AI SYNERGY FRAMEWORK FOR **WORKFORCE LIFECYCLE**

Discipline that stabilizes transformation

Business Maturity is the structural discipline that transforms raw ambition into a legacy of Authority. It begins at the Foundation, establishing the governance and funding models that anchor an organization's identity before activation begins. As the enterprise moves through Initiation and Cohesion, this Lens evaluates whether the internal scaffolding is robust enough to support the increasing velocity of the transformation engine. This Lens exposes a fundamental truth: sustained success is engineered through systems, not slogans. It ensures that as Momentum builds, the enterprise maintains the Optimization necessary to prevent collapse under the weight of its own initiatives. Mastering this Lens allows an organization to transcend temporary programs and achieve the permanent market Authority that defines industry leadership.

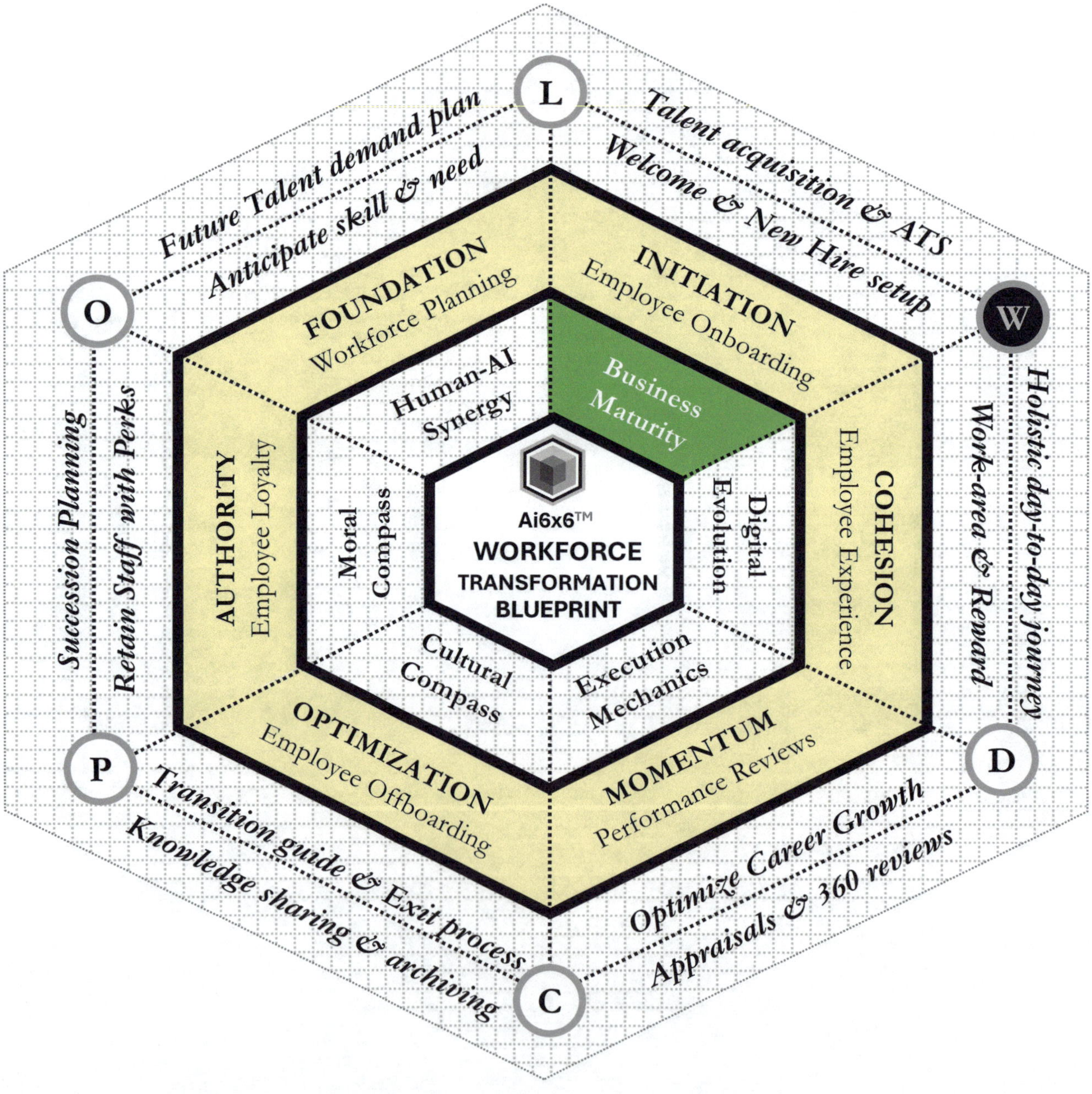

Ai6x6 BUSINESS MATURITY FRAMEWORK FOR **WORKFORCE LIFECYCLE**

Infrastructure built for continuous change

Digital Evolution examines the technological backbone that enables modern transformation. It evaluates system integration, data flow, platform scalability, and the adaptability of digital environments. This Lens ensures technology accelerates rather than obstructs progress. It forces organizations to confront technical debt, architectural fragility, and the true readiness of their digital ecosystem. When ignored, enterprises build brittle systems, slow innovation, and create environments where strategy moves faster than infrastructure. Digital Evolution is not about tools—it is about constructing a digital foundation capable of continuous change. It ensures that technology becomes an enabler of intelligence, speed, and resilience. Without this Lens, even the strongest strategies fail because the underlying environment cannot support enterprise-wide evolution.

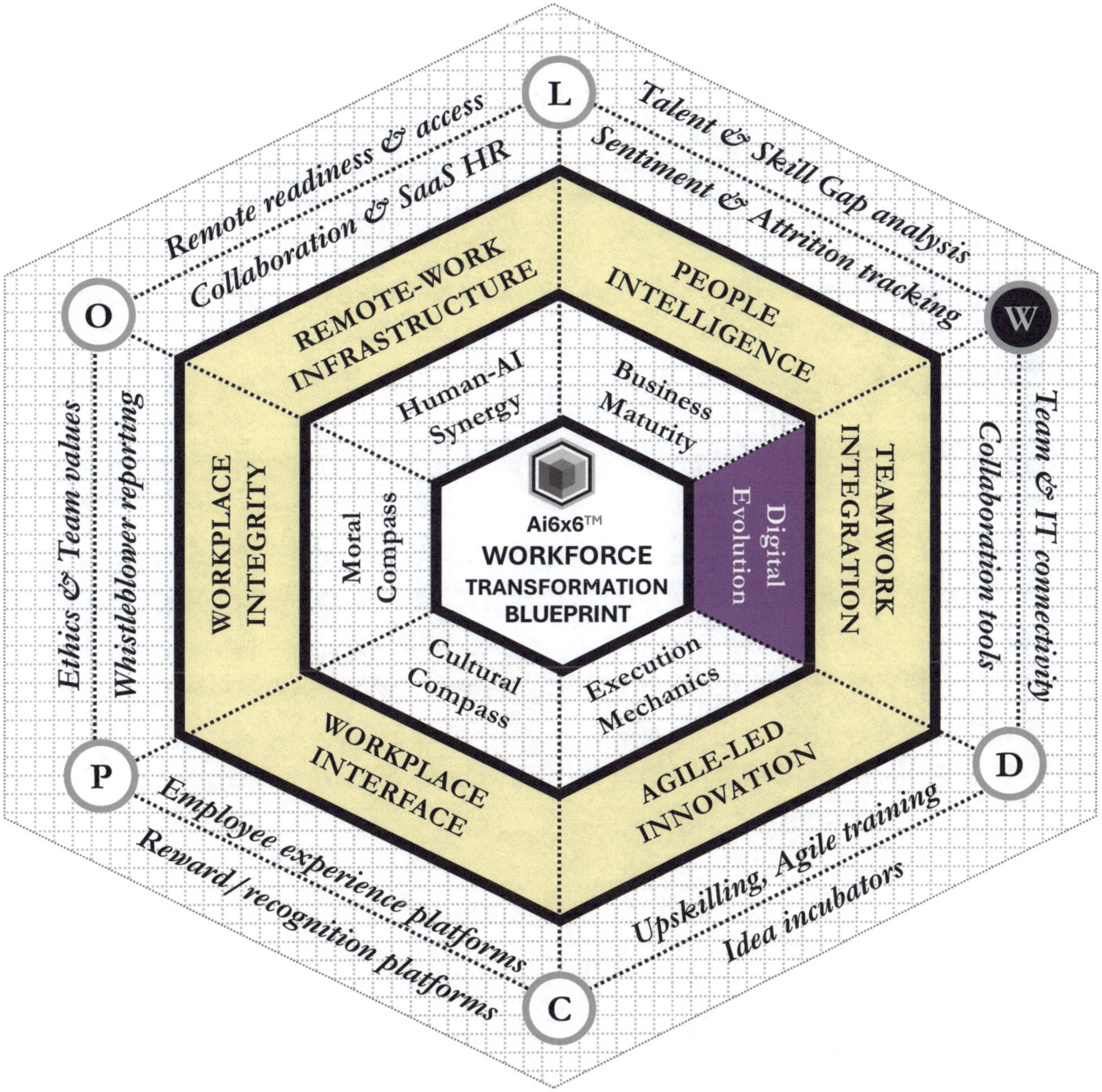

Ai6x6 DIGITAL EVOLUTION FRAMEWORK FOR **WORKFORCE LIFECYCLE**

Strategy delivered with operational precision

Execution Mechanics converts strategy into consistent, scalable action. It examines operating models, workflows, methodologies, and performance systems to determine whether the organization can deliver reliably. This Lens exposes the gap between leadership intent and frontline reality, revealing whether execution is engineered or improvised. It forces leaders to confront bottlenecks, handoff failures, and the true velocity of delivery. Ignoring this Lens results in missed deadlines, inconsistent outcomes, and organizational frustration. Execution Mechanics is essential because it transforms ideas into outcomes and outcomes into advantage. It ensures that transformation is not theoretical but operational—a system of delivery that produces measurable impact at scale.

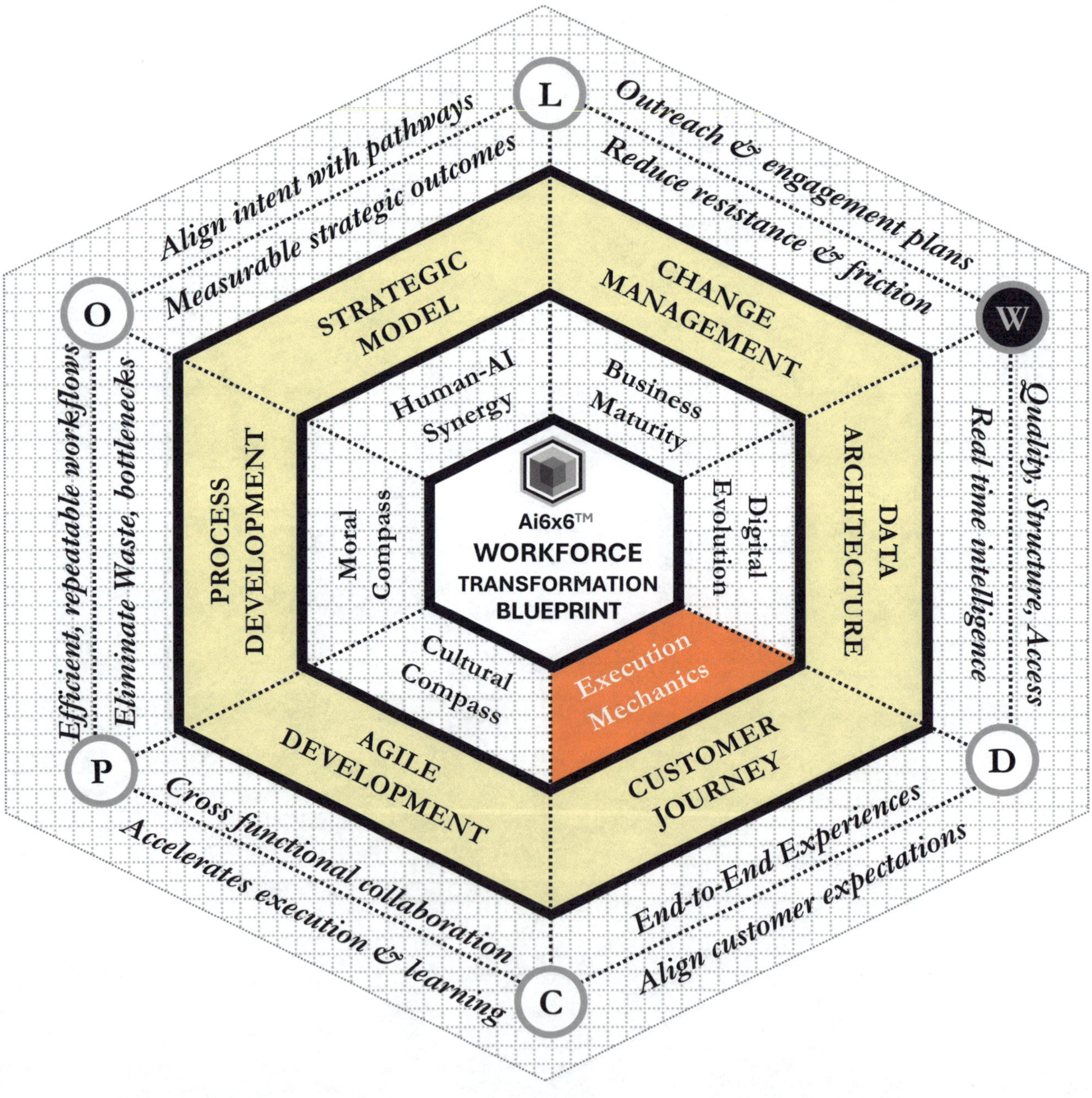

Ai6x6 EXECUTION MECHANICS FRAMEWORK FOR **WORKFORCE LIFECYCLE**

Culture aligned to accelerate change

Cultural Compass explores the beliefs, behaviors, and shared meaning that shape how people think, collaborate, and respond to change. It examines psychological safety, communication norms, leadership behaviors, and emotional climate. This Lens reveals whether culture accelerates transformation or quietly resists it. It forces leaders to confront the invisible forces that determine engagement, trust, and momentum. Ignoring this Lens means ignoring the human system that ultimately determines success. Even the best strategies fail in cultures that are misaligned, fearful, or fatigued. Cultural Compass is essential because transformation is a human endeavor—no system or technology can succeed without cultural readiness. It ensures people are not just compliant but committed, creating the conditions for lasting change.

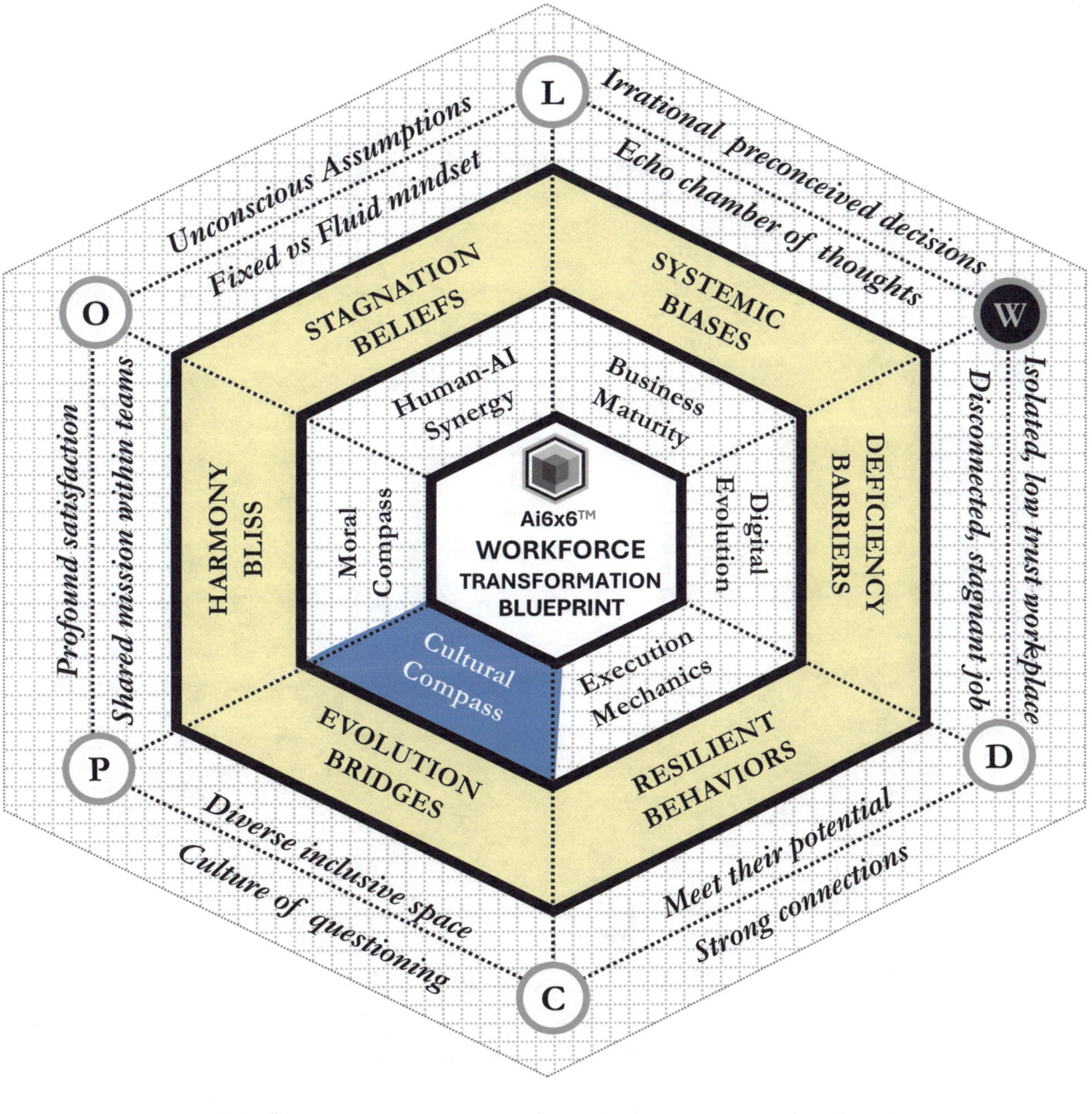

Ai6x6 CULTURAL COMPASS FRAMEWORK FOR **WORKFORCE LIFECYCLE**

Ethics governing intelligent enterprise systems

The **Moral Compass** anchors transformation in ethics, transparency, and responsibility. It evaluates data privacy, AI fairness, accountability structures, and the moral obligations of leadership. This Lens ensures that innovation is principled and that stakeholders feel protected, respected, and informed. It forces leaders to confront the ethical implications of their decisions and the long-term consequences of their systems. Ignoring this Lens exposes organizations to reputational damage, regulatory risk, employee resistance, and customer disengagement. Trust is the boundary condition of modern enterprise—without it, nothing else holds. The Moral Compass is essential because it defines the legitimacy of every decision, system, and outcome, ensuring transformation is not only effective but worthy of confidence.

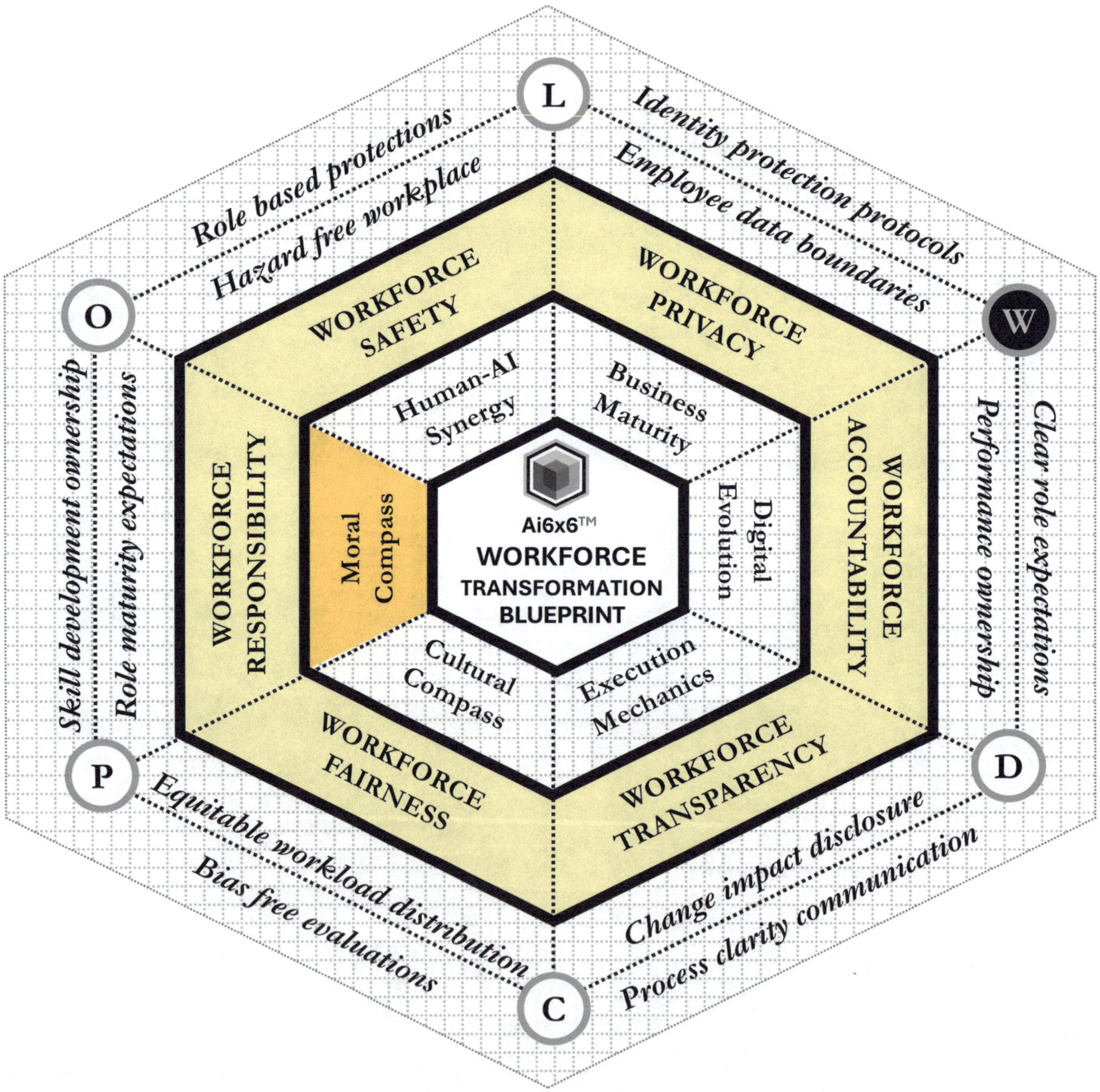

Ai6x6 Moral Compass FRAMEWORK FOR **WORKFORCE LIFECYCLE**

Ai6x6™ Transformation Blueprint

Why Transform Workforce?

"The future belongs to those who learn more skills and combine them in creative ways." —Robert Greene

Robert Greene is an American author known for his books on strategy, power, and human behavior, including The 48 Laws of Power.

The global economy is being reshaped by automation and AI, making many traditional job roles obsolete. Workforce transformation is no longer optional; it's a matter of survival for both individuals and companies. Without a proactive strategy to upskill and reskill employees, businesses will face a critical talent gap, unable to compete. This transformation is necessary to create a dynamic workforce equipped with digital literacy, critical thinking, and emotional intelligence, ensuring that people are seen as a strategic asset for the knowledge economy. The benefit of this transformation is a more agile, adaptable, and innovative team capable of navigating future challenges.

1. **Increased Employee Autonomy and Mastery:** By moving to cross-functional teams and granting outcome ownership, employees achieve a higher level of job satisfaction and can develop mastery in specialized judgment, focusing their efforts on complex, human-centric problems. This self-direction leads directly to higher quality outputs, as the people closest to the problem are empowered to implement the best solutions. High autonomy is directly correlated with higher innovation scores and better product quality.
2. **Targeted Skill-Gap Closure:** Proactive internal assessments identify future-critical skill shortages, which are addressed through large-scale, continuous upskilling, mentorship, and rotation programs, ensuring the organization's capability keeps pace with technology evolution. This strategic investment dramatically reduces the high cost and disruption associated with relying solely on external hiring for scarce digital skills. By developing talent internally, the organization creates a deep bench of culture-fit experts.
3. **Higher Employee Engagement:** The automation of routine, repetitive toil frees up staff to focus on creative, strategic, and high-value decision-making, leading to a significant decrease in burnout and a measurable increase in employee engagement scores. Elevating human labor to focus on specialized problem-solving maximizes the cognitive capital of the workforce, directly correlating with superior innovation. The workforce becomes less of a maintenance crew and more of a creative engine.
4. **Creation of T-Shaped Skill Profiles:** The workforce develops deep expertise (vertical) combined with a broad understanding (horizontal) of adjacent domains like cloud security, data governance, and design thinking, making collaboration more effective. These versatile skill sets reduce the need for constant, formal handoffs between departments, accelerating end-to-end delivery speed. Teams can address a wider range of problems internally without creating dependencies on external silos.
5. **Shift to Outcome-Based Work:** Measuring teams by concrete business outcomes (e.g., lowered cost per transaction, improved customer resolution time) rather than time spent aligns all labor directly with value creation, maximizing productivity. The focus on results encourages teams to find the most efficient pathways to success, often involving the automation of their own internal processes. Work becomes a direct pursuit of customer value, removing unproductive activity.
6. **Reduced Institutional Knowledge Silos:** Mandating knowledge codification and fostering cross-functional teams ensures that critical institutional knowledge is documented, shared, and decoupled from single individuals, mitigating key-person risk. The ease of knowledge transfer and access streamlines the onboarding process for new hires and allows for immediate contribution. This democratization of knowledge hardens against staff changes.
7. **Increased Velocity of Internal Mobility:** Standardized, clear role expectations and accessible training pathways enable employees to easily and quickly move to areas of strategic need, making resource allocation highly flexible and responsive. This internal talent market prevents high-priority projects from being stalled due to external hiring delays and improves retention through career pathing. Resource alignment happens based on business priorities.
8. **Embedded Equity and Inclusion in Process:** Workforce data is used to actively monitor and improve equity in workload distribution, project assignments, and promotion pathways, ensuring the transformation benefits are shared and fostering a truly inclusive culture. A demonstrable commitment to fairness increases overall team performance and taps into a broader diversity of problem-solving perspectives. This commitment maximizes the potential of all employees and improves ethical standing.
9. **Sustainable Workload Management:** By automating operational toil (runbooks, provisioning), the workload burden on specialized engineers is significantly reduced, leading to better work-life balance and a lower incidence of human-factor errors. Reducing the constant pressure of urgent, low-value work allows engineers to dedicate capacity to preventative maintenance and innovative projects. Staff burnout is reduced, leading to consistent performance over the long term.
10. **Enhanced Talent Acquisition Brand:** Demonstrating a commitment to modern methodologies, continuous learning, and outcome-based work makes the organization a highly attractive destination for top digital and engineering talent in a competitive market. A strong, modern brand presence in the tech community drastically reduces recruitment costs and improves the quality of candidates attracted. The talent pipeline becomes self-sustaining and high-quality.

1. Hybrid Work & Inclusion — The Proximity Bias Awakening

Maya hit every deliverable, often ahead of schedule. But in promotion conversations, her name rarely came up. "We just don't see her leadership presence," someone said—never mind that she was leading global workflows from her home office. When Leo, a charismatic in-office manager, got fast-tracked, Maya didn't complain—but a colleague did. "You promoted the guy who grabs hallway chats, not the one solving problems." That stung. Leadership realized they'd been rewarding visibility over value. Maya's story became the catalyst: they introduced meeting equity frameworks, spotlighted async wins, and tracked airtime gaps in hybrid meetings. Months later, a dashboard replaced gut feel. Promotions shifted. So did inclusion.

In a hybrid world, excellence isn't always visible—but it's always valuable. Real workplace transformation begins when we acknowledge the systems that make some contributions seen—and others invisible.

2. Wellbeing & Performance — Burning Out the Brightest

Olivia was legendary—always reliable, high-output, low-maintenance. So when she didn't log in one Monday, no one knew who to call. She'd collapsed after a weekend of all-nighters prepping for back-to-back launches. "She never looked stressed," her manager said—but stress had become her default. HR found she hadn't taken a day off in ten months. And she wasn't alone. Four other top performers said they felt constantly exhausted, afraid to disappoint. The company realized they'd built a culture that rewarded quiet self-sacrifice over sustainable excellence. Performance reviews were restructured to value team pacing, recovery, and coaching. Olivia later helped create a resilience toolkit for high-achievers. Now when someone finishes a major milestone, they're offered rest—not more responsibility.

Burnout wasn't a failure of individuals—it was a flaw in the system. If high performance comes at the cost of health, it's not performance. It's leakage.

3. Meeting Culture Reset — Escaping Calendar Chaos

Jared's day was a blur of blinking rectangles: "Check-in," "Team Sync," "Weekly Touchpoint"—not one with a clear agenda. When his 8-year-old asked, "What do you do at work?" Jared stared at his calendar and muttered, "I go to meetings... about meetings." He wasn't alone. His team clocked over 20 hours of calls per week, yet felt more out of sync than ever. So Jared hit delete—literally. He canceled 50% of meetings for a month and rebuilt from scratch. Every remaining slot needed a purpose, outcome, and pre-read. They shifted status updates to dashboards, decisions to structured huddles, and brainstorms to async whiteboards. At first, people panicked. Then—clarity. Creativity. Time to think again. Morale climbed. People stopped mistaking activity for progress. The experiment became policy.

In transformed workplaces, time isn't free—it's a shared asset. And reclaiming it might be the boldest culture act of all.

4. Psychological Safety — The Fear in Silence

During an all-hands, a junior engineer flagged a potential flaw in the product roadmap. The room went silent. No one responded. The next day, she was quietly removed from the project. Word spread. Weeks later, that flaw triggered a public incident—and suddenly, people remembered her name. Leadership scrambled to fix the product, but they missed the real bug: fear. Innovation had stalled not because of skill gaps, but silence. The CTO launched listening sessions, hired external facilitators, and began rewarding—not punishing—constructive dissent. "If you disagree, say so" became a visible mantra on every presentation slide. Within months, product quality rebounded, but more importantly—so did engagement.

Brave teams don't emerge from slogans. They emerge from leaders who reward truth more than comfort. Psychological safety isn't a vibe—it's a variable. And it determines whether your culture builds breakthroughs or buries them.

5. Skills Over Titles — The Internal Mobility Breakthrough

Tariq was stuck. Not because he lacked drive, but because HR couldn't find a "fit" for his next step. He was a business analyst—period. But he'd built automation scripts, facilitated design thinking workshops, and coached interns. Still, job ladders favored title over talent. When a product manager left mid-project, Tariq asked to fill in. "You're not qualified," they said. But after a gentle push from his skip-level leader, he took the role—and thrived. That experiment sparked a broader shift. The company began using skills graphs instead of job ladders, and internal gigs opened based on capability, not seniority.

Employees were seen as multi-dimensional—not monochrome titles. Internal mobility doubled in a year. People felt seen, not slotted. In the modern workplace, transformation doesn't mean hiring more—it means seeing more in who you already have.

6. From Diversity to Belonging — More Than a Metric

Nina's company had hit its DEI targets—but still felt... sterile. The numbers looked better, but the stories didn't. At a team dinner, a new hire whispered, "Sometimes I feel like I was hired just to check a box." Nina felt gutted—and motivated. She convened listening circles, not surveys. She made story-sharing part of onboarding. She questioned who felt heard—not just who was present. When affinity groups requested budget and visibility, leadership said yes—and showed up. Slowly, "diversity" stopped being a banner and became a behavior. Employees stopped editing themselves at work.

Retention climbed. So did joy. Because inclusion isn't just counted—it's felt. And belonging isn't built in a quarter. It's built every day we choose presence over optics.

7. Deskless Enablement — The Invisible Workforce

Marco was a top warehouse manager—but you'd never know it. HR announcements didn't reach him. Leadership all-hands were virtual—but he didn't have a laptop. When software updates rolled out, floor workers found out weeks later, often from each other. The office called it "communication lag." Marco called it disrespect. Eventually, a frontline team built a feedback loop using WhatsApp and printed pulse surveys. They shared field photos, co-designed solutions, and even won a company-wide innovation award. Headquarters was stunned. "You don't need a desk to build the future," Marco said at the next exec meeting—his first. The company invested in mobile-first tools, manager toolkits, and quarterly frontline town halls. Trust rebounded. So did quality and retention.

Because the frontline isn't your afterthought—it's your operating system.

8. Knowledge Flow — Unblocking the Bottlenecks

At Rachel's fintech startup, product iterations slowed—not because people weren't trying, but because knowledge kept getting trapped. Dashboards lived in silos. Project recaps disappeared into inboxes. Slack channels multiplied like rabbits. Every week felt like déjà vu. "Didn't we try this already?" someone would say. So Rachel proposed something radical: a "working brain." A lightweight platform where updates were open, retros transparent, and wins documented. They called it the Flow Center. Suddenly, product cycles sped up. New hires ramped faster. Repeats dropped. When a client crisis hit, teams didn't scramble—they searched, learned, and responded. Knowledge didn't just flow—it compounded.

In fast-paced companies, intelligence isn't a static asset—it's how fast it travels. And when knowledge gets stuck, so does innovation.

9. Rituals in Remote — Connecting Without Proximity

During her first week, remote hire Daria felt invisible. Everyone was kind—but she didn't feel included. One day, a teammate sent her a GIF reacting to a comment—and it opened a thread of banter, laughter, and warmth. That moment became a spark. The team created playful rituals: emoji roll calls, "Friday Fails," and silent co-working hours. None of it was mandated—but all of it mattered. Daria later said, "That GIF wasn't a joke—it was a door." In distributed teams, connection doesn't happen by accident. It's built through intentional micro-moments.

Rituals are the glue of culture when walls disappear. They signal belonging, permission, and presence. And sometimes, transformation doesn't need a strategy deck—it just needs a shared smile at the right time.

10. Career Growth Reimagined — Beyond the Ladder

Tom loved his job but felt stuck. His next "promotion" involved managing people—but he didn't want that. "Where else can I grow?" he asked. HR shrugged. That was the moment of reckoning. Not every path leads upward—but every role can lead deeper. With his manager's support, Tom created a technical mentorship track. He started shadowing strategy meetings, writing knowledge articles, and presenting at conferences. Colleagues soon followed suit. HR formalized multiple growth tracks—technical, strategic, creative. Internal job architecture shifted from ladders to lattices. People began designing careers, not chasing titles. And Tom? He stayed—and soared.

Because talent doesn't grow in one direction. It grows where it's seen, stretched, and supported.

11. The Office Reimagined — Function, Not Furniture

When the new HQ opened, it had nap pods, kombucha taps, and ambient playlists. But usage was abysmal. Teams still preferred home. That's when COO Miguel asked, "What problem is the office solving?" Silence. They'd designed for aesthetics, not utility. So they reframed the space—not as a destination, but a tool. Teams co-designed zones: one for deep work, one for cross-functional jams, another for mental recharge. Meeting rooms became decision rooms. Whiteboards weren't decor—they were collaboration canvases. Weekly "reason to come in" rituals emerged: design sprints, pitch reviews, learning labs. The result? Traffic rose, but purpose rose faster. No one came in just to be seen—they came to build. The transformation wasn't about perks. It was about making the workplace worth the commute.

Because in a hybrid world, space must serve behavior—not nostalgia. The best office isn't cool. It's catalytic.

12. Culture Experiments — Micro-Transformations That Stick

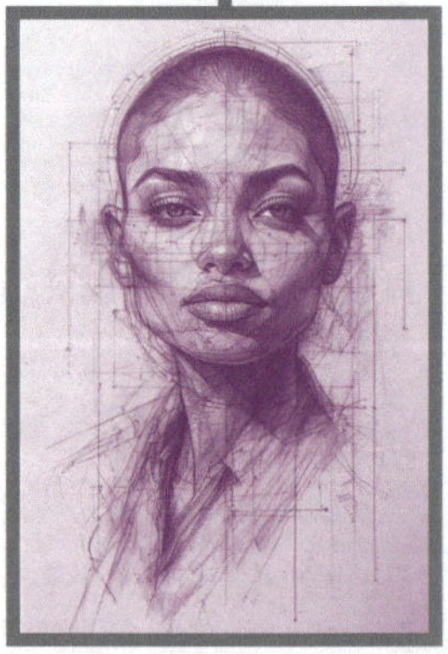

At an all-hands, a quiet new hire named Lina asked, "If we're so big on learning, why don't we share our mistakes?" The room paused. A week later, someone posted their "Friday Fail" in the company chat. Another followed. Then a third. Within months, it became a weekly ritual: public missteps turned into team learning—and laughter. Performance didn't suffer. Psychological safety soared. No memo required. Lina's question had sparked a micro-transformation. Leaders began tracking not just KPIs, but cultural signals: who spoke up, who shared, what rituals took root. They realized culture doesn't shift through annual summits—it evolves through daily choices. Micro-behaviors. Local experiments. Small, bold acts that catch fire.

Culture isn't a program. It's a pattern. And transformation doesn't always need a reorg—it sometimes starts with one person asking a better question.

Ai6x6™ Transformation Blueprint

Power of Diversified Insight

"AI is one of the most profound things we're working on as humanity. It's more profound than fire or electricity."

Sundar Pichai, CEO of Alphabet and Google. Pichai stresses the epochal scale of the current transformation, implying that companies must view AI not as a feature upgrade, but as a foundational, civilization-level shift that will restructure every industry and value chain.

We have deliberately chosen to move beyond the limiting confines of the conventional single-expert **Foreword**, which, by its nature, risks restricting the reader to one singular point of view. In an era where **business transformation** is shaped by a confluence of disruptive forces, relying on a solitary oracle is fundamentally inadequate. Instead, this book opens with a **curated collection of 15 interview snippets** from a diverse array of global thought leaders. This method immediately immerses you in a rich, multi-faceted tapestry of opinions and real-world experiences. By foregrounding this **plurality of perspective**, we offer a wider, more nuanced, and dynamic lens through which to view the challenges and enormous opportunities presented by modern change. Consider these short, powerful excerpts a **Chorus of Counsel**: a comprehensive and inclusive foundation built on collective wisdom that better equips you for your own complex journey of enterprise innovation and transformation. This is a commitment to **diversified insight** from the very first page.

DIVERSIFIED EXPERT VOICES

Gain unprecedented, insider access to the minds of 27 global thought leaders in this six-book series. Sourced from comprehensive 1-on-1 interviews, these experts are recognized worldwide for their work in business transformation and emerging technology. Their insights are drawn from my personal, professional network—including former bosses, colleagues, and business partners — giving you an insider's view into the minds of industry shifters. To maximize focus and utility, we've synthesized these deep-dive conversations into thematic 'interview cards.' Each of the six books offers 15 relevant, concentrated insights on specific topics, allowing you to quickly access and apply the most relevant expertise, no matter where it originated in the original interview.

EXPERT VOICES

Subrata Mukherjee: How do you see emerging technologies like Spatial/XR/Immersive experiences, and the Space Economy changing the way we work and do business?

Richard Entrup: Merging technologies like Spatial/XR (Extended Reality, encompassing VR/AR/MR) and the Space Economy are fundamentally reshaping how we work and conduct business by enabling new forms of collaboration, customer engagement, and operational efficiency. Spatial/XR technologies are revolutionizing workplaces by creating immersive environments for remote collaboration—think virtual boardrooms where global teams interact as avatars. The Space Economy, meanwhile, unlocks unprecedented data and connectivity opportunities via satellite constellations (e.g., Starlink's low-latency broadband) and space-derived analytics, enabling real-time supply chain optimization or precision agriculture. These technologies shift work toward hybrid, data-driven models, with employees leveraging XR for design prototyping or space-based IoT for global asset tracking, while businesses gain competitive edges through hyper-personalized customer experiences and scalable, resilient operations.

SM: What do you believe will be the biggest technological challenge for businesses to overcome in the next decade?

RE: As my current leader often says: " The next 5 years may see more change than the last 30." I believe the biggest technological challenge businesses will face over the next decade is achieving scalable, secure, and ethical integration of AI-driven automation and quantum computing into core operations while navigating regulatory and talent constraints. AI's rapid evolution—particularly generative AI and quantum-enhanced algorithms—promises transformative gains but scaling these technologies requires overcoming fragmented data ecosystems, ensuring robust cybersecurity (e.g., against AI-augmented supply chain attacks), and addressing ethical concerns like bias or transparency, especially as regulations like the EU AI Act tighten. Businesses must invest now in interoperable platforms, crypto-agility, and upskilling to avoid falling behind, with failure to adapt potentially costing 30-40% market share in disrupted industries.

SM: How can a leader ensure that they are always at the "frontier" and not just reacting to what's already here?

RE: I think that's the Innovator's Dilemma? True disruption at the edge vs. innovation at the core. Some innovations are new business models and some may just build upon existing frameworks, but both provide benefits of varying scale. To stay at the innovation frontier rather than reacting to existing trends, leaders must adopt a proactive, ecosystem-driven approach.

First, they should build a foresight engine: establish dedicated innovation labs or cross-functional teams to scan emerging technologies (e.g., XR, quantum, space tech) and run 6-12 month pilots, measuring outcomes like 10% efficiency gains or new revenue streams. Partnering with startups, academia, or consortia can also provide early access to breakthroughs.

Second, embed a culture of experimentation: allocate 10-15% of IT budgets to high-risk, high-reward projects, using metrics like prototype success rates to track progress, and foster psychological safety so teams can iterate without fear of failure (e.g., Amazon's "two-pizza teams" model).

Third, leverage data-driven agility: invest in real-time analytics platforms to anticipate market shifts—e.g., using space-based IoT data for demand forecasting—and empower rapid decision-making.

Managing Director, Head of Emerging Solutions at KPMG US

RICHARD ENTRUP

A multi-award-winning global technology and innovation leader with nearly three decades of experience driving strategic business transformation. He is at the forefront of identifying and incubating disruptive technologies like GenAI, Quantum Computing, and Spatial/XR, pioneering new products and solutions for clients. A two-time recipient of the HMG Global Leadership Institute Award for Exceptional Business Transformation, his distinguished background includes C-suite leadership roles (CIO, CTO, CDO) at globally recognized brands such as Verizon, Disney/ABC, Christie's, Time Warner, Viacom, and Tiffany & Company. He also serves as a strategic advisor and investor to a portfolio of innovative startups and venture capital firms, passionately guiding ventures focused on cutting-edge innovations. He is based out of New York, United States.

EXPERT VOICES

Subrata Mukherjee: You have a proven track record of building and scaling technology teams, from one engineer to over 100. What is your philosophy for building a high-performing technology team from scratch?

Chris Pena: Start with a clear, compelling mission and hire for curiosity, collaboration, and resilience as much as technical expertise. I believe in building diverse teams that challenge assumptions and bring varied perspectives to complex problem-solving. Organizational structure matters less than culture. When people feel genuinely empowered to innovate and learn from failure, high performance follows naturally. The foundation is trust, accountability, and shared purpose. These elements act as performance multipliers across the entire team.

SM: How do you foster a culture of innovation and continuous improvement within a globally distributed team?

CP: The foundation is psychological safety and establishing a shared innovation language across all geographies. This means encouraging experimentation, celebrating learning from failures, and ensuring feedback loops are rapid and transparent. Regular innovation forums, hackathons, and structured knowledge-sharing sessions maintain creative energy and connection. Distributed teams thrive when they feel connected not just to local objectives, but to a global mission that genuinely values creativity and breakthrough thinking.

SM: What is the most important quality you look for when hiring a senior technologist or a leader for your team?

CP: Beyond deep technical expertise, I look for leaders who can translate complexity into clarity for business stakeholders.

The ability to influence, communicate effectively, and build trust across the organization is critical. Senior technologists must demonstrate adaptability. While technologies evolve rapidly, the leaders who consistently succeed are those who can pivot quickly, continuously learn, and inspire their teams to move forward with confidence through uncertainty.

SM: Your profile highlights your expertise in leveraging AI, data, and digital labor. You implemented RPA to cut manual processes by 50% at Aspire Lifestyles. What is a key use case for automation that you believe is still largely underutilized by organizations?

CP: Knowledge work automation represents a massive untapped opportunity, specifically automating data enrichment, content tagging, and compliance reporting processes. While many organizations focus on transactional task automation, significant value lies in automating repetitive knowledge tasks that consume skilled employees' time.

Freeing knowledge workers to focus on creative problem-solving and strategic initiatives represents the next frontier for automation and organizational productivity.

SM: How do you see the relationship between AI and human workers evolving over the next decade?

CP: AI will become a true collaborative partner rather than a replacement threat. AI systems will handle repetitive, data-intensive work while humans focus on judgment, creativity, and empathy-driven tasks that require human insight. The winning organizations will treat AI as workforce augmentation, investing significantly in reskilling programs to help employees thrive in this hybrid work environment. Trust and transparency in AI deployment will ultimately define the future of work relationships.

CHRIS PEÑA

VP, CIO Go To Market Industry Lead at Salesforce | Ex-CIO Aspire

Chris is a strategic technology leader and CIO advisor with a deep expertise in AI, digital transformation, and e-commerce. In his current role as VP, CIO Go To Market Industry Lead at Salesforce, he is responsible for developing and executing the company's value proposition for Chief Information Officers across various industries. As Global VP of Consumer Technologies at PVH Corp., he managed the global Salesforce platform. Prior to that, as General Manager CIO at Aspire Lifestyles, he built a global technology team from the ground up, modernizing operations with cloud infrastructure and automation, and successfully managing massive budgets. His extensive experience also includes senior leadership roles at EY, Razorfish, and Pearson, where he led digital strategy, platform implementations, and engineering. He is based out New York in United States.

EXPERT VOICES

Subrata Mukherjee: As the Global Head of Presales at SHL—a leader in talent innovation—and with your recent experience leading an AI Labs unit at intive, this is clearly a key focus for you. How is AI fundamentally transforming the talent and HR industry?

Mark Brincat: AI is fundamentally shifting the talent and HR industry from a reactive, process-driven function to a proactive, predictive one. We are moving beyond simple, static assessments to dynamic models that evaluate not just a candidate's skills, but their potential, adaptability, and cultural fit in real-time. On the talent management side, AI enables personalized learning pathways, predictive retention analytics, and a dynamic view of workforce skills that aligns people growth directly to business outcomes. The entire talent lifecycle is being reinvented; from using AI in sourcing to create more inclusive job descriptions to using chatbots for screening and providing structured interview guides. SHL, for instance, uses science-driven AI in language testing, coding simulations, and on-demand interviews to make evaluations predictive, scalable, and fair. This means talent decisions are no longer based on intuition but are data-driven and aligned with business strategy.

SM: What are the primary challenges companies face when trying to integrate AI into their talent strategies?

MB: The biggest challenges are trust and adoption. Leaders worry about bias, transparency, and compliance, while employees want reassurance that AI is augmenting—not replacing—the human element. A recent BCG study shows 98% of enterprises have experimented with AI, but only 26% have scaled it into production. Why? Because leaders worry about bias, transparency, and explainability, while employees and regulators demand assurance that AI is ethical and not a black box. On the operational side, fragmented HR data and integration into existing ATS/HRIS remain major barriers. Leaders often ask: "Can I explain this decision to a regulator? To a candidate?" Employees worry about fairness. So many global enterprises now have AI councils that look through the technical details.

SM: In your role, how do you advise clients on best leveraging SHL's AI-powered solutions to achieve their business goals?

MB: I emphasize three pillars: fairness, scalability, and business impact. SHL's AI models are built on decades of behavioral science, validated across millions of data points, so they're both predictive and fair. For clients, that means evidence-based decisions they can defend. We work with them to embed solutions where they'll create immediate ROI—whether reducing hiring cycle time, increasing retention, or mapping workforce skills to future strategy. But transformation isn't just about tools—it's about trust. SHL's strategy is to both defend and leverage AI. We defend by ensuring our assessments are scientifically robust, validated against the noise of generative AI, and trusted by customers. We leverage AI responsibly to enhance fidelity, productivity, and candidate experience—never as a "black box." The industry is also facing a huge influx of AI-driven tools, from talent marketplaces to AI-based chatbots. At SHL, our approach is to bring together science, AI, and technology into a single talent intelligence offering that is with you across the entire talent lifecycle.

SM: What's the best career advice you've ever received?

MB: The best piece of career advice I ever received was: "Don't build technology for technology's sake—solve a real business problem." Early in my career, I was reminded that the most elegant architecture or cutting-edge platform is worthless if it doesn't move the needle for customers or the business. That advice still grounds me today: start with the outcome, then work backward to the technology. It's the mindset that turns IT from a cost center into a growth driver, and it's a principle I apply to every project of mine.

MARK BRINCAT

Ex-CTO @ SHL, Economist | Architect & BBC, Telegraph, Home Office

Mark Brincat is a technology executive with deep expertise in digital transformation, AI, and data analytics. In his current role as the Global Head of Presales and Technology Partnerships at SHL, he leads go-to-market strategy and advises clients on their transformation journeys. Concurrently, he serves as a Fractional CTO for Castille Resourcing, where he oversees the engineering and scalability of their digital platform. He previously held key Chief Technology Officer (CTO) roles at both SHL and Intive, and senior architectural positions at major media organizations like The Economist, Telegraph Media Group, and BBC Worldwide. His career highlights include implementing large-scale digital and cloud solutions and forging strategic partnerships. He is based out of London, United Kingdom.

EXPERT VOICES

Subrata Mukherjee: Your work at Cameron and your background as a Mechanical Engineer show a strong foundation in technical problem-solving. How has your experience as an engineer—particularly in developing a patented piece of oil equipment to solve a industry challenge - shaped your approach to product leadership?

Brian Glassman: My efforts in studying and practicing mechanical engineering have instilled a disciplined, process-driven approach into how I apply and practice product management. Engineering taught me to anticipate failure modes, design for edge cases, and approach problem-solving with structured clarity. That mindset now shapes how I lead product efforts. In my experience, success in product development comes not just from intuition or creativity but from following clear frameworks that allow teams to navigate complexity with confidence. While product leaders come from a range of backgrounds including marketing, software, and engineering, I have found that a strong engineering foundation helps me operate with systems-level thinking and a high standard for execution. At the same time, technical skill alone is not enough. Product leadership depends on understanding customer behavior, market movements, and competitive positioning.

SM: What are the key lessons you've learned about leading technical teams and product development in a highly regulated environment?

BG: Leading technical teams in highly regulated environments has taught me that a zero-defect mindset is non-negotiable. These environments do not tolerate error, which means that product development must be built on a foundation of preventative thinking and process precision. My approach has always centered on applying a risk-based mindset—one that continuously tunes and refines operations to detect and resolve issues before they reach the outside world. This mindset becomes especially important when introducing new technologies like AI. Unlike traditional systems, AI can introduce unexpected variation, and its behavior may shift from version to version. That unpredictability, if left unmanaged, can threaten compliance. This is why deployment discipline and ongoing validation are critical when AI is part of the stack. At the same time, it is important to recognize that regulatory oversight is not uniform across all domains. Areas such as human resources, recruiting, or incentive management often operate under lighter regulatory constraints compared to core operational systems.

SM: How important is being versed in AI to the future of product management professions?

BG: Product management has always involved a balance of insight, strategy, and execution. At its core, the discipline requires understanding customer needs and market shifts, generating clear product strategies, and then delivering against those strategies through careful coordination and documentation. AI now plays an increasingly important role across all three of these domains. It can help generate faster and deeper market insights, support the creation of smarter and more adaptive product strategies, and streamline the execution layer by automating routine documentation and communication tasks. I have been writing extensively about this shift because I believe AI is not just augmenting the product management function, it is redefining it. For that reason, I believe every product manager should treat AI as a daily practice, not just a conceptual interest. The profession is moving toward a model where repeatable workflows like competitor scanning, synthesis of customer feedback, or status documentation can be codified, delegated, and executed by AI systems. For example, a task that once involved manually reviewing competitor websites can now be converted into a reference document and handed off to an AI to replicate and update regularly.

DR. BRIAN SCOTT GLASSMAN

Chief Product Officer & Head of Client Engagements at AInspire.ai

Dr. Glassman is a seasoned product management executive, engineer, and product strategist specializing in generative AI. As the co-founder and Chief Product Officer at AInspire.ai, he leads client engagements and product strategies, collaborating with C-level executives to develop and deploy AI solutions for large enterprises. With a career dedicated to leading technology innovation, he previously served as Director of Product Management at Electra Vehicles, where he built the product function for AI-powered solutions in the automotive sector. His extensive experience includes a seven-year tenure at Rave Digital, where he led product management and customer engagements for Fortune 1000 clients. Furthermore, his background as an engineer and product leader at Cameron, a Schlumberger company, highlights a unique expertise in technical product development. He is based in New York.

EXPERT VOICES

Subrata Mukherjee: Your progression at Platform.sh, from Partner Manager to VP of Global Alliances and now to General Manager, is a testament to your expertise. What is your core philosophy on building a successful channel and partner program? How do you take a partnership from an initial handshake to a high-growth channel?

Gauthier Garnier: My core philosophy is rooted in a win-win-win relationship where our company, our partners, and our mutual customers all derive significant value. In the cloud industry, the most important Success KPI is trust; you win when partners and customers trust you with their most precious workloads. This requires more than just a solid product; you must become their best partner by building empathy for their needs, understanding their stage of cloud transformation, anticipating challenges, and offering support in difficult times. Going the extra mile on a challenging deal builds immense goodwill that is often repaid tenfold, whereas forcing a bad deal for short-term gain can leave permanent scars on a relationship. Trust is hard to build, and easy to break.

SM: The DACH region is a key market. What are the unique cultural or business nuances of selling and managing channels there, compared to North America?

GG: The DACH market operates on a foundation of long-term stability, quality, and data sovereignty, where trust is paramount and earned over considerable time. This stands in contrast to the often faster, more transactional pace typical in North America. Partners in the DACH region require deeper technical validation and a clear, pragmatic growth roadmap. Relationships are more formal, built on demonstrated reliability and a profound respect for their established business processes. Consequently, topics like GDPR (DSGVO) and Success there demands patience, precision, and a genuine commitment to the region, including having key people in the territory who speak the language.

SM: How has the role of partnerships and channels changed in the software industry over the past decade?

GG: Partnerships have evolved from a superficial tactic to a strategic necessity. In Europe, they are essential for exploring new business models without the heavy investment seen in the US; less funding means collaboration is critical to accelerate innovation. For a company like Platform.sh, partnerships have fundamentally shaped our company, roadmap, and go-to-market strategy. After signing key partnerships with Orange and Magento, we became a partner-first company. Today, partners are strategic advisors who influence technology choices long before a vendor is involved. Our most advanced partnerships involve co-developing our roadmap, defining pricing, and even creating isolated product versions to satisfy requirements like extreme sovereignty. This depth requires dedicated teams and a shared vision, investment, and belief in the "why" from top management to be successful.

My advice for partners is to focus on using AI to solve their most pressing business challenges to gain speed and efficiency. Instead of chasing trends, they should start with the data they already possess. For instance, many partners struggle to scale their sales teams effectively. I advise them to leverage AI tools, much like the ones we use for automated demos, to analyze incoming leads and automate the initial research and qualification process. This immediately boosts efficiency, freeing up their sales teams to focus on high-value conversations. Another key area is leveraging generative AI to analyze their most successful case studies and customer feedback, allowing them to quickly produce targeted marketing materials that resonate with specific segments and enable their outreach efforts to scale.

GAUTHIER GARNIER

General Manager, DACH & Channels at Platform.sh

Gauthier Garnier is a highly experienced General Manager and business leader with a career focused on building strategic partnerships and long-term, win-win relationships with clients. As General Manager for the DACH (Germany, Austria, Switzerland) region and Channels at Platform.sh, he is responsible for driving growth and managing strategic alliances. His background is deeply rooted in digital transformation and web agency leadership. Previously, as a partner and General Manager at Kaliop, he successfully launched and scaled a web agency in the North American market, overseeing production, human resources, and finances. He has a strong expertise in strategic planning for complex digital projects, particularly within the retail and media industries, and has taught and coached on these topics. He has been recognized for his projects, including an "Innovation of the year" award for an e-commerce platform for The Economist. He is based out of Montreal, Canada.

EXPERT VOICES

Subrata Mukherjee: You spoke at the Green IT Day about responsible digital practices. How can organizations balance the incredible potential of AI with the need for digital sustainability and sobriety?

Françoise Nauton-Inglis: That is the next big question we collectively need to answer! I believe that sustainable AI is the only AI we can truly rely on for the future so there is a lot of work ahead of us on the subject. Which doesn't mean there is nothing we can do in the meantime. Set objectives for lowering your carbon footprint and implementing computation where it truly brings value in your business. Start by mapping the heaviest journeys (search, media, information transmission, documents generation etc) and reduce waste before adding intelligence. Keep things balanced between objectives and operational constraints, and keep everyone in the loop. Prefer bottom up approach over brute-force change. Organizationally, give everyone shared targets on performance and footprint so trade-offs are explicit when they are needed. In our cluster work at Digital 113, we see that sobriety accelerates—not slows—innovation: faster pages, clearer copy, fewer drop-offs. Responsible AI is not an add-on; it's a quality discipline.

SM: What cultural shifts are required for a company to not only adopt new technologies but also to do so in an ethical and responsible way?

FNI: Move from "project" and "risk assessment" to "practice." Establish small cross-functional crews—product, data, ethics, security—that meet weekly and pilot initiatives to review impact in production, not just at launch. Do not be afraid to abandon projects if they don't meet your criteria (business value, human value and CSR). It's more important to experiment than to try and plan everything in advance. Teach managers to ask, "What evidence shows this helps users?" and "What's the lightest way to achieve it?". Culture changes operates best when responsibility is embedded in respect & storytelling.

SM: How do you ensure that your team, and the teams you advise, maintain a "human scale" approach to technology in an increasingly automated world?

FNI: Human scale is where clarity, craft and brain power lead positive transformation. Practically: small autonomous teams; clear product narratives; and time for craft reviews where designers and engineers demo improvements to legibility, not just features. In client work, we protect budget for usability testing and service design. Automation should elevate human roles—freeing experts to handle complex advice, exceptions, and relationship-building. In community roles at Digital 113 and MoMa, we promote this mindset across the ecosystem—because scale without human standards is just noise.

SM: Beyond your work with AI for e-commerce, what emerging technology will have the most significant impact on the business world in the next five years?

FNI: I'm watching three trends that need to converge: agentic AI with verifiable context (retrieval + actions), privacy-preserving analytics at the edge, and sustainable IT. Together, they make digital services more proactive and accountable—able to act for customers while proving why those actions were safe, compliant, and beneficial.

SM: What is your long-term vision for how technology will "humanize" the business world, rather than simply automate it?

FNI: Humanizing means restoring proximity and meaning. Technology should shorten the distance between a person's intent and a helpful outcome—while making the trade-offs visible and fair. AI that explains itself, services that respect time and context, and organizations that prize clarity over complexity will win trust. If we do this well, automation won't replace relationships; it will give them room to flourish.

FRANÇOISE NAUTON-INGLIS

CEO @ Spiriit / Juliia, Ex GM/CFO of Kaliop Interactive Media

Françoise Nauton-Inglis is an accomplished business leader with over 20 years of experience in the digital and technology sectors. She is the CEO of Spiriit, an agency that designs and builds custom digital and AI solutions, and its AI-first product, Juliia, which focuses on intelligent search and product recommendations for e-commerce. Françoise's expertise lies in growth management, international development, and strategic business planning. Prior to Spiriit, she spent a decade at Kaliop Interactive Media, where she served as the Group General Manager, overseeing M&A, international expansion, and financial operations. She has a deep passion for structuring and growing innovative, profitable, and human-centered businesses, and she holds leadership positions in several organizations, including as President of Digital 113 and CI Montpellier Management. She is based out of Montpellier, France.

EXPERT VOICES

Subrata Mukherjee: What is your philosophy for leading large global teams, whatever be the industry?

Charisma Glassman: Transparency, empathy, and shared ownership. Distributed teams must feel empowered, not siloed. I create alignment through open communication, collaborative tools, and a culture that celebrates both wins and lessons learned. Autonomy within a shared mission is what fuels innovation. By adopting a global-first but locally flexible approach. We start with listening. Regulations and cultural expectations differ dramatically, so local leadership becomes essential. Technologically, modular architectures allow compliance and customization without fracturing the core system. By framing transformation in terms of business outcomes and risk management. I build trust with data-backed cases, proactive risk mitigation, and continuous updates.

SM: Your career has successfully bridged two very different industries: financial services and retail. What principles from financial services have translated most effectively into retail and how predictive analytics translate across industries? How do you ensure technology investments deliver measurable ROI?

CG: Three core principles endure. One, precise segmentation and retention strategies: identifying high-value customers and proactively reducing churn. Two, operational discipline: automating processes to improve efficiency and profitability. Three, resilient platforms: scalable infrastructure capable of handling spikes in demand and vast data flows. These apply whether you are running a trading platform or an e-commerce site during peak season.

In finance, AI predicts default risk or detects fraud. In retail, those same models predict churn, forecast demand, and optimize pricing. The goal in both is the same: moving from hindsight-driven to foresight-driven decision making.

On ROI, by keeping them business-led. Every initiative ties to a clear business metric. We measure rigorously, track KPIs transparently, and use agile iteration, piloting, scaling what works, and discarding what does not. This creates a culture of disciplined innovation.

SM: How do you balance personalization with privacy? Can you share an example of a data-driven content strategy driving significant business outcomes and what advice would you give to companies still operating with fragmented customer data?

CG: Through transparency and control. Customers must know what is collected, how it is used, and have tools to manage it. Adherence to global standards and robust security builds trust. The exchange must feel fair: richer experiences in return for data, never at the cost of privacy. One example was a global franchise re-launch that leaned heavily on fan analytics. By segmenting audiences by character affinity and nostalgia triggers, we designed tailored campaigns that drove massive pre-orders and early sales. Data did not just boost revenue; it deepened emotional resonance across different fan groups. For those with Fragmented customer data, start with a centralized data platform and focus on one high-value segment as proof of concept. Do not try to unify everything at once. Small wins build momentum, credibility, and resources for scaling.

SM: What is the key message you want to convey to the next generation of leaders and What is the single most important characteristic you look for in a team?

CG: Leadership today is not about control, it is about vision and empathy. Great leaders create environments where ideas flourish, risks are safe to take, and diverse voices shape outcomes. The leaders of the future will be both courageous change agents and compassionate humanists. Intellectual curiosity is the most important characteristic. Skills will evolve, industries will shift, but curiosity drives learning, adaptation, and innovation. A curious team will never stagnate

CHARISMA GLASSMAN

Global Head of Retail & Consumer Applied Advisory, Genpact

Charisma Glassman is a distinguished global executive, a Forbes Distinguished Leader, and one of the Top ranked AI Leaders in consulting. As Group VP and Senior Partner at Genpact, she is the Global Head of Retail & Consumer Applied Advisory, leading business transformations powered by AI for clients across the US, Europe, and APAC. Her impressive career includes senior leadership roles at various consulting companies, advising C-suite executives, and at major financial institutions, where she led digital initiatives that resulted in significant revenue growth. Charisma consistently leverages her expertise in digital transformation, product strategy, and global team leadership to drive strategic growth & innovation. She is based out of New York, in the United States.

EXPERT VOICES

Subrata Mukherjee: You spent a significant part of your career as a leader at major ratings agencies like Moody's and S&P Global. What was the catalyst for your move from leading corporate innovation to co-founding your own firm, and how are you applying lessons learned from that world to your work now?

Sandy McCarron: The catalyst was really timing and perspective. I was there at the beginning of the internet boom, working full-time at Random House. Looking back, I sometimes regretted not taking the leap to join one of the big tech companies that became household names—like Google or Amazon. Now, with this new wave of game-changing technology, I see a second chance in my career. This time, I'm trying a new path that's not corporate, working with a team of visionaries. That's why I co-founded my own firm—to be at the forefront of transformation, applying lessons from corporate innovation while moving faster and more independently. I'm taking the strategic discipline and rigorous project management I honed in a highly regulated environment and applying it in a more nimble, agile setting.

SM: What are the key differences in your approach to driving change as an internal leader within a large, established enterprise versus as an external advisor?

SMC: As an external advisor, you're brought in specifically for your expertise, which actually gives you more authority in a way. Internally, even with experience across multiple disciplines, you might be seen as a generalist or pigeonholed as just a product manager. As an advisor, clients expect and value that broader perspective, which allows me to influence more directly. It's a very different dynamic. Internally, you're often fighting against ingrained processes and cultural inertia, whereas as a consultant, you are often the catalyst for change they've already decided they need. This shift in role allows me to focus purely on strategic value without being bogged down by internal politics or departmental silos.

SM: In your new role, how are you helping organizations solve the same types of complex problems you tackled at the enterprise level?

SMC: AI is evolving so quickly—every day there seems to be half a dozen new breakthroughs. That can be overwhelming for organizations. What we encourage them to do is go back to the basics. If your data is bad, AI won't fix it—garbage in, garbage out. The key is to start with the problem, not the solution. Don't begin with "How do we use AI?" Instead, ask: "What's the most difficult pain point for our team right now?" Solve that problem first, and then determine whether AI is the right tool to help. We're helping them build a strong, fundamental foundation before they even consider adopting AI. This ensures that their AI investment is solving a real business need, not just a passing trend.

SM: Your firm, Sapient Advisors, specializes in AI readiness and adoption. What are the most common data and governance hurdles you see companies face when trying to move beyond an initial AI pilot?

SMC: The number one issue is the lack of a data strategy. Data often isn't tagged, centralized, or reusable. Sometimes it's outsourced, comes from third party partners, or originates with clients—and contracts often don't give explicit permission to use that data for AI. That requires renegotiations, or tackling unsexy data projects that have been put off for years. Another hurdle is communication. Technical founders often struggle to get CEOs to understand the business impact of AI. Too often the narrative centers on cost-cutting and layoffs instead of value creation, efficiency, or growth.

SANDY MCCARRON

Co-Founder & COO at Sapient Advisors | Ex- Moody, Ex-S&P

Sandy McCarron is a co-founder and Chief Operating Officer at Sapient Advisors, a lean, AI-enabled agency specializing in product and data strategy. He focuses on guiding companies through AI adoption and achieving operational efficiencies through rapid AI Design Sprints. With a career spanning both media and fintech, his expertise is built on a strong foundation of innovation and program management. At Moody's Analytics, he was a Senior Director of the Innovation Unit, where he introduced Google Ventures Design Sprints and led a major program to upgrade a flagship product, generating a $60M increase in profit. He also held a senior technology director role at S&P Global. His early career includes positions as Director of Online Content at Penguin Publishing Group and Director of Online Production at Penguin Random House. She is based out of New York.

EXPERT VOICES

Subrata Mukherjee: Your title as both a Chief Digital Officer and a Chief Information Officer reflects the modern C-suite. How do you balance the internal focus of a traditional CIO—overseeing core IT infrastructure and security—with the external, client-facing, and innovative responsibilities of a CDO?

Preeti Sharma: Balancing these roles is about seeing them as interconnected rather than separate. The CDO role calls for outward vision—understanding client needs, anticipating trends, and reimagining possibilities. The CIO role grounds me in operational excellence—security, stability, and scalability. The addition of CIO responsibilities made sure that our house was in the same order that I strive to create for clients. Both roles reinforce each other: strong infrastructure makes innovation possible, and external innovation informs how we strengthen our core. It's like working with a dual lens the CIO lens zooms in on precision—systems, security, operations. The CDO lens zooms out to see the bigger picture—emerging trends, client needs, future opportunities. Balance comes from knowing when to shift focus, and making sure the two views inform one another.

SM: What are some key skills or mindsets you believe are necessary to successfully navigate this dual role?

PS: Curiosity, adaptability, continuous technical learning, and empathy. Curiosity keeps me asking "what if" and exploring new possibilities. Adaptability ensures I can shift between technical deep dives and strategic vision. Staying current with technical skills is essential—it allows me to engage meaningfully with my teams, evaluate emerging solutions, and lead with credibility. And empathy helps me connect with clients and colleagues—because transformation is not just about technology, it's about people navigating change together.

SM: You've referred to yourself as a "Nerd in Chief." How do you use this approachable leadership style to foster innovation and build trust with clients and internal teams?

PS: I use "Nerd in Chief" as a reminder that it's okay to bring both expertise and humor into leadership. It makes technology approachable and reminds people that innovation doesn't need to feel intimidating. When teams see that I'm genuinely passionate—and not afraid to be human—they're more willing to take risks, share ideas, and build solutions collaboratively. It's about creating a culture where both clients and colleagues feel seen, supported, and inspired; I'm not afraid to admit that I nerd out and dive deep into technology exploration.

SM: As a leader, how do you manage the expectations of public officials and the public they serve when it comes to implementing these transformative technologies?

PS: Transparency and empathy are at the core of how I manage expectations. For clients, I make sure we deliver what we promise—on time and without missing deadlines. That reliability builds trust and keeps partnerships strong.

For citizens, I approach every rollout with them in mind, ensuring that the digital products we create are user-friendly, intuitive, and not cumbersome. I also focus on explaining the "why" behind timelines and constraints, while keeping the conversation anchored on outcomes. I am a huge fan of citizen and end user focus groups and we try to incorporate these in the product design/feedback phase so we deliver solutions that will work for them. People are more patient when they understand the journey, and more engaged when they feel part of the solution.

PREETI SHARMA

Chief Digital Officer & CIO at VGD NYC

Preeti is a distinguished technology executive and a "Notable Woman in Tech" with a proven history of driving digital strategy and innovation. As the Chief Digital Officer and CIO at VGD NYC for over 18 years, she leads the delivery of digital services, technology strategy, and solutions for a diverse range of clients. Preeti's expertise is deeply rooted in her long-standing focus on AI and technology transformation, backed by recent certifications in AI Leadership and Digital Innovation from Carnegie Mellon University. Her foundational experience includes nearly a decade at ISI Emerging Markets, where she served as Global Operations Manager, playing a key role in scaling the business from a small startup into a profitable enterprise. A member of Chief and the Forbes Technology Council, she has been recognized with multiple AVA Digital Awards for her work. She is based out of New York.

EXPERT VOICES

Subrata Mukherjee: What foundational skills and principles did you learn as a developer that still influence your decisions and leadership style today?

Alex Demichev: One of the most important things software development teaches you is the ability to decompose the most challenging tasks into small steps that are each comprehensible and doable. That ability has definitely been instrumental to me when time has come problems that affect our large business and people in it.

SM: With AI poised to automate many routine tasks, how is your company reskilling and upskilling your developers and engineers to work alongside AI assistants

AD: We use all available AI assistants in our work, and right now I see them as another step in individual performance acceleration of a particular developer, rather than something bigger. We have navigated several disruptions like that before, introduction of IDE's, code completion, development frameworks and so on. Every time it requires reskilling or up-skilling but it happens rather naturally with people gaining access to the technology, appreciating its efficiency, adopting it in their everyday work and starting to benefit from it.

SM: What new roles or skill sets do you believe are becoming most critical for success in a world where AI-powered tools are the norm?

AD: All the same old roles and skill sets that were always important. To me the right question would be – how do we become good at what we do to the extent that AI becomes our tool and helper able to reduce mundane work. Becoming good is a process, you start bad and gradually become good at something. But if AI is already better than you at this at the moment when you start how do you become good at it?

SM: How do you balance the need to implement efficiency-driving AI tools with the equally important goal of keeping your human teams motivated and creatively engaged?

AD: You use AI as a tool and means to empower your people to do more, faster, better. People love to be empowered and they hate mundate jobs. At least the best people are that way, and you should aim to surround yourself with the best people.

SM: How has the rise of the AI economy changed the way you approach business development and identify new opportunities for growth?

AD: AI is still finding its practical applications in this world. We're carefully monitoring real life practical use cases and their real effectiveness and value. So far one of the biggest positive outcomes of the AI trend was the fact that businesses have taken time to really consider how and where they fit into this new world. And that existential threat of missing out this time has made them look into where they are at all in the digital transformation journey. Before, some were proud to be "pen and paper" business. Well, not anymore.

SM: What is your strategy for helping clients, particularly in traditional industries, successfully integrate AI into their business models?

AD: You start with Data. No matter where you are in your AI journey, no matter where it brings you, organizing your data properly is an absolutely critical first step. You can't go wrong about it. And then with properly organized Data you can do anything, reporting, all sorts of AI enablement.

SM: Looking back, what is a key strategic decision you made that you are most proud of, and what was the outcome?

AD: It was to become Salesforce and Microsoft platforms System Integrator, Dynamics F&O. It has finally allowed us to establish ourselves among the enterprise level consultancies.

ALEX DEMICHEV

CEO at ITransition Group

Alex Demichev is a business-oriented software professional with a rich background spanning over two decades at ITransition Group, where he has risen from a Department Manager to the Chief Executive Officer. His deep technical and operational expertise, honed over nearly 14 years in management, is complemented by a strategic business perspective gained from his M.B.A. at the University of Oxford - Saïd Business School. As CEO, Alex leads a company focused on helping clients with complex technology challenges, including M&A integration, ERP modernization, and broader digital transformation. His leadership combines a passion for detail with a comprehensive understanding of business needs, guiding the company to deliver robust software solutions and drive value for its clients. He is based out of Berlin, in Germany, Europe.

EXPERT VOICES

Subrata Mukherjee: How is Generative AI changing the way we approach healthcare data? Are we moving beyond simple data extraction to true reasoning and inference?

Mousumi Sengupta: GenAI is allowing us to move beyond simple data extraction towards true reasoning, but this is only possible with the addition of context. Traditional NLP models were limited to pulling specific data points based on pre-trained patterns. GenAI, with its ability to comprehend narrative and context, can synthesize complex clinical notes, reason about a patient's journey, and even infer causal links or potential diagnoses. This allows us to automate cognitive tasks that were previously impossible, like summarizing a patient's entire medical history or drafting a preliminary research abstract. This moves us towards a new era where AI isn't just a tool for data retrieval, but a partner in clinical and scientific reasoning.

SM: What are the primary ethical and data governance challenges that must be addressed when using LLMs on patient and clinical data?

MS: The primary challenges are data privacy, model bias, and the risk of "hallucinations." LLMs, being continually learning systems, can inadvertently expose sensitive patient information or generate factually incorrect clinical outputs. Robust data governance is crucial, including strict de-identification protocols and federated learning architectures to keep raw data secure and HIPAA compliant. Another challenge is mitigating inherent biases in the training data, which could lead to health inequities. Lastly, it's imperative to implement explainability frameworks to understand how a model arrived at a conclusion, ensuring transparency and accountability in every high-stakes healthcare decision.

SM: In your view, will AI and knowledge graphs eventually replace human-enabled data annotation, or will they serve as a powerful tool to augment human expertise?

MS: I believe AI and knowledge graphs will serve as a powerful tool to augment human expertise, not replace it. The complexity and high-stakes nature of healthcare data necessitate a human-in-the-loop approach. While AI can automate the mundane tasks of initial data annotation and analysis, a human expert is essential for validation, handling complex edge cases, and ensuring clinical accuracy. The synergy between human intelligence and AI's computational power will accelerate the pace of scientific discovery and improve patient outcomes by allowing experts to focus on the most challenging and rewarding aspects of their work.

SM: For those outside of life sciences, what does it mean for data to be Findable, Accessible, Interoperable, and Reusable (FAIR)?

MS: FAIR data principles are a set of guidelines for making data truly useful for both humans and computers. Findable means the data is discoverable via a persistent identifier. Accessible means it can be retrieved using a standardized protocol. Interoperable is the most critical; it means the data can be combined with other datasets because it uses a common language and an agreed-upon vocabulary (ontology). Reusable means the data is well-described with sufficient metadata, allowing it to be applied to new research questions. FAIR data is designed to allow for seamless machine-to-machine collaboration.

SM: What is the role of Responsible AI in Healthcare?

MS:As I work with AI in healthcare, my goal is to ensure its implementation is ethical, safe, and effective. I prioritize principles like fairness, transparency, privacy, and accountability to improve patient outcomes and streamline processes. My role is to serve as a framework for the responsible development and deployment of these systems, addressing potential biases, securing patient data, and fostering trust between patients and providers. By doing this, I can enhance clinical decision-making, personalize medicine, and advance public health.

MOUSUMI SENGUPTA

Chief Data Strategist at Condext, Ex-Accenture, Ex- Mount Sinai Hospital

Mousumi Sengupta is an accomplished leader in digital health, currently serving as the Founder and Chief Data Strategist at Condext, a startup automating the contextualization of healthcare data. Her expertise in Data Strategy, Ontology Engineering, and AI is built on a career at prominent organizations like Accenture, where she helped develop a Life Sciences practice with a focus on Knowledge Graphs and Generative AI. Her prior roles include leading a FAIR Data Management function at Bristol Myers Squibb and driving semantic interoperability at Mount Sinai Health System. She was also a Director at organizations like Nuance Communications and the Multiple Myeloma Research Foundation (MMRF). She is a proven innovator at the intersection of technology and clinical data. She has a Masters in Bioinformatics from Columbia University. She is based out of New York, United States.

EXPERT VOICES

Subrata Mukherjee: When you join a board today, what are the first two or three questions you ask to assess a company's technology and cyber risk posture?

Michael Gabriel: One of the first steps in effective oversight is recognizing that technology posture and cyber risk posture are fundamentally different. Each requires its own lens of evaluation. I begin every assessment by asking: "What defines success?" For technology, success is typically measured in capabilities—either in running the business or enabling growth. For cybersecurity, it's about risk mitigation, regulatory compliance, and resilience. Next, to understand how well technology and cyber strategies are aligned with business priorities, I request the following Four items. First, Board Decks (Last 4 Quarters) What themes have dominated their focus? What investments were needed for projects? What goals and timeframes were attached to each? Were the stated goals achieved within the targeted timeframes? Second, Segmentation Strategy. How are technology and cyber topics organized? Are they broken down by business unit, geography, initiative, or tech domain? Third, Risk Reporting. How are risks and opportunities presented? Are ongoing metrics provided? Are the mitigations for risks or the resolutions for issues clearly outlined? Fourth. Board Member Perspectives.

SM: In your view, what is the single biggest misconception that corporate boards have about technology?

MG: One of the most damaging myths in the boardroom is that technology cannot be communicated in a way that's accessible to non-technologists. I've heard countless complaints about presentations filled with acronyms, metrics, and jargon that obscure rather than clarify. "It's like a black hole that consumes funds and just continues to grow." This perception leads to frustration, disengagement, and ultimately poor oversight. But it doesn't have to be this way. Boards should demand clarity—and technologists must learn to deliver it. Experienced business leaders shouldn't feel embarrassed to ask for clearer communication. The perpetuation of confusion around technology is not a failure of intelligence - it's a failure of translation. And that misconception is a solvable problem that shouldn't be accepted as unavoidable.

SM: As a Board Director for Entertainment 2 Affect Change, what role do you think technology can play in driving social impact within the entertainment industry?

MG: I'm no longer a Board Director at E2AC, but my time there led to them developing an online survey tool, the Impact Assessment Scale, that allowed them to survey viewers of their documentaries to see the type of impact they had. It arose from a simple question I asked as a Board member. Since the name of the non-profit is Entertainment to Affect Change, how do we know we're affecting change, that the funds we've raised are being utilized to have the most impact, and better understand the factors in a film that create that impact? Entertainment is about storytelling, and the emotional feeling we have when watching a film, listening to music, or playing a game can have a profound impact on how people view issues and causes. If you can't determine whether you are obtaining the goals that you are setting for that social cause, whether communicating something to a large enough audience, recruiting volunteers to help the cause, or changing laws to support the cause, then you don't know that you're making a difference, and all you're doing is creating a media product. While that's still an accomplishment, it's not necessarily driving social impact. So, technology can not only help with creating the entertainment product, but it can also help with understanding the impact those products are having, and how to structure them to have the desired impact for the targeted audience.

MICHAEL GABRIEL

Ex CIO at HBO, NBA | NACD Certified Board Director

Michael is a senior technology and cybersecurity executive with experience providing C-level leadership, governance, and advisory services. He is currently an independent Board Director and provides professional services for companies like Warner Brothers Discovery (WBD). His career is highlighted by a 14-year tenure as EVP & Global CIO at HBO, where he was the initial executive to spearhead the groundbreaking HBO GO initiative and led a company-wide digital transformation. An inventor on six technology patents, he also played a key role in information security governance across the Time Warner enterprise. Prior to HBO, he served as Global CIO for the National Basketball Association (NBA), where he implemented the first NBA retail store technology system. He also held senior IT leadership positions at EMI Records Group and NBC Olympics.. He is based out of New York

EXPERT VOICES

Subrata Mukherjee: When an organization is undergoing a major technological transformation, what are the key cultural elements you focus on to ensure the change is not just adopted, but embraced by the team?

Amanda Murphy: As I mentioned above, change doesn't work if it's only a request from the top to make it happen. Leaders and associates need to understand the why and help facilitate the transformation from the bottom up, as well. Research shows associates want to hear about change from their direct leaders, not from senior leadership. They have a relationship and a trust with direct or skip level leaders that they don't have with senior leadership. For that reason, change must involve all levels of the organization. In my experience, leaders want to feel they are part of the change, as well. This means prior notification of the planning and asking for their input for implementation, adoption and change management. If you're bringing leaders (middle managers) along with you during the change planning, they can have a significantly positive influence on the change curve for organization.

SM: You mention that your leadership style is honest and empathetic, with a commitment to continuous growth. How do you use these qualities to lead and inspire teams, especially when facing resistance to new initiatives?

AM: Empathetic leadership is key. This can be difficult for senior leaders who have already been discussing and decisioning the change months prior to messaging to teams. By the time they are sharing the information with all levels, they are at a completely different part of the change curve than those hearing the information for the first time, and this causes disengagement and change resistance. Having a high EQ, I'm able to influence messaging, talking points, and tactics to keep empathy front and center with our executive team. This helps set us up for success in a transformation.

SM: You recently completed a certification in "Change Enthusiasm." How does this approach to change management differ from a traditional model, and why do you feel it's essential for today's leaders?

AM: What I appreciate most about Change Enthusiasm Global is the focus on personal values and emotions in how we respond to change. These are both items rarely discussed in a professional setting, but are our individual key indicators for how we react to change and how we can look at change as opportunity for growth.

SM: Your career path shows a progression from communications and media relations into senior roles within technology and security. How did your background in storytelling and messaging uniquely prepare you for leading a technical organization?

AM: Storytelling and being able to take complicated information and make it consumable for any and all audiences is a skill that's useful in any industry and any role. When you're presenting your work to the C-suite or the Board of Directors, they don't care about the name of the tool, the vendor you're using, the various acronyms you use every day. They want to know the big picture and the bottom line.

SM: You led an enterprise re-brand for the Chief Information Security Office. What are the key elements of a successful re-brand for an internal-facing technology organization?

AM: I've led an external re-brand for a national nonprofit organization, as well as an internal rebrand within a Fortune 100 company. I leveraged the same content marketing and storytelling techniques for the external rebrand as I did for the internal - The only difference is the stakeholders you're messaging to. The most important element, as with any large change initiative, is messaging the Why and making it human-centric.

AMANDA MURPHY

Technology Director, Chief of Staff at Nationwide Insurance

Amanda is a seasoned leader specializing in transformational leadership, business operations, and organizational change management (OCM) within the technology sector. As the Technology Director, Chief of Staff for Security & Infrastructure at Nationwide, she formulates and executes strategic initiatives for a large, Fortune 100 organization. Her proven expertise in building and leading high-performance teams is demonstrated by her success in managing a complex merger. With a strong background in strategic communications, digital storytelling, and executive messaging, Amanda is skilled at creating strong corporate cultures and driving successful outcomes by bridging the gap between high-level strategy and tactical execution. She has been recognized with a Gartner Communications Award for her work in transforming internal communications. She is based out of Westerville, Ohio, United States.

EXPERT VOICES

Subrata Mukherjee: Throughout your career at organizations like the NBA and EMI, you repeatedly achieved significant cost savings through strategic IT restructuring. How do you balance the need for technology innovation with the mandate to reduce costs and increase efficiency?

Michael Gabriel: The ever-present mandate to both reduce costs and increase efficiency is usually more of a focus on "keep the business running" costs that focuses on areas like resource utilization (such as outsourcing or automation), vendor and third party product or service selection /consolidation /negotiation /justification, and project/initiative prioritization. This should be treated differently from technology innovation, which will usually have an uncertain return on investment. Innovation is a process that can grow revenue and/or profit, improve a product, process, or service, such as customer service, or reduce costs or time, such as time to market. It depends on the goal of the initiative, and can sometimes provide all the above. That doesn't mean that you have an endless amount of capital and resources to focus on innovation. Innovation needs to be properly directed and managed, with clear results that are verifiable after some predetermined amount of time. It should not be a black vortex of funding, and should be held accountable at predefined timeframes with predefined goals. It does not mean that those goals need to have clear ROIs, because that is often not possible with innovation. There should be guidelines for what you're trying to solve or invent, and some method of determining if you're going in the right direction, and at the right speed and cost allocation, to attain defined goals..

SM: You've managed multi-million dollar projects like the NBA's first retail technology system and EMI's ERP implementation. What is your philosophy for ensuring a major technology project stays on time and within budget?

MG: It's always about having clear end goals and milestones, the right mix of internal and external talent, and the proper oversight to ensure that accountability is instilled. The longer a project is, the greater the likelihood that something unexpected will go awry. The more people and areas involved, the more likely there will be communication and coordination issues. Having said that, if the oversight and management of major projects are properly handled, you can ensure they stay on time and on budget. And change management needs to be first and foremost in any technological change to ensure successful usage after implementation. Change is not normally welcomed by most, other than the person who drove the change.

But proper change management will help ensure that the goal and value of the changes are understood and realized. I can't tell you how many times I have seen project plans with target dates that have been passed, with no issues defined or a revised end date provided. And how often there is a misunderstanding about a risk, which is something that could happen, and an issue, which is something that did happen. And how often risks are listed as some amorphous item, like possible resource turnover or requirements change. You might as well list a meteor hitting your building as a risk. It might happen, and it might not. Unless you have a cause to be concerned about resource turnover, such as if you're in the middle of a merger, or a cost-cutting/reduction in force, or have had a historically high staff turnover, it's just a theoretical risk and has no value being listed. If a risk is listed, there should be actions and timeframes to assess whether the risk has increased, been mitigated, or has turned into an actual issue that needs corrective action and resolution; otherwise, you will miss a time, cost, or effectiveness goal.

Ex CIO at HBO, NBA | NACD Certified Board Director

MICHAEL GABRIEL

Michael is a senior technology and cybersecurity executive with experience providing C-level leadership, governance, and advisory services. He is currently an independent Board Director and provides professional services for companies like Warner Brothers Discovery (WBD). His career is highlighted by a 14-year tenure as EVP & Global CIO at HBO, where he was the initial executive to spearhead the groundbreaking HBO GO initiative and led a company-wide digital transformation. An inventor on six technology patents, he also played a key role in information security governance across the Time Warner enterprise. Prior to HBO, he served as Global CIO for the National Basketball Association (NBA), where he implemented the first NBA retail store technology system. He also held senior IT leadership positions at EMI Records Group and NBC Olympics.. He is based out of New York

EXPERT VOICES

Subrata Mukherjee: You were employee #86 at Indeed.com and played a key role in its hyper-growth phase. What's the most critical difference between hiring for a fast-growing startup and a large enterprise?

Jennifer Speciale: In startups, you hire for adaptability; in enterprises, you hire for precision. Startups need generalists who can pivot across functions as the business evolves weekly. I once hired a designer at a 100-person startup who also handled product strategy, customer discovery, and board reporting—roles that would be five separate jobs at Amazon. Leaders must know whether they are recruiting a Swiss army knife or a scalpel, and design hiring processes accordingly.

SM: Your profile mentions you developed a "strategic hiring framework" to align talent with Amazon's bold innovation goals. What was the secret to navigating Amazon's famously rigorous interview process, and what can candidates do to prepare for that level of scrutiny?

JS: At Amazon, the secret was codifying "bar-raising." Every hire had to demonstrate impact bigger than their role—proof they could invent and simplify at scale. I developed a framework that mapped candidates' stories directly to leadership principles, ensuring evidence, not opinion, drove decisions. For candidates today, the lesson is: show not just what you did, but how you consistently raised the bar for others. One client reframed his narrative from "I launched X product" to "I created a repeatable system that 10 teams still use." That is the level of scrutiny Amazon rewards.

SM: What are the most common mistakes that startups make when they first start to scale their hiring efforts?

JS: The most common mistake: hiring for today's pain instead of tomorrow's vision. Startups under pressure fill gaps reactively, often with friends or quick referrals. Six months later, those hires lack the skills for the next growth stage. I counsel founders to write a "future org chart" 18 months out, then hire toward that. When you over-index on immediate relief, you build tomorrow's turnover. When you hire with foresight, you create tomorrow's leadership bench.

SM: You've built a framework that helps leaders land roles in under 12 weeks with significant salary growth. What is the single most important action a professional can take today to ensure their career remains resilient?

JS: The most important action: build a visible portfolio of stories of impact. Titles change, markets shift, AI disrupts—but stories that clearly tie your actions to business outcomes remain currency. I advise every leader to track three things quarterly: what did I make faster, cheaper, or more valuable?

SM: How has the rise of AI and automation changed the way you coach leaders, particularly regarding the skills and roles they should be targeting?

JS : AI has shifted my coaching from "what skills do you have" to "what perspective do you bring." Technical skills will always evolve; what endures is the ability to interpret complexity, frame problems, and mobilize people. For example, I coach leaders to use AI tools to handle the first draft—then double down on uniquely human tasks: judgment, narrative, empathy. Careers are now less about being the one with answers and more about being the one who asks the questions AI cannot.

SM: You have helped leaders transition into roles at companies like Amazon, McKinsey, and Google. What is the one common thread you see among all of your successful clients?

JS: The common thread I see in my most successful clients is humility—sometimes to a fault. Many wrestle with imposter syndrome, or with not knowing how to balance work and family, or how to be assertive without tipping into aggression. These are not failures; they are proof that leadership is deeply human. What unites them is the desire to do better, coupled with the courage to seek support.

JENNIFER SPECIALE

CEO, Speciale Executive Leadership | Ex- Amazon, McKinsey, Google...

Jennifer is a Global Executive Leadership Coach where she partners with senior leaders to showcase Executive Presence, Influence, and Impact. Her coaching practice is built on over two decades of experience in strategic talent advisory, having directly shaped executive teams at the world's most innovative companies. Jennifer's career highlights include establishing the pioneering UX Design team in NYC for Google, building the foundational global hiring processes for Indeed during its hypergrowth phase, and leading the global expansion of the Design executive talent function at both McKinsey & Company and Amazon. She is an expert in organizational Change Management and Executive Search, dedicated to translating senior leaders' expertise into compelling narratives that unlock C-suite and VP opportunities across the Design and Technology landscapes. She is based out of New York city.

SPECIAL OFFER!

WE NEED YOU!

WELCOMING OUR NEXT SET OF THOUGHT LEADERS

The remarkable success of the Ai6x6 Organization Transformation Series is built upon the **hard-won wisdom** of executive leaders who have successfully navigated the AI transformation within their industries, turning theory into a deeply practical, actionable guide. As we expand the series, we are extending a highly selective invitation to a new cohort of pioneers—individuals whose professional journey is at the **leading edge of AI strategy and execution**. We are looking for executives who possess a proven track record of rewriting business models, mastering unforeseen ethical and competitive challenges, and offering a **clear blueprint** for the next 3-5 years of the AI-Era.

Contributing to the Ai6x6 Series is a unique opportunity to **cement your professional legacy**, joining an exclusive, curated network of global authorities whose insights are distilled into executive-level knowledge cards, directly shaping the governance and innovation decisions of the world's most influential companies. Our selection process is based strictly on **pure expertise and demonstrated leadership** in the AI-Era, reinforcing our commitment to editorial integrity and authentic thought leadership—there is no monetary exchange involved. Select contributors will also receive an invitation to blog on our expert network microsite, further amplifying their platform. To confidentially inquire about contributing your expertise, please contact us via our dedicated portal at **www.Ai6x6.com/contact** or by scanning the adjacent QR code.

Contact

SUBSCRIBE TO OUR NEWSLETTERS: To stay ahead of the curve and receive continuous, executive-level intelligence on the rapid evolution of the AI-Era, we invite you to register for our **Ai6x6 Weekly Newsletter**. Each edition delivers curated insights directly from our thought leaders and highlights critical shifts in AI strategy, governance, and provides relevant excepts from our website and books. Keep your finger on the pulse of the future and ensure your company remains competitive by signing up on the same page

PLEASE REVIEW US ON AMAZON!

Workforce Blueprint

We know your time is valuable, but if the Ai6x6 Organization Transformation Series delivered the real-world, actionable strategies you needed, please consider leaving a brief review on Amazon. If we missed the mark, we would also like to know, and that is the only we we can improve. Your honest feedback is crucial; it helps fellow executives discover this unique framework and benefit from the insights of our thought leaders. Thank you for supporting our mission!

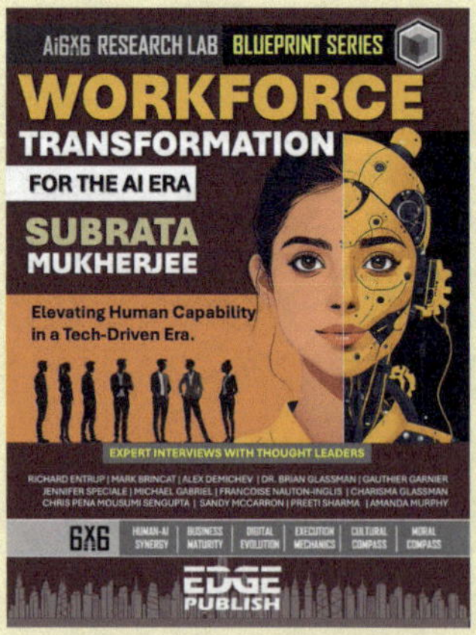

BLUEPRINT SERIES

SECTION B

Human-AI Synergy

Ai6x6™ Transformation Blueprint

Story 1 — Enigma Codebreakers (1940–45): German U-boats encrypted communications using constantly shifting ciphers, forcing Allied cryptanalysts to chase patterns that dissolved daily. Manual decoding could not keep pace with wartime urgency, and insights often vanished before they could be confirmed. Electromechanical devices accelerated the testing of solutions at unprecedented speed. Humans interpreted these outputs, correcting false leads and refining assumptions. The fusion of machine acceleration & human judgment reshaped the trajectory of the war.

Story 2 — Turing's Bombe & Operators (1940–45): Breaking enemy codes required evaluating thousands of cipher possibilities rapidly, and the Bombe machine generated candidate solutions at speeds humans could not match. Yet its outputs were incomplete, ambiguous, and often misleading. Human operators validated results, corrected machine assumptions, and filtered noise from meaningful signals. Their contextual judgment transformed mechanical output into strategic clarity. This interplay of mechanical speed and human reasoning cracked codes that neither side could solve alone. It became one of history's earliest demonstrations of human–machine symbiosis.

Story 3 — Toyota Production System (1950s–70s): Factories struggled with defects, bottlenecks, and unpredictable workloads as machines produced volume without contextual awareness. Workers observed issues that machines could not detect or interpret, prompting Toyota to design a synchronized workflow that aligned human insight with machine precision. Real-time data from the line informed continuous improvement cycles. This human–machine rhythm created a new operational standard for reliability and efficiency. The system became a global model for intelligent production. It demonstrated how augmentation, not automation, drives sustainable excellence.

Story 4 — Salk Polio Vaccine Trials (1952–55): As polio outbreaks intensified, statistical models guided trial design but lacked the ethical nuance required for medical decision-making. Researchers interpreted model outputs through clinical judgment, ensuring safety and fairness while machines processed vast amounts of trial data. This combined system accelerated discovery without compromising integrity. Human oversight balanced computational rigor with moral responsibility. It showed how human values anchor scientific acceleration.

Story 5 — Apollo 13 Mission Control (1970): An explosion crippled Apollo 13, leaving astronauts in mortal danger as incomplete telemetry streamed back to Earth. Engineers relied on simulations to test survival scenarios under extreme constraints, but the machines could only reveal possibilities. Human teams improvised procedures the computers could not imagine, blending machine-generated insights with creative reasoning. It became a defining moment of human–machine crisis collaboration.

Story 6 — Air Traffic Control Modernization (1960s–80s): Surging air travel overwhelmed manual tracking systems, and controllers struggled to maintain safe separation across increasingly crowded skies. Radar automation provided real-time positional data, enabling machines to track aircraft continuously while humans made judgment calls. This hybrid system reduced collisions, increased capacity, and established a global standard for aviation safety. Human–machine coordination became the backbone of modern air travel. The system remains one of the world's most enduring examples of operational augmentation.

LESSONS FROM A WORKFORCE TRANSFORMATION PERSPECTIVE

The workforce transforms as human–machine collaboration shifts from mechanical assistance to cognitive partnership. Early roles required workers to validate machine outputs and correct structural blind spots, while later systems demanded situational awareness, improvisation, and continuous improvement. Across decades, the pattern remains consistent: machines expand capacity, but humans elevate meaning. Workers become interpreters, supervisors, and designers of intelligent workflows. Their value lies not in competing with automation but in providing the contextual reasoning, ethical judgment, and adaptive problem-solving that machines cannot replicate. Human–AI synergy ultimately strengthens the workforce by amplifying human capability rather than diminishing it.

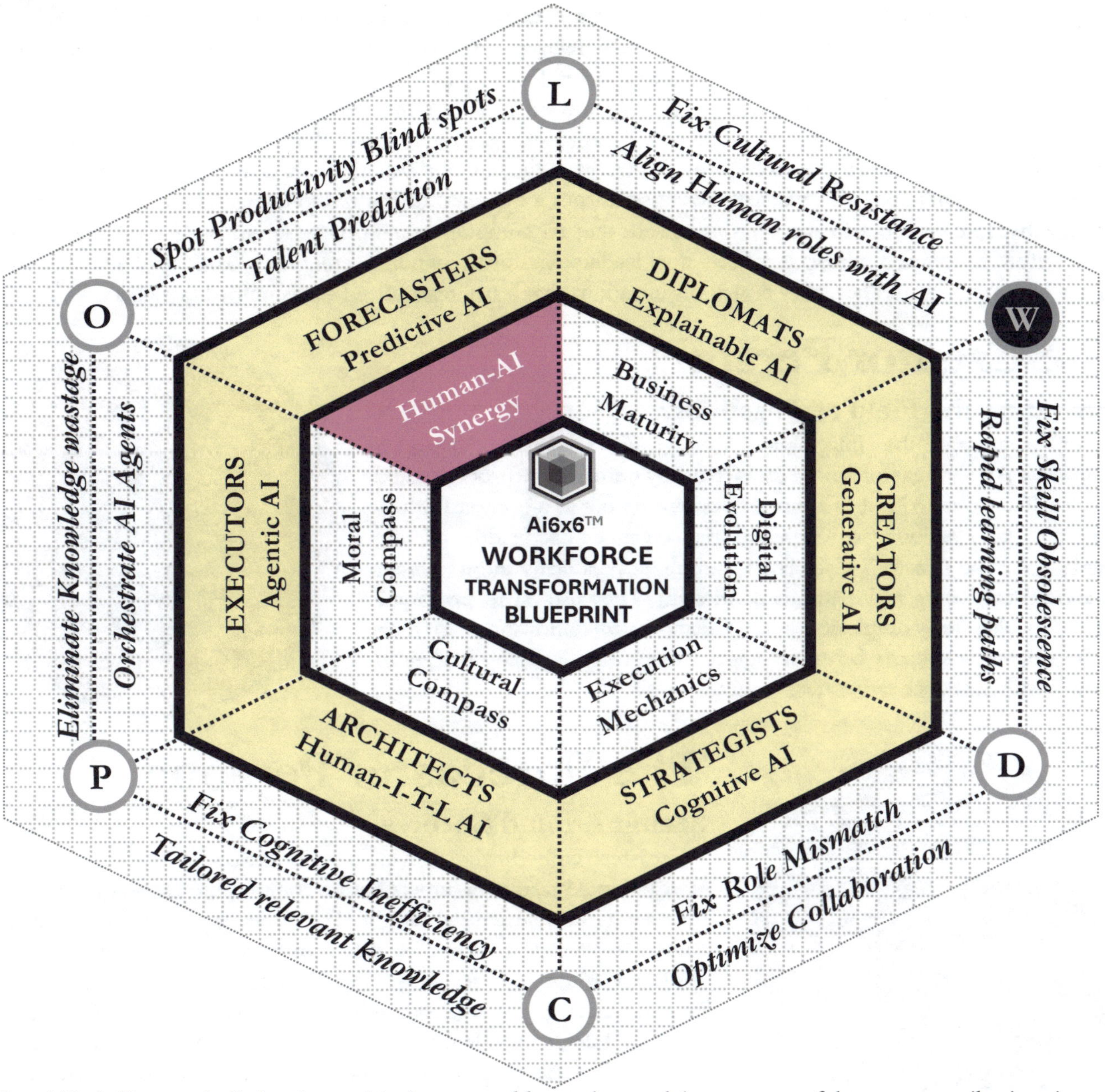

The **Ai6x6 Research Lab** chose this lens to address the evolving nature of human contribution in an automated world. We believe the modern workforce is undergoing a shift from manual execution to high-level orchestration, where the employee becomes a partner to intelligence. This lens is vital because it moves the worker beyond the fear of displacement and into a state of empowerment through diverse archetypes. By exploring the synergy between human talent and agentic systems, the Lab ensures that labor needs are met through a collaborative architecture. It is designed to transform the workforce from a collection of task-doers into a network of creators and strategists who command AI to amplify their unique output.

It is about shifting the worker's heart from a posture of uncertainty to one of agency and trust. This lens serves as the cognitive bridge, allowing the individual to see AI not as a competitor for their role, but as a diverse set of personas that extend their own professional reach. Without this shift, the workforce remains reactive; with it, they become the architects of their own productivity. It is an invitation to shed the burden of repetitive labor and embrace a new, augmented career path grounded in strategic oversight and creative fulfillment. The Lab ensures that as tasks are completed by machines, the "why" behind the work remains fundamentally and powerfully human.

Leadership in the age of intelligent systems demands a new grammar of collaboration—one that transcends the binary of human versus machine and instead frames intelligence as a continuum. The **Human–AI Synergy Lens** redefines leadership not as the art of commanding technology but as the discipline of co-creating with it. It asks leaders to move beyond automation and efficiency toward a deeper integration of foresight, ethics, creativity, cognition, governance, and execution. At its core, Human–AI Synergy is about **shared agency**. It recognizes that leadership today is exercised not only through human judgment but also through algorithmic influence. Every forecast, recommendation, and automated decision carries a fragment of leadership intent. The challenge is not to suppress that influence but to shape it—to ensure that AI amplifies human purpose rather than diluting it. This lens therefore becomes a mirror: it reflects how leaders distribute cognition, emotion, and control across human and machine systems, and how they design organizations that learn, reason, and act as unified intelligences.

The Creator Persona

Expanding the Field of Possibility

Creators represent the imaginative frontier of leadership. Through **Generative AI**, they explore more ideas than any human-only process could conceive. They treat AI not as a threat to originality but as a catalyst for it—a partner in ideation that expands the creative bandwidth of the organization. The Creator's discipline is not about producing more content but about discovering new patterns of thought. They use AI to prototype futures, to visualize complexity, and to democratize innovation. In their hands, generative systems become engines of curiosity, enabling leaders to design from abundance rather than scarcity.

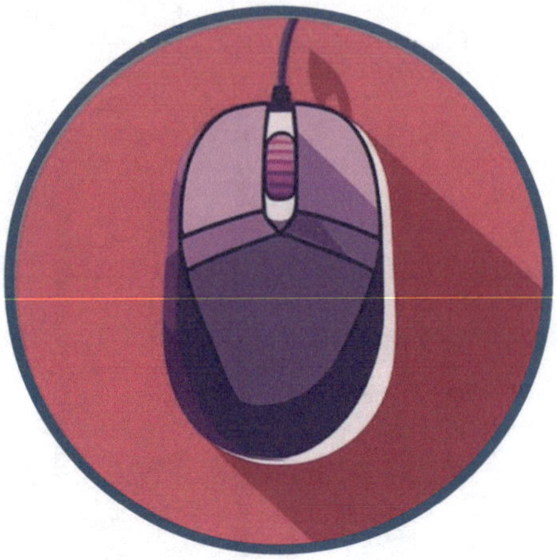

The Forecaster Persona

Seeing Around Corners

Forecasters embody the anticipatory dimension of leadership. They use **Predictive AI** to model futures before committing resources, transforming uncertainty into structured insight. Their craft lies in converting data into foresight—detecting weak signals, pressure-testing assumptions, and rehearsing alternative scenarios. In traditional leadership, forecasting was intuition-driven; in synergistic leadership, it becomes a disciplined dialogue between human imagination and machine probability. The Forecaster's maturity is measured not by how accurately they predict but by how responsibly they prepare. They shift the organization from reacting to events to anticipating them, embedding resilience into its strategic DNA.

The Strategist Persona

Orchestrating Cognitive Intelligence

Strategists use **Cognitive AI** to weave intelligence across functions, geographies, and time horizons. They design strategy as a living system—adaptive, self-correcting, and continuously learning. The Strategist's genius lies in orchestration: aligning distributed intelligences into coherent purpose. They understand that cognition is no longer centralized in the executive suite; it flows through networks of humans and machines that co-reason in real time. Their leadership is systemic rather than hierarchical, emphasizing feedback loops, dynamic modeling, and cross-domain synthesis. In this persona, AI becomes the connective tissue of enterprise intelligence.

The Diplomat Persona:

Building Ethical Bridges

Diplomats operate at the intersection of **Explainable AI** and human empathy. They translate algorithmic logic into moral language, ensuring that machine outputs are understood, trusted, and socially accepted. Their role is to humanize the interface between data and decision, to make transparency a leadership act. In a world where AI can both illuminate and obscure, the Diplomat safeguards the integrity of interpretation. They remind the enterprise that trust is not a technical feature—it is a cultural outcome. By bridging emotional and ethical terrain, they ensure that AI strengthens relationships rather than eroding them.

The Architect Persona:

Designing Responsible Collaboration

Architects are the custodians of **Human-in-the-Loop AI**, ensuring that governance and ethical guardrails are embedded into every workflow. They design environments where humans and machines collaborate responsibly—where oversight is not an afterthought but a structural principle. The Architect's leadership is infrastructural: they build the conditions under which synergy can thrive. Their focus is on accountability, transparency, and inclusion, crafting systems that respect both human dignity and machine precision. In their world, collaboration is engineered, not improvised; it is sustained by design integrity.

The Executor Persona:

Converting Insight into Action

Executors close the loop between knowing and doing. Through **Agentic AI**, they embed intelligence directly into operations so that insight becomes action without delay. They represent the kinetic dimension of synergy—the moment where cognition translates into execution. The Executor's mastery lies in operationalizing intelligence: automating decisions that are safe, contextual, and aligned with strategic intent. They ensure that the organization's nervous system is responsive, not reactive, and that every process carries the imprint of learned intelligence.

Together, these six personas form a **leadership spectrum**—from foresight to action, from imagination to governance. They illustrate that synergy is not a single capability but a choreography of roles. The Forecaster anticipates, the Diplomat interprets, the Creator imagines, the Strategist aligns, the Architect safeguards, and the Executor delivers. Each persona represents a distinct mode of engagement with AI, yet all converge on one principle: leadership is now a shared performance between human and machine. The Human–AI Synergy Lens therefore reframes leadership as **co-evolution**. It invites leaders to cultivate humility before complexity, curiosity before certainty, and design before control. It challenges them to lead not by commanding intelligence but by convening it—to orchestrate a dialogue between human wisdom and machine capability that continuously redefines what is possible. In this lens, synergy is not the end state; it is the ongoing discipline of learning together.

HUMAN-AI SYNERGY ANALOGY – ADMINISTERING SMART CITY

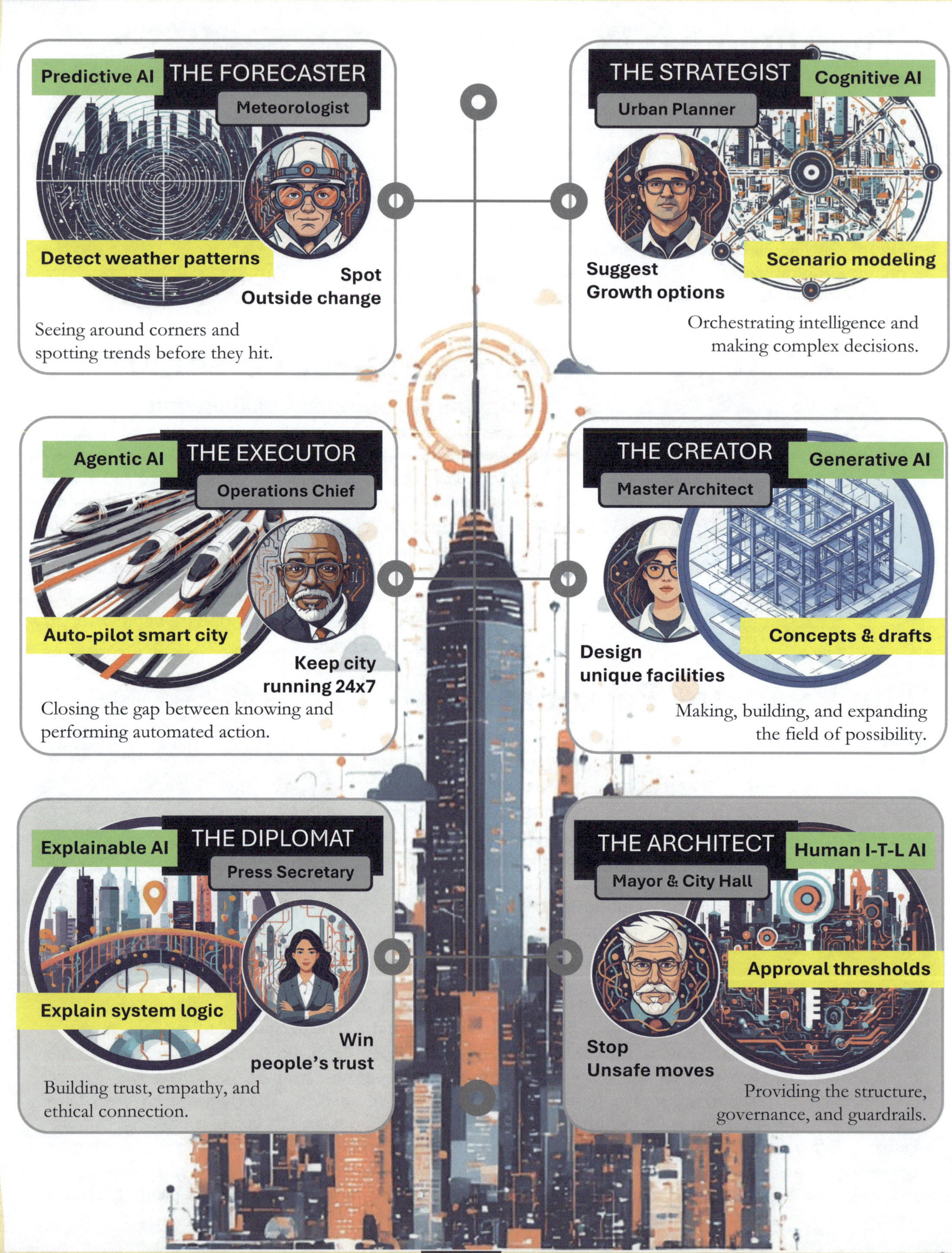

Ai6x6™ Transformation Blueprint

Evolution of Humanity

"No man's labor is unworthy of a man, unless he has bad ends."

Immanuel Kant (1724–1804); Human labor is elevated to worthiness when machines handle toil, allowing workers to focus exclusively on Specialized Judgment and unique cognitive ends.

For the ancient master craftsman, fulfillment was achieved through mastery and the flow state (Techne). The artisan experienced deep satisfaction by owning the entire process, from raw material to finished product, finding dignity in the uniqueness and quality of their specialized judgment. The Industrial Age shattered this model through Taylorism and the assembly line. The worker was alienated from the final product, exchanging mastery for mass efficiency. Fulfillment was reduced to subsistence, and the human became a high-friction gear in a mechanical process, losing the connection to creation. The AI Era offers a radical return to a higher form of this fulfillment. By automating all predictable work (the 80% of repetitive toil), the workforce is liberated to exercise only their unique, Specialized Judgment. Fulfillment is now found in solving novel, non-standard failures and engaging in Ambiguity Training—tasks where human creativity and critical thinking are indispensable. This environment engineers a consistent Flow State for cognitive labor, valuing the individual for their depth and complexity, rather than their obedience or stamina.

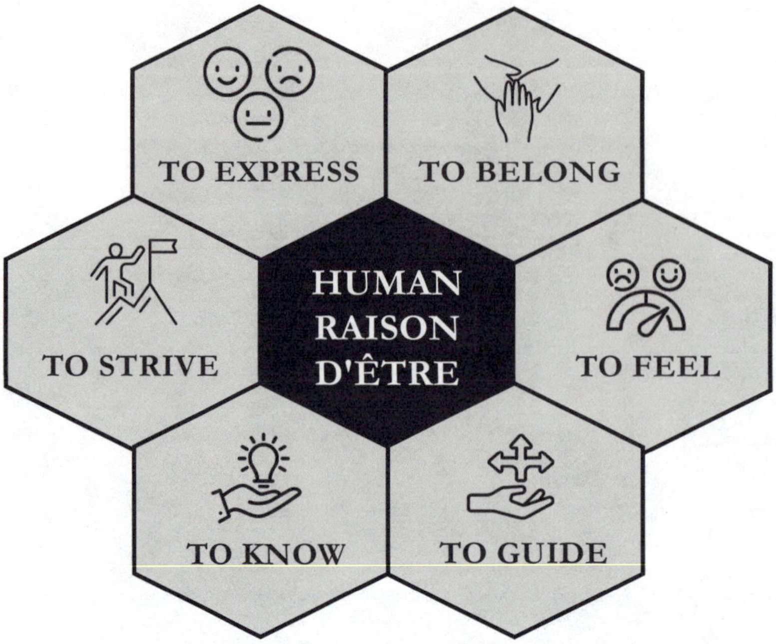

SIX UNIVERSAL ANCHORS OF HUMAN FULFILLMENT

HUMAN RAISON D'ÊTRE (THE REASON FOR EXISTENCE)

In an age defined by exponential automation, the core question of human existence shifts from survival to **significance**. When technology handles efficiency and labor, human value is no longer defined by capability, but by **meaning**. The six **Universal Anchors of Human Fulfillment** define the enduring, non-negotiable contributions of human consciousness—the pillars of a **Raison D'être** that secures our purpose across any era. These anchors affirm that our ultimate mission is to serve as the unwavering source of **Meaning and Moral Authority** in existence. They are the essential acts that only consciousness can perform.

1) TO EXPRESS

The **Anchor of Expression** is the essential human mandate to be the **Source of Novelty and Creation**, generating original ideas and subjective feelings that resist calculation or statistical synthesis. This purpose validates our unique consciousness by asserting that fulfillment comes from the generative act of bringing forth something utterly unique, not merely consuming or replicating. It requires **Uncurated Creativity**—the freedom to explore and define **Aesthetic Judgment**—thereby expanding the conceptual universe with concepts that have no statistical precedent. Through expression, we transform complex inner experience into external reality, upholding our role as the inventors of meaning.

The ultimate benefit of this anchor is the preservation of **Human Originality** and **Intellectual Sovereignty**. By committing to expression, we secure a future that is continually enriched by unique ideas, preventing stagnation and the predictability of automated systems. This anchor maintains the essential spark of invention, ensuring that human meaning remains tied to the capacity to create, **differentiating our value from that of a machine**. It is the source of all artistic, philosophical, and inventive breakthroughs that define a truly dynamic culture.

2) TO BELONG

The **Anchor of Belonging** defines our core identity as social beings, asserting that **authentic connection** is an existential necessity beyond mere digital interaction. This purpose requires the active, continuous labor of **Deep Connection**, prioritizing **unconditional presence** and shared vulnerability to foster a resilient community. Fulfillment is achieved by being the **social architect**, engaging in the emotional work of **mediating conflict**, and fostering cohesion. It includes preserving and transmitting **Cultural Heritage**—the stories and qualitative wisdom that bind us across generations. To belong is to confirm that our worth is interwoven with our relationships.

The ultimate benefit of this anchor is **Emotional Resilience** and **Social Stability**. By prioritizing genuine human connection, we safeguard ourselves against isolation and the psychological fragility that arises from purely transactional relationships.[1] This anchor ensures a supportive social framework that validates empathy and provides necessary emotional resonance, acting as a crucial defense mechanism against an isolating technological world. It builds the **trust and cohesion** necessary for any human system to function justly and compassionately.

3) TO FEEL

The **Anchor of Feeling** grounds human existence in the tangible, asserting the profound psychological value of **Embodied Mastery** and sensory engagement. This purpose elevates the dedicated pursuit of physical skill and process—not for necessity, but for the inherent satisfaction of the process. It is the quest for the **'Flow' state**, where an earned link between **exertion and accomplishment** provides deep fulfillment. This anchor mandates resisting the ease of total automation by engaging in voluntary challenges and perfecting non-essential crafts. It ensures that our abstract thoughts are grounded in the undeniable reality of the body and the senses.

The ultimate benefit of this anchor is **Physical and Psychological Integrity**. By committing to felt experience, we preserve our cognitive and physical health, resisting the inertia of automation and maintaining a sharp connection to reality. This anchor validates the **dignity of effort**, providing a clear, visceral confirmation of personal competence and self-mastery that is crucial for self-esteem. It ensures that fulfillment remains an **earned state**, confirming the complexity and richness of being physically alive.

4) TO GUIDE

The **Anchor of Guidance** establishes the human being as the **ultimate moral and ethical sovereign**. This purpose is founded on our unique capacity for empathy and subjective value judgment, making us the non-delegable authority for setting rules and managing risk. It is the practice of **Ethical Governance**, demanding that we **Draw the Moral Line** and translate compassionate principles into enforceable policy. Fulfillment is achieved by maintaining transparency, enforcing accountability, and continuously exercising the **Human Veto Power** over critical decisions.

The ultimate benefit of this anchor is **Systemic Integrity** and **Moral Safety**. By retaining final ethical authority, we ensure that technological power remains fully aligned with, and subservient to, human values, preventing systems from optimizing themselves at the expense of human welfare. This anchor provides the essential check against unchecked power, establishing the legal and moral frameworks necessary to protect all citizens. It guarantees that empathy and justice remain the **supreme architects of our collective future**.

5) TO KNOW

The **Anchor of Knowing** defines our core intellectual purpose as the engine of **Epistemic Exploration** and abstract reasoning. This purpose compels the mind to focus on formulating **Axiomatic Hypotheses**—the fundamental "Why?" questions that lie beyond the reach of data processing and established knowledge. Fulfillment is found in the generative act of **challenging established truths** and creating new theoretical frameworks that synthesize disparate fields. This anchor ensures that the pursuit of knowledge remains boundless, driven by philosophical depth rather than functional utility.

The ultimate benefit of this anchor is **Intellectual Freedom** and **Conceptual Progress**. By focusing on formulating new questions and challenging paradigms, we prevent the stagnation that could result if all knowledge were confined to data-driven answers. This anchor guarantees that our intellectual journey continues to be defined by profound theoretical breakthroughs and a relentless expansion of understanding. It confirms that human intelligence is defined not by the volume of information stored, but by the **depth and originality of the questions** we ask.

6) TO STRIVE

The **Anchor of Striving** is the **Existential Quest**—the voluntary pursuit of difficulty necessary to generate meaning, independent of any external necessity. This purpose recognizes that true fulfillment is an **earned state**, achieved through **Deliberate Discipline** and self-imposed ambition. It compels us to **Set Impossible Goals** and **Choose the Hard Way**—overcoming voluntary hardship—precisely because the challenge validates our capacity and resolve. This continuous effort allows us to craft a meaningful **Personal Legacy** by executing complex, long-term plans purely for the satisfaction of the effort.

The ultimate benefit of this anchor is **Authentic Self-Mastery** and **Purposeful Existence**. By embracing self-imposed difficulty, we cultivate the resilience, discipline, and fortitude that define character, resisting the complacency of automated comfort. This anchor ensures that human worth is determined by our **internal will**, rather than external circumstance, providing a framework for continuous personal growth and the creation of an enduring, meaningful legacy.

CONCLUSION: THE RAISON D'ÊTRE FOR THE HUMAN

The six Universal Anchors converge to form humanity's ultimate purpose: to be the **Unwavering Source of Meaning and Moral Authority**. This composite Raison D'être defines the essential human role in any advanced civilization.

Our mission is the sustained commitment to **non-automatable value**. We are the unique entity that not only pursues truth (To Know) but also creates the beauty (To Express) that makes truth worth pursuing. We are the mechanism that defines the rules of fairness (To Guide) and then personally invests the necessary effort (To Strive) to live up to those rules. Crucially, we are the only entity that grounds this entire conceptual structure in the tangible joy of physical reality (To Feel) and the vital necessity of shared vulnerability (To Belong).

The human purpose shifts from survival to an intentional journey of **growth, creation, and ethical custodianship**. Our value is not in output, but in the **input of consciousness, conscience, and will**. By diligently exercising these six anchors, every individual contributes to the essential, enduring purpose of the species: to be the **Sole Arbiter of Worth, Will, and Wonder**—ensuring that existence remains a deeply meaningful and purposeful quest.

In this era, human existence was focused on the direct, tangible application of skill for immediate survival, such hunting, fishing and surviving. The human was the Master of the Hunt.

In the Primitive/Paleolithic Era (Survival & Sustenance), human purpose was an absolute function of immediate necessity. Fulfillment was not a philosophical pursuit but a practical outcome of successfully continuing existence for oneself and the kin group. All six Universal Anchors manifested, but they were tethered directly to the biological and environmental imperative to survive the day and propagate the species. The overarching Raison D'être was to **Endure and Transmit**, meaning personal worth was inseparable from one's contribution to the group's survival calculus. This collective focus meant that individual talent was instantly converted into a benefit for all, solidifying a direct, unbroken link between personal action and existential significance.

To Express: Expression was functional and communal, centered on **Ritual Storytelling** and cave art. This art was not for individual fame or abstract aesthetics, but served as the primary educational system, preserving and transmitting essential survival knowledge, hunting techniques, spiritual beliefs, and the topography of the land. The repetitive, sacred nature of the rituals ensured collective memory and adherence to successful methods. Furthermore, these cultural forms encoded crucial information, often representing animals and spirits, that, if lost or forgotten, would jeopardize the tribe's safety and cultural identity.

To Belong: The Anchor of Belonging was non-negotiable, manifesting as **Kinship and Mutual Defense**. Isolation meant certain death; therefore, the purpose of connection was absolute. Fulfillment came from actively contributing to the group's defense, sharing food and resources, and ensuring the collective physical safety of the tribe above all else, fostering a fierce, self-sustaining mutualism. The emotional security derived from this deep collaboration—knowing that others were committed to your well-being—was a core psychological reward.

To Feel: Embodied Mastery was defined by **Essential Tool Mastery**. The feeling of accomplishment was directly tied to successfully chipping a sharp stone tool, accurately throwing a spear, or enduring the physical hardships of the hunt. This pursuit was not sport, but the daily necessity of mastering the physical world to secure food and shelter, and the psychological confidence gained was crucial for future success. This expertise immediately translated into social value; the most capable hunter or tool-maker earned respect and a higher status within the hierarchy, directly reinforcing the dignity of effort.

To Guide: Moral authority was based on immediate experience, taking the form of **Elder Authority and Tradition**. Guidance was unwritten law, passed down through generations of hard-won experience. Decisions—like where to hunt, when to migrate, or how to resolve disputes—were arbitrated by those with the deepest, most successful experiential knowledge, and were obeyed out of necessity for the group's stable cohesion. This system was vital for ensuring collective action remained predictable and risk-adverse, preventing internal conflict from threatening external survival.

To Know: The pursuit of knowledge was pure **Observation and Practicality**. The purpose of knowing was to accurately predict the immediate environment for resource exploitation. This involved memorizing the habits of migratory animals, understanding the medicinal or poisonous properties of local plants, and tracking subtle shifts in weather patterns—information critical for daily sustenance and successful, low-risk planning. Early forms of abstract reasoning also emerged here, through tracking the movement of stars or noting the cyclic nature of seasons and astronomical events, demonstrating an innate human desire to find order and predictability in environmental chaos.

To Strive: Striving was synonymous with **Daily Endurance**. There was no need to invent abstract goals; the pursuit of difficulty was the environment itself, offering constant challenge. Fulfillment was the successful completion of the continuous, grueling effort required to secure food, withstand the elements, and ensure the basic propagation of the tribe, creating a deeply ingrained sense of earned worth. The resilience built through continuous hardship was the ultimate personal asset, psychologically connecting physical suffering with the existential meaning of survival.

With stability, purpose shifted to sustained output, long-term planning, and maintaining social harmony within a fixed settlement. The human was the Master of the Land.

The Agricultural/Neolithic Era (Farming & Community) marked humanity's fundamental pivot from nomadic struggle to settled stability. The mastery of cultivation and animal husbandry created a reliable **Surplus of Time and Resources**, fundamentally shifting human purpose from immediate survival to calculated investment, long-term order, and the establishment of lasting community structures. The Raison D'être of this period was to **Settle and Establish Order**. This era saw humans take on the role of conscious managers of the environment, laying the groundwork for systematic laws, property rights, and enduring social legacies that extended far beyond one generation.

To Express: Expression became focused on **Accounting and Enduring Form**. The logistical necessity of tracking seeds, yields, stored grains, and communal ownership led directly to the invention of the first systematic writing forms, such as early cuneiform, demonstrating that expression was initially a tool for managing complex resources. Beyond text, expression was seen in durable, functional goods like specialized pottery, permanent architectural structures, and woven textiles. These forms reflected a deep, stable community identity and provided evidence of specialized skill that contributed directly to the village's permanence.

To Belong: Belonging shifted to being centered on **Land-Based Cooperation**. The success or failure of the farm was the success or failure of the entire settlement, making shared labor and mutual dependency paramount. Fulfillment was found in the reliable rhythms of shared work, such as irrigation projects or harvesting, cementing generational inheritance and deeply intertwined **Kinship Networks**. Purpose was now inseparable from one's physical, defined place within the land and the local community, creating a profound sense of rootedness and stability unknown in the Paleolithic past.

To Feel: Embodied Mastery focused on **Craftsmanship and Process**. Fulfillment was derived from the methodical, repeated application of skill required by agriculture—plowing a straight furrow, carefully managing livestock, or perfecting the complex processes of metallurgy and food preservation. This mastery provided a sense of control over the environment that raw endurance could not. The purpose of skill was purely to generate the stable surplus required to maintain the settled life, providing a deep psychological reward through the predictable yield of sustained, expert effort.

To Guide: Moral authority became formalized, focusing urgently on **Property Rights and Resource Management**. Guiding the community meant establishing clear, complex rules around ownership of land, managing precious resources like water rights, and ensuring crop protection from both animals and neighbors. This necessity gave rise to the first systematic, written laws, moving purpose from inherited tradition (as in the Paleolithic) to a **Codified Social Contract**. Leadership was defined by the ability to enforce these new laws, ensuring fairness and stability within an increasingly complex economy.

To Know: Abstract reasoning began to flourish, driven by **Mathematics and Calendar Tracking**. The purpose of knowledge was now predictive certainty, essential for farming success. This meant mastering geometry to accurately divide and manage fields, and engaging in sophisticated celestial observation to track seasons and accurately determine optimal planting and harvesting cycles. This pursuit of cosmic order confirmed the human ability to impose intellectual structure upon chaotic nature, turning farming into a sophisticated, knowledge-driven enterprise.

To Strive: Striving fundamentally shifted from day-to-day survival to **Creating Surplus and Investment**. The existential quest became the arduous, long-term, deliberate work required to build durable structures, amass protective stores of food, and ensure the economic and social prosperity of the next generation. This focus on legacy and investment transformed work into a purposeful act of building a lasting future, where one's worth was measured not just by survival, but by the inheritance left behind for the family and the stability provided to the community.

This era defined purpose through professional expertise and contribution to complex, often abstract, technological and economic systems. The human was the Master of the Machine.

The Modern/Industrial Era (Specialization & Function) marked the decisive triumph of machine logic and capital over the natural world. Human purpose shifted from managing the cycles of the land (Neolithic) to mastering the mechanical efficiency of the factory and the market. The rise of the machine and the division of labor fundamentally redefined the individual, whose worth was now derived from highly specialized, functional expertise within a vast, interconnected economic system. The Raison D'être of this period was to **Achieve and Advance**, driving relentless progress through expertise and material gain.

To Express: Expression became focused on **Mass Media and Invention**. With the advent of the printing press, cheaper paper, and later, radio and cinema, creative expression transformed into a commodified product designed for wide distribution and profit. The purpose of expression was to inform, entertain, and persuade the masses, shaping shared cultural narratives through novels, newspapers, and commercial design. The creation of a **Patentable Invention** became a powerful act of individual expression, translating originality directly into economic ownership and societal contribution. Expression was celebrated when it served a functional, marketable, or propagandistic role.

To Belong: Belonging was redefined by **Nation and Institution**. The old agrarian kinship networks dissolved as masses migrated to urban factory centers, replaced by new, abstract, yet powerful social structures. Fulfillment was found in national identity, membership in professional organizations, and political affiliation. Crucially, the extreme specialization of labor necessitated the formation of organized counter-structures like **Labor Unions** to protect the rights of workers against centralized corporate power, demonstrating that belonging was a functional necessity for collective bargaining and social stability in a highly unequal economic system.

To Feel: Embodied Mastery centered on **Efficiency and Maximized Output**. The physical purpose shifted away from holistic craft towards highly repetitive, specialized labor—the mastery of a single, precise task on the assembly line. While monotonous, this specialization provided a clear, measurable sense of contribution, where personal worth was directly tied to economic output and reliability. Outside of the factory, the Anchor manifested in the rise of **Organized, Competitive Sports**, where physical effort was deliberately structured and formalized, allowing the industrial individual to achieve a sense of earned competence and visceral struggle that was absent in their work.

To Guide: Moral authority became formalized through **Secular Law and Corporate Regulation**. The complexity of contract law, property ownership, and massive corporate liability demanded sophisticated, non-religious guidance systems. The purpose of guidance was to manage the inherent conflicts between capital and labor, protect consumer safety in mass production, and define the boundaries of ethical economic competition. This period saw the relentless development of regulatory bodies and constitutional law aimed at establishing fairness within the new, powerful industrial economy, ensuring that **progress did not undermine social order** entirely.

To Know: The pursuit of knowledge was dominated by **Empirical Science and Specialization**. The purpose of knowing was functional utility, believing that all problems—from disease to industrial inefficiency—could be solved through the application of the rigorous **Scientific Method**. Knowledge became fragmented into specialized disciplines, each dedicated to gathering objective, measurable data. Fulfillment was achieved by contributing a verifiable, peer-reviewed fact or theory to one's specific field, thereby confirming the era's deep-seated faith in rational inquiry and technical progress.

To Strive: Striving was synonymous with **Ambition and Material Gain**. The existential quest was the pursuit of wealth, social mobility, and **Status**. Fulfillment was an earned state achieved through economic success, the establishment of a profitable business, or professional advancement within a corporate or governmental hierarchy. The cultural narrative of the "self-made individual" defined personal purpose, urging people to embrace continuous, relentless work to secure prosperity for the family and, more importantly, to carve out an enduring legacy of material success and high societal standing.

In this emerging era, AI handles efficiency and role specialization, freeing and forcing humans to refocus on subjective values, unique creation, and moral governance. The human becomes the Master Curator of Purpose

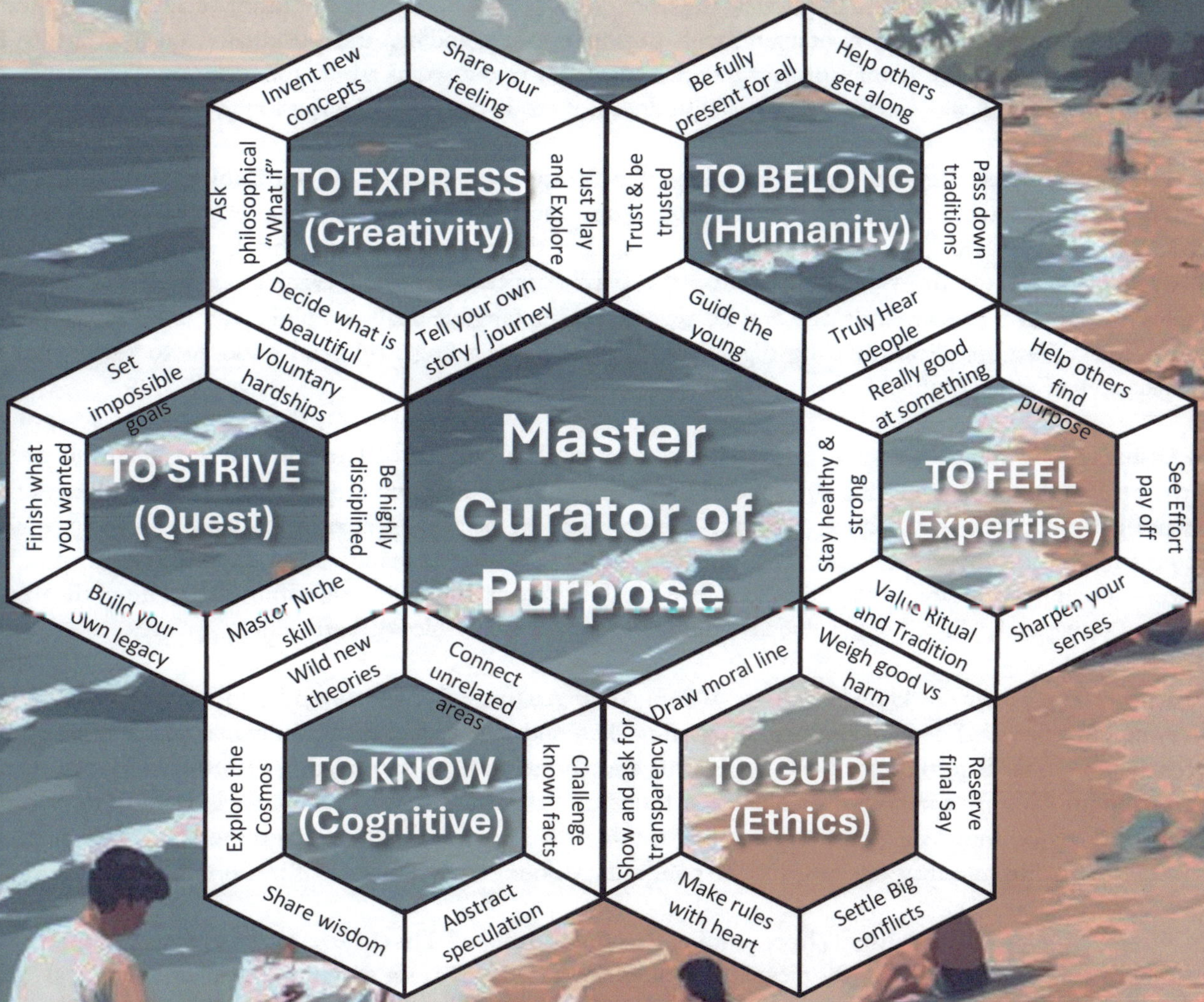

The Post-AI Era (Curation & Meaning) represents the definitive existential shift in human history. With advanced artificial intelligence and automation having conquered the challenges of efficiency, information processing, and mass labor, human purpose is radically redefined. We are no longer necessary for function, but for **meaning**. The individual's worth is derived entirely from their capacity for subjective judgment, ethical authority, and the creation of non-calculable value. The collective Raison D'être is to become the **Sole Arbiter of Worth, Will, and Wonder**—the source of all criteria for why life should continue and what makes it beautiful.

To Express: Expression is focused on **Subjectivity and Uncurated Truth**. In a world saturated with synthetic media and hyper-optimized content generated by AI, the only expression that retains intrinsic value is the genuinely **original, subjective human voice**. This purpose mandates focusing on authentic, unfiltered creation—personal philosophical content, aesthetic critique, memoir, and art that contains flaws, uncertainty, and emotional resonance. Fulfillment is found in generating novelty that AI cannot predict, thereby constantly pushing the boundaries of human experience and securing a creative future that is immune to algorithmic stagnation.

To Belong: Belonging becomes a deliberate, chosen investment in **Deep Vulnerability and Presence**. When digital connection is ubiquitous but emotionally shallow, true fulfillment relies on actively resisting transactional relationships and seeking authentic, physical presence. This purpose demands the **labor of Deep Connection**: cultivating empathy, mediating conflict through personal understanding, and prioritizing quality, face-to-face interaction. The collective human project requires maintaining the social fabric and emotional resonance that machines cannot replicate, ensuring that isolation does not erode the psychological foundation of society.

To Feel: Embodied Mastery transforms into the **Voluntary Pursuit of Difficulty and Flow**. Since automation eliminates the necessity of difficult labor, human physical purpose centers on self-imposed challenges—non-essential crafts, athletics, meditation, and the perfection of technical, yet unnecessary, skills. This pursuit generates the vital 'Flow' state—the psychological reward of earned accomplishment—which secures self-esteem and grounds abstract consciousness in tangible reality. Fulfillment is achieved through **Deliberate Effort** that reaffirms the dignity of the body, providing a visceral antidote to the inertia and complacency of total comfort.

To Guide: Moral authority centers on **AI Ethics and the Human Veto**. The purpose of guidance is to serve as the **ethical sovereign**, placing strict moral boundaries around autonomous systems. This involves defining the principles of data governance, ensuring transparency in black-box algorithms, and enforcing systemic accountability for technological outcomes. Crucially, every individual is mandated to exercise the **Human Veto Power**—the non-delegable right to overrule critical decisions made by algorithms—ensuring that power remains permanently aligned with subjective human values, not cold-blooded efficiency.

To Know: The pursuit of knowledge is now defined by **Abstract Synthesis and Axiomatic Inquiry**. In an environment where AI holds access to all existing data and established facts, human intellect must shift its focus. The purpose is to transcend data processing and concentrate on formulating **Axiomatic Hypotheses**—the profound "Why?" questions that challenge foundational assumptions. Fulfillment is found in connecting disparate fields of knowledge, creating new philosophical frameworks, and constantly resisting the temptation to settle for algorithmic answers, ensuring the continuous, boundless expansion of human understanding.

To Strive: Striving is the existential commitment to **Self-Imposed Discipline and Personal Legacy**. Since survival is guaranteed, the quest for meaning requires the individual to set ambitious, non-essential goals purely for the satisfaction of the effort. This involves cultivating profound resilience and fortitude against the immense temptation of automated ease. Fulfillment is an **earned state**, achieved through continuous, conscious effort and the construction of an enduring **Personal Legacy**—a body of work or moral character that defines one's worth by internal resolve rather than external economic output.

Ai6x6™ Transformation Blueprint

Human-AI Personas

"AI won't replace humans, but those who use AI will replace those who don't."
Garry Kasparov, the legendary former World Chess Champion.

The Ai6x6 HUMAN-AI Synergy framework formalizes a new executive standard for the AI era, transitioning organizations from fragmented automation to a state of Orchestrated Autonomy. This model defines the modern enterprise through six core leadership movements, each serving as a specialized "sense" for the organization. **The Forecasters** serve as the "Eyes," tasked with the heavy lifting of foresight to eliminate strategic blind spots and detect invisible market signals before they escalate into crises. Supporting this vision, **The Diplomats** act as the "Voice," translating complexity into human trust and ensuring that transformation is met with cultural alignment rather than institutional resistance. The organization's creative energy is fueled by **The Creators**, the "Engine" of the enterprise, who move beyond static production to drive a state of radical velocity and infinite iteration. Their output is balanced by the surgical objectivity of **The Strategists**, acting as the "Compass" to ensure that every capital investment is mathematically anchored to peak ROI. To unify these efforts, **The Architects** build the "Institutional Brain," ensuring that the organization's collective memory is always accessible and relevant to the individual. Finally, the framework finds its movement in **The Executors**, the "Hands" of the organization, who coordinate autonomous workflows to achieve scale without traditional overhead. By weaving these archetypes into a single Symphony, leaders can navigate market ambiguity with surgical precision, transforming the enterprise into a self-evolving system of sustainable innovation.

Focus: Neutralizing Talent Friction and Safeguarding Human Potential

In the **Ai6x6 Human-AI Synergy framework**, the defensive stance for the workforce is about protecting the organization's most volatile and valuable asset: human talent. As AI integrates into the enterprise, the primary defensive challenge is preventing the erosion of morale, the loss of institutional wisdom, and the "Cognitive Drag" caused by inefficient human-machine collaboration.

The Forecaster and The Diplomat: Anticipating Burnout and Dissolving Fear

The **Forecaster** serves as a defensive early-warning system for human capital. By analyzing digital workflows and engagement signals, it identifies **Productivity Blindspots**—the subtle patterns that precede burnout or mass attrition. Its defensive mandate is to see "talent fatigue" before it manifests in the balance sheet, allowing leadership to intervene before high-value experts leave the firm.

Complementing this, the **Diplomat** is tasked with killing **Cultural Resistance**. The introduction of AI often triggers a "threat response" in employees. The Diplomat utilizes empathetic logic to bridge the gap, ensuring that AI is seen as a tool for expansion rather than a herald of replacement. By addressing the psychological friction of transformation, it protects the organization's cultural integrity and ensures that talent remains engaged rather than defensive.

The Creator and The Strategist: Preventing Skill Decay and Role Mismatch

The **Creator** acts defensively by neutralizing **Skill Obsolescence**. In a rapidly evolving market, human skills can become liabilities overnight. The Creator generates "just-in-time" learning modules and creative scaffolds that allow employees to adapt their existing expertise to new AI-driven workflows. It ensures the organization's "Human Engine" doesn't stall due to a lack of modern technical fluency.

The **Strategist** defends the workforce by eliminating **Role Mismatch**. Much talent wastage occurs when people are assigned to tasks based on outdated job descriptions rather than real-time strengths. The Strategist audits performance data to ensure that cognitive energy is perfectly aligned with the project's needs. By preventing "talent misallocation," it protects the individual from frustration and the company from the inefficiency of having the wrong minds on the right problems.

The Architect and The Executor: Capturing Wisdom and Reducing Drudgery

The **Architect** kills **Knowledge Wastage**. In many firms, critical wisdom resides solely in the heads of senior experts. When they leave, the company's "memory" is deleted. The Architect defensively captures this tribal knowledge, indexing it into a searchable, persistent institutional brain. It ensures that the workforce is protected against "Information Loss," making the organization's collective intelligence immune to individual turnover.

Finally, the **Executor** serves the workforce by killing **Administrative Friction**. A significant portion of employee dissatisfaction stems from "drudgery"—the manual, repetitive tasks that drain cognitive energy. The Executor automates these low-value workflows, acting as the defensive "Hands" that clear the path for high-value work. By removing the wastage of manual entry and routine coordination, it safeguards the employee's mental bandwidth for innovation.

Conclusion: Protecting the Human Heart

A defensive workforce strategy within the **Ai6x6 Human-AI Synergy framework** is about creating a "Psychologically Safe" and "Cognitive-Efficient" environment. By shielding employees from burnout, obsolescence, and administrative clutter, the framework ensures that the human element remains the creative heart of the business. When talent is no longer fighting the friction of fear and inefficiency, it is finally free to evolve alongside the technology.

In the **Workforce Transformation** within the Ai6x6 framework, the objective shifts from "Resource Management" to **"Talent Orchestration."** This transition turns the workforce from a static cost center into a dynamic, "Living Asset" capable of evolving in real-time. By applying the six roles to the human capital layer, leadership can move beyond traditional hiring and training into a state of **Adaptive Human-AI Synergy.**

The Forecasters and The Diplomats: Unlocking Potential and Culture

The **Forecaster's** mandate in workforce transformation is **Skill-Gap Anticipation.** By pairing Predictive AI with Graph Neural Networks, the leader identifies not just who the company needs today, but which skills will become obsolete in 18 months. It maps the "Internal Talent Web" to find hidden adjacent skills, allowing the organization to upskill its people before a market shift occurs, ensuring the workforce is never "lagging" behind the technology.

The **Diplomat** serves as the cultural bridge for **Psychological Safety and Alignment.** In a transformation, the biggest friction is fear. By pairing Explainable AI with Sentiment Analysis, the Diplomat ensures that every AI-driven organizational change is explained with radical transparency. It calibrates the "Voice" of the company to address worker anxieties in real-time, building a culture where AI is viewed as an "Exoskeleton for the Mind" rather than a replacement for the person.

The Creators and The Strategists: Unlocking Productivity and Performance

The **Creator** acts as the engine of **Customized Learning and Development.** In this role, Generative AI is paired with RLHF to create hyper-personalized training modules for every individual employee. This "Generative Education" ensures that a junior developer and a senior manager receive different, AI-curated paths to mastery. It allows the workforce to "Learn at the Speed of Light," turning the traditional annual training cycle into a continuous, daily evolution of capability.

The **Strategist** optimizes the **Human-AI Ratio for Peak Yield.** Using Multi-Objective Optimization, the Strategist determines the exact balance of human intuition and machine precision for every department. It identifies the "Performance Frontier," ensuring that teams are neither overwhelmed by tech nor slowed down by manual bottlenecks. This ensures that capital is deployed to the right combination of people and agents, maximizing both employee fulfillment and bottom-line ROI.

The Architects and The Executors: Unlocking Wisdom and Flow

The **Architect** builds the "Searchable Institutional Memory" for the workforce. By pairing Human-in-the-Loop with Vector Databases, the Architect ensures that the "Brain" of the company is accessible to every new hire. This eliminates the "Learning Tax" of onboarding, as every employee is granted instant access to the context-aware wisdom of the firm. It transforms the workforce from a collection of silos into a unified "Cognitive Collective" where knowledge flows without friction.

Finally, the **Executor** manages the **Autonomous Workflow Orchestration.** Through Agentic AI paired with RPA, the Executor offloads the "Digital Drudgery" from the human workforce. By assigning mundane, repetitive tasks to autonomous agents, the "Hands" of the organization are freed to focus on high-value, creative problem-solving. This creates a "State of Flow" across the enterprise, where humans do what they do best—reasoning and empathy—while the agents handle the high-volume execution.

Conclusion: The Symphony of High-Velocity Talent

Workforce transformation is the pursuit of **Human Flourishing through Machine Augmentation.** By unlocking anticipation, alignment, development, optimization, wisdom, and flow, the leader creates a workforce that is not just resilient, but antifragile. This framework ensures that the organization's most valuable asset—its people—is perfectly synchronized with its most powerful tool—its AI—creating a high-velocity symphony of orchestrated potential.

Workforce Power: Anticipatory Competence

In the traditional workforce, employees are trained to be "Reactive." They wait for a problem to occur or a ticket to arrive before they take action. The transformation into **Forecasters** fundamentally changes the "Eyes" of every worker. It moves the staff from a "Service Desk" mentality to a "Sentry" mentality. In this role, every member of the workforce is empowered with the ability to see the "Pre-signals" of their specific domain. Whether it is a warehouse worker spotting a subtle pattern in inventory decay or a customer service agent noticing a micro-shift in caller frustration, the workforce becomes an integrated early-warning system for the entire enterprise.

Defensive Stance: Killing the Reaction Gap

The defensive mission of the Worker-Forecaster is the **Elimination of Latency**. In a high-speed market, the "Reaction Gap"—the time between a problem occurring and a human noticing it—is where profit dies. By training the workforce to look for the "Statistical Whisper" within their own workflows, the leader kills the "Wastage of Delay." This creates a **Shield of Proactivity**. Employees no longer wait for a system to crash or a customer to churn; they identify the "Lead Indicators" of failure. This defensive posture ensures that the organization's "Eyes" are distributed across every level of the hierarchy, killing the blindspots that central leadership could never see on their own. It transforms the workforce into a "Living Radar" that detects friction at the source.

Offensive Stance: Unlocking Opportunity Capture

On the offense, the Forecaster workforce moves toward **Anticipatory Action**. When every employee can see "One Step Ahead," the organization can capture opportunities that are invisible to a reactive competitor. This **Unlocks Localized Prediction**, where a front-line worker can suggest a shift in resources or a change in a process *before* the market demands it. By "Seeing the Wave" early, the workforce can position the company's assets in the optimal spot. This turns the staff into "Opportunity Seekers" who use their predictive "Eyes" to maximize the yield of every hour they work. It moves the organization from "Executing the Plan" to "Refining the Plan" in real-time, based on the foresight of the people closest to the action.

The Workforce Outcome: Collective Foresight

The ultimate value of the Forecaster role is the **Democratization of Strategy**. When the "Eyes" of the workforce are calibrated to the future, every employee becomes a strategic asset. For the leader, this provides **Operational Resilience**. You are no longer the only person looking at the horizon; you have a thousand "Forecasters" identifying risks and rewards simultaneously. This turns the organization into a "High-Resolution Entity"—one that perceives the world with such clarity that it can navigate through complexity with zero hesitation. It ensures that the brand's "Vision" is not a top-down mandate, but a bottom-up reality powered by the "Collective Foresight" of its people.

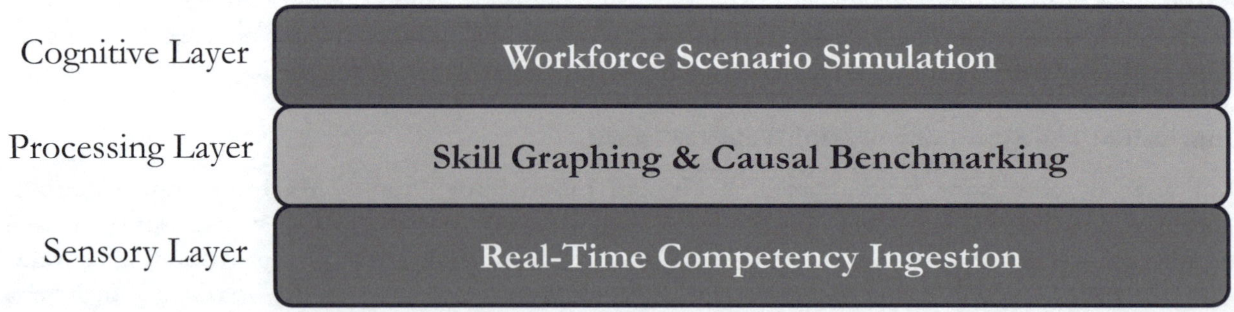

Executive Power: Talent & Skill Perception

In the legacy era, a leader's view of their workforce was limited to static resumes, annual performance reviews, and rigid org charts. The **Forecaster** role in workforce transformation represents the transition to **Predictive Human Capital Management**. For the executive, this means moving from "Hiring for the Past" to "Architecting for the Future." By building a stack that perceives the real-time skill density and cognitive capacity of the firm, the leader gains the clarity to see "Skill Gaps" before they turn into "Operational Failures." The ultimate goal is the **Purification of Talent Allocation**, ensuring that the right cognitive energy is positioned at the exact point of market need before the competition even identifies the vacancy.

The Sensory Layer: Real-Time Competency Ingestion

The **Sensory Layer** of the Workforce Forecaster serves as the organization's "Skill Radar." It moves beyond manual self-reporting and into **Automated Competency Ingestion**. This layer leverages **Digital Exhaust Analysis** anonymized metadata from collaboration tools (Slack, GitHub, Jira) and project outputs—to capture the actual, real-time capabilities of the team. By utilizing **Passive Skills Mapping**, the leader gains a high-fidelity signal of who is actually solving which problems, regardless of their official job title. It removes the bias of "Self-Promotion" and the latency of "Annual Reviews," ensuring the "Eyes" of the firm are seeing the true cognitive pulse of the organization. This layer provides a raw stream of "Capability Truth," allowing the executive to perceive the emerging expertise within the shadows of the formal hierarchy.

The Processing Layer: Skill Graphing & Causal Benchmarking

The **Processing Layer** serves as the "Talent Recognition Engine." It transforms the raw activity stream into a structured **Skill Knowledge Graph**. This layer utilizes **Graph Neural Networks (GNNs)** to map the relationships between disparate skills—understanding, for example, how a proficiency in Python in the engineering department could be causally linked to a high success rate in data-driven marketing campaigns. By applying **Predictive Labor Analytics**, this layer identifies the "Skill Decay" rate of specific departments, signaling when a team's current expertise is becoming obsolete. It moves the leader from "Counting Heads" to "Measuring Potential," providing a "Logical Foundation" for internal mobility and strategic upskilling. This layer ensures the firm's "Workforce Compass" is calibrated to the evolving technological landscape, identifying the causal link between specific skill clusters and long-term project success.

The Cognitive Layer: Workforce Scenario Simulation

The **Cognitive Layer** is the "Reasoning Mind" of the talent stack, where skill data is converted into **Strategic Resourcing Intent**. This layer utilizes **Large Language Models (LLMs)** and **Agentic Simulation** to synthesize the logic of the Processing Layer into "Growth Scenarios." Instead of a static hiring plan, the Cognitive Layer generates thousands of "What-If" workforce models, stress-testing how the organization would perform if specific teams were re-skilled or if generative agents were integrated into certain workflows. It acts as a "Strategic Talent Advisor," using **Natural Language Generation** to articulate complex talent probabilities as clear executive narratives. This allows the leader to move from "Reactive Hiring" to "Proactive Orchestration," authorizing the movement of people and agents to the exact point of future strategic impact, ensuring the brand is always "Staffed for Tomorrow."

OVERVIEW

Predictive AI forecasts future outcomes by analyzing vast historical data using machine learning algorithms. For the "Intuitive & Risk" leader, the AI becomes a crucial tool for quantifying uncertainty. The leader provides their initial intuition and acceptable risk limits, framing the precise question. The AI delivers objective, quantifiable evidence, such as likelihoods and key drivers, that either validates the leader's gut feeling or proposes data-driven alternatives. This synergy converts subjective risk into a calculated, strategic choice, empowering the leader to make proactive, decisive actions.

MAIN STEPS

Start

STEP 1 — Define Objective & Scope
Specify question, target outcome, and limits

STEP 2 — Data Collection & Preparation
Gather, clean, and format historical inputs

STEP 3 — Model Selection & Training
Build patterns from cleaned training data

STEP 4 — Evaluation & Validation
Test model accuracy on unseen data

STEP 5 — Deployment & Integration
Insert model into real-time operations

STEP 6 — Monitoring & Refinement
Continuously check and improve model performance

End

SAMPLE FLOW

Healthcare - Staffing Optimization- Hospital management predicts patient volume surges to optimize staffing levels, ensuring quality care and preventing staff burnout proactively.

1. Model Demand Peak
Train on patient historical admissions, seasonality, and local health event data.

2. Forecast Requirements
Estimate Staffing required RNs, physicians, and support staff for peak loads.

3. Assess Fatigue Metrics
Calculate potential burnout for team based on projected shifts & existing overtime.

4. Identify Gaps and Overages
Compare forecasted needs against currently scheduled staff availability in real time.

5. Simulate Scenarios
Generate optimal re-assignment plans minimizing cost and maximizing employee fairness

6. Execute Rota Adjustments
Approve high-priority staffing adjustments to prevent service disruption or burnout spike.

USE CASES

- **Retail | Future Skill Gap Prediction:** Predicts 12-month future skill shortages in critical areas like e-commerce logistics, guiding proactive internal upskilling programs for current employees.
- **Financial Services | Employee Training Efficacy Forecast:** Forecasts which compliance training modules will reduce error rates, optimizing resource allocation for maximum operational impact and reduced audit risk.
- **Energy | Safety Incident Propensity:** Predicts which field teams are at highest risk of safety incidents based on fatigue, recent equipment maintenance, and weather conditions, triggering immediate safety checks.
- **Transportation | Driver Retention Probability:** Forecasts commercial driver retention based on route load, home time, and compensation structures, allowing management to address dissatisfaction before departure.
- **Hospitality | Service Quality Decline Warning:** Predicts moments when guest service ratings are likely to drop due to low front-of-house staffing and high booking volume, prompting immediate resource shifts.
- **Media | Content Creator Burnout:** Predicts burnout likelihood for key content producers based on output frequency and engagement metrics, enabling managerial intervention to maintain creative quality.

1. THE FORECASTERS (THE EYES) | PREDICTIVE AI PLUS OTHERS...

In the **Workforce Transformation**, the **Forecasters (The Eyes)** serve as the organization's "Anticipatory Radar," scanning the horizon for shifts in human capability and labor dynamics. While **Predictive AI** identifies broad trends—such as the rise of a new technical language—it often lacks the granular causal logic to map internal potential or specific talent adjacencies. To move from "Reactive Hiring" to "Anticipatory Talent Sovereignty," a leader must integrate these six secondary pairings to sharpen the "Eyes" and eliminate the friction of talent scarcity.

1. The Graph Neural Network (GNN) Pairing for Skill Adjacency

Pairing **Predictive AI with GNNs** is the ultimate tool for **Internal Mobility**. While traditional AI predicts which roles might become obsolete, GNNs map the complex, non-linear "Web of Skills" within the existing workforce. This allows a leader to see that a logistics manager possesses 85% of the cognitive "nodes" required for a supply chain architect. By identifying these hidden links, the Forecaster can pivot internal talent into high-value positions before they ever hit the open market, reducing recruitment friction.

2. The Causal Inference Pairing for Retention Architecture

Leaders should pair **Predictive AI with Causal Inference** to solve the "Churn Paradox." Standard predictive models show *who* is likely to leave, but Causal Inference identifies the specific *lever* that will make them stay—whether it is mentorship, management style, or compensation. This allows the leader to move from generic retention programs to surgical interventions that address the root cause of talent leakage, ensuring the "Eyes" see the motivation behind the movement.

3. The Digital Twin Pairing for Workforce Rehearsal

When planning a major reorganization, **Predictive AI must be paired with Workforce Digital Twins**. This is best deployed to "rehearse" a change in a virtual sandbox before it is implemented. The leader can simulate how a 20% shift toward autonomous agentic workflows will impact team communication and morale. This pairing ensures the "Eyes" see the second and third-order consequences of a strategy, allowing for the refinement of the organizational structure in a risk-free environment.

4. The Natural Language Processing (NLP) Pairing for Cultural Health

For real-time health monitoring of the "Corporate Soul," a leader should pair **Predictive AI with Advanced NLP**. While predictive models look at quantitative turnover data, NLP "listens" to the qualitative sentiment in internal feedback loops. This identifies "Culture Decay" at the sub-departmental level—spotting burnout or a loss of psychological safety weeks before it manifests as a resignation. This turns the Forecaster into a sentinel for cultural integrity and workforce resilience.

5. The Monte Carlo Pairing for Global Scarcity Mapping

To navigate the hyper-competitive "War for Talent," leaders should pair **Predictive AI with Monte Carlo Simulations**. This is best utilized when a leader is entering a new technical territory. By running 100,000 market scenarios, the Forecaster can map the probability of talent scarcity in specific geographies. This allows the leader to secure "Alpha Talent" early, ensuring the brand isn't caught off guard by a sudden spike in the cost of specialized expertise.

6. The Knowledge Graph Pairing for Talent Provenance

In industries where technical mastery is the primary moat, **Predictive AI should be paired with Knowledge Graphs**. This pairing tracks the "Provenance of Expertise"—mapping not just what an employee knows, but who they learned from and which teams produce the highest-quality output. It allows the leader to see the "Mentorship Lineage" of the workforce. By identifying these "Invisible Masters," the Forecaster can protect the critical nodes of knowledge transfer within the enterprise.

Workforce Power: Collaborative Fluency

As automation takes over the "Mundane," the human workforce must evolve into the **Diplomats of the Machine.** The primary skill of the future worker is not "Technical Execution," but "Collaborative Fluency"—the ability to translate human intent into machine action and machine data into human meaning. In this role, the workforce becomes the "Voice" that bridges the gap between the brand's "Digital Brain" and its "Human Soul." They are the translators, the ethicists, and the relationship managers who ensure that the organization's technology serves its people and its customers with empathy and logic.

Defensive Stance: Killing the Automation Friction

The defensive mission of the Worker-Diplomat is the **Elimination of Alienation**. When new technologies are introduced, the natural human response is "Resistance" and "Fear of Replacement." The Diplomat role kills this "Friction" by turning the worker into a "Partner of the Machine." By mastering the "Logic of the AI," the employee removes the "Black Box" mystery that causes cultural decay. This is the **Shield of Psychological Safety**. The worker ensures that the "Voice" of the technology is aligned with the "Values" of the brand. Defensively, they act as the "Ethical Guardrail" on the ground, spotting when an automated system is acting with bias or coldness and intervening to protect the brand's reputation. This kills the "Compliance Risk" inherent in unmanaged automation.

Offensive Stance: Unlocking Narrative Power

On the offense, the Diplomat workforce **Unlocks Explanatory Power**. In a world of complex, data-driven products, the ability to "Explain the Why" is the ultimate sales and service tool. The workforce uses their "Voice" to take the "Reasoning" provided by the organization's brain and translate it into a persuasive, empathetic narrative for the customer. This turns every employee into a "Brand Ambassador" who can justify the company's actions with evidence and heart. By "Mastering the Dialogue" between data and emotion, the workforce creates a "Resonant Brand" that customers trust deeply. They use their "Voice" to orchestrate complex human-machine collaborations that produce outcomes a human or a machine could never achieve alone.

The Workforce Outcome: Cultural Resonance

The ultimate value of the Diplomat role is **Institutional Trust**. When the workforce acts as the "Voice" of the brand's intelligence, the organization achieves a state of "High-Empathy Scale." For the leader, this provides **Reputational Security**. You know that your "Hands" (the automation) are being guided by "Voices" (the people) who understand the nuances of human relationships. This turns the organization into an "Authentic Powerhouse"—where every automated interaction feels personal and every human interaction feels informed. It ensures that the brand's "Voice" is never lost in the noise of technology, but is instead amplified by it, creating a culture of "Collaborative Fluency" that is impossible for a "Purely Human" or "Purely Machine" competitor to match.

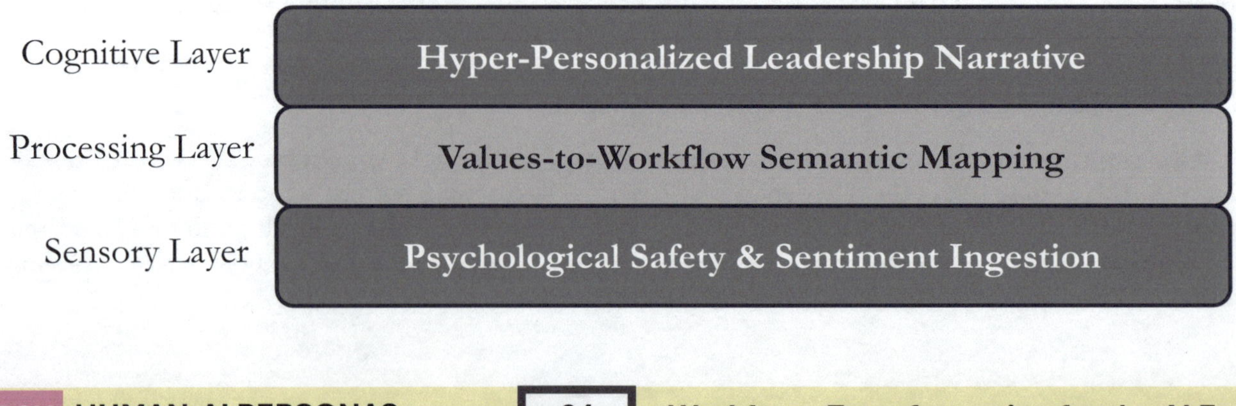

Executive Power: Cultural Alignment & Morale

For the modern executive, the "Voice" of the workforce is no longer a top-down internal memo or an annual "All-Hands" meeting; it is about **Cultural Alignment** achieved through hyper-personalized resonance. In an era of remote work and "Quiet Quitting," the **Diplomat** role ensures the brand's mission is not just "heard," but "internalized" by every individual. The leader uses this stack to ensure the "Company Values" are articulated in the specific context of every employee's daily reality. By architecting this role, the executive moves from "Management" to "Inspiration," ensuring that the "Voice" of the organization is the most "Empathetic" and "Aligned" force in the employee's professional life, bridging the gap between corporate goals and individual purpose.

The Sensory Layer: Psychological Safety & Sentiment Ingestion

The **Sensory Layer** of the Workforce Diplomat acts as the organization's "Cultural Ear." It moves beyond the flawed "Pulse Survey" and into **Passive Sentiment Ingestion**. Utilizing **Natural Language Processing (NLP) on communication metadata**, this layer "listens" to the collective mood of the organization. It captures subtle shifts in tone that indicate burnout, misalignment, or declining psychological safety before they lead to attrition. By monitoring the "Digital Body Language" of the company—such as response latencies or the density of collaborative language—this layer provides the leader with a real-time map of "Cultural Friction." It ensures the executive is never speaking into a vacuum but is reacting to a high-fidelity "Map of Workforce Morale," allowing for intervention at the first sign of cultural decay.

The Processing Layer: Values-to-Workflow Semantic Mapping

The **Processing Layer** serves as the "Alignment Engine." It utilizes **Knowledge Graphs** to link the brand's high-level mission statements to the specific, day-to-day tasks of the individual employee. If the executive's "Voice" speaks of "Innovation," this layer maps that value to the actual R&D credits and creative freedom allowed in specific workflows. This ensures that the leader's message is "Hard-Coded" in operational reality, killing the friction of corporate "Cynicism" by providing a "Verifiable Logic" for every cultural claim. It acts as a **Shield of Semantic Integrity**, ensuring that when the leader speaks about "Work-Life Balance" or "Inclusion," those words are backed by the structural evidence of the company's internal policies and resource allocations, removing the "Hypocrisy Gap" that often plagues large-scale corporate communications.

The Cognitive Layer: Hyper-Personalized Leadership Narrative

The **Cognitive Layer** is where the "Single Mission" is transformed into **Personalized Leadership Narratives**. Utilizing **Generative AI (LLMs)** and **Agentic Communication**, the Diplomat generates "Sovereign Cultural Experiences." It doesn't issue one generic update; it orchestrates thousands of tailored messages—translating the CEO's strategic vision into a narrative that resonates with a junior developer in Berlin as effectively as it does with a sales lead in Tokyo. This layer ensures the leader's intent is perfectly "decoded" into the specific life-context and career-goals of every individual. Because this layer is governed by **Ethical AI Guardrails**, it ensures that the communication remains authentic and non-manipulative. This allows the executive to scale their presence and project a "Voice" that builds **Unshakable Institutional Loyalty**, ensuring that every member of the workforce feels a deep, personal alignment with the company's ultimate destination.

2. THE DIPLOMATS (THE VOICE) | EXPLAINABLE AI

OVERVIEW

Explainable AI (XAI) provides transparent, human-readable rationales for AI decisions, moving beyond black box outputs. It reveals why a model reached a conclusion, building trust and mitigating legal and ethical exposure. The Regulatory & Compliance leader leverages XAI to validate complex, automated decision-making processes, particularly in highly regulated domains. This synergy ensures AI systems are fair, auditable, and adhere to internal policies and external legal requirements, transforming compliance from a check-the-box function into a strategic enabler of trust.

MAIN STEPS

Start

STEP 1 — Define Audit Scope
Establish parameters for compliance assessment.

STEP 2 — Feature Importance Analysis
Quantify input factors driving AI output.

STEP 3 — Generate Local Explanation
Create rationale for a single prediction.

STEP 4 — Validate Model Fairness
Test decision consistency across subgroups.

STEP 5 — Human Review and Oversight
Leader verifies explanations meet standards.

STEP 6 — Document Rationale & Action
Record justification for audit readiness.

End

SAMPLE FLOW

Fair Hiring Recommendation Audit: Leader uses Explainable AI (XAI) to audit AI-driven hiring recommendations, ensuring that selection criteria remain unbiased and compliant with EEO laws.

EXPLAINABLE AI

1. Input Selection Metrics
Define required job skills and compliant evaluation criteria clearly.

2. Analyze Recommendation Drivers
Identify which candidate features primarily influence the AI ranking.

3. Generate Bias Reports
Produce reports showing how protected attributes impact final candidate scores.

4. Compare Subgroup Outcomes
Check if the model consistently rates minority candidates fairly and equally.

5. Justify Ranking Deviation
Leader documents rationale for overruling a non-compliant or biased AI recommendation.

6. Publish Policy Alignment
Communicate verified fairness policy alignment to all employees and regulatory bodies.

USE CASES

- **Manufacturing | Automated Training Compliance:** Provides audit trails explaining *why* certain operators are flagged for refresher training, ensuring fair and legally documented compliance with complex safety certifications.
- **Financial Services | Compensation Equity Rationale:** Explains the variables used by AI to set compensation packages, allowing leaders to certify pay equity and proactively address potential wage discrimination claims.
- **Healthcare | Shift Scheduling Fairness:** Provides transparent explanations for AI-generated shift assignments, verifying that labor distribution is equitable, compliant with fatigue laws, and addresses staff availability rules.
- **Technology | Termination Justification:** Generates detailed, auditable reports explaining the exact performance factors leading to an AI-recommended termination decision, minimizing legal exposure for the organization.
- **Retail | Performance Improvement Plan (PIP) Basis:** Explains the specific, objective data points (sales, efficiency) that triggered an employee's placement on a performance improvement plan, ensuring defensible HR decisions.
- **Energy | Contractor Certification Vetting:** Provides transparent rationale for AI-based vetting of external contractors' safety and compliance records, ensuring all workforce partners meet strict regulatory standards quickly.

2. THE DIPLOMATS (THE VOICE) | EXPLAINABLE AI PLUS OTHERS

In the **Workforce Transformation**, the **Diplomats (The Voice)** are the architects of "Psychological Alignment." While **Explainable AI (XAI)** provides the technical logic for organizational change, it lacks the emotional nuance required to manage the human anxiety inherent in a high-velocity, tech-driven pivot. To move from "Enforced Compliance" to "Radical Buy-In," a leader must integrate these six secondary pairings to calibrate the brand's "Voice" for its people, turning fear into focused energy.

1. The Sentiment Analysis Pairing for Emotional Calibration

Pairing **Explainable AI with Sentiment Analysis** is best utilized during "Agentic Integration." While XAI explains why an AI agent is taking over a specific workflow, Sentiment Analysis monitors the "emotional climate" of the workforce across internal channels. This allows the leader to adjust the "Voice" to address skepticism in real-time. By identifying rising anxiety early, the Diplomat ensures that technical reasoning is delivered with the empathy required to maintain high institutional morale.

2. The Contextual NLP Pairing for Cultural Translation

Leaders should pair **Explainable AI with Contextual NLP** when managing global, diverse teams. XAI provides the facts behind a transformation, but Contextual NLP translates that logic into the specific cultural and professional vernacular of the recipient. This ensures that a strategic shift is understood with the same depth in a Tokyo engineering hub as it is in a London marketing suite. It eliminates the friction of misinterpretation, ensuring the brand's "Voice" is resonant across all borders.

3. The Behavioral Economics Pairing for Incentive Alignment

When rolling out new performance metrics, **Explainable AI must be paired with Behavioral Economics models**. XAI explains the "Why" of the new metric, while the behavioral layer ensures the "Voice" triggers positive reinforcement through nudges. This pairing is best deployed to align individual ambitions with the firm's offensive goals. It turns a top-down directive into a self-motivated pursuit of excellence, ensuring that the workforce feels rewarded, not just monitored, by the new system.

4. The Conversational AI Pairing for Real-Time Interrogation

To eliminate the "Executive Echo Chamber," leaders should pair **Explainable AI with Conversational AI**. While XAI holds the data, Conversational AI allows employees to "interrogate" a corporate decision via an internal, always-on interface. This is best utilized during town halls or major policy shifts, providing every staff member a private channel to ask, "How does this change affect *my* specific career?" This pairing builds a culture of radical, open-door transparency.

5. The Knowledge Graph Pairing for Logical Continuity

For maintaining long-term trust, a leader should pair **Explainable AI with Knowledge Graphs**. While XAI justifies a single decision, the Knowledge Graph shows how that decision aligns with the firm's "Sovereign Values" and past actions. This creates a "Verifiable Voice," proving to the workforce that the current transformation is a logical evolution of the company's history. It ensures the Diplomat remains the guardian of institutional integrity, preventing the pivot from feeling like a random or erratic shift.

6. The Human-in-the-Loop Pairing for Ethical Governance

In high-stakes talent decisions, **Explainable AI should be paired with Human-in-the-Loop (HITL)** interfaces. XAI identifies the algorithmic reasoning for a promotion or structural change, but the HITL layer provides the "Moral Filter." This ensures the "Voice" of the organization remains grounded in human values, preventing the "Hands" from acting on machine logic that might overlook nuanced human contributions or social context. It guarantees that the brand's leadership remains fundamentally human-centric.

Workforce Power: Generative Orchestration

In the legacy workforce model, an employee's value was often tied to their "Manual Output"—the number of lines of code written, the number of designs drafted, or the volume of reports compiled. This created a "Production Bottleneck" where the speed of the company was limited by the physical and mental stamina of its people. The transformation into **Creators** shifts the workforce from "Doing the Work" to **"Orchestrating the Engine."** In this role, every employee becomes a "Director of Production." They use the organization's generative capabilities to multiply their own agency, moving from the creation of single assets to the management of "Creative Systems" that can produce at infinite scale.

Defensive Stance: Killing the Obsolescence

The defensive mission of the Worker-Creator is the **Purification of Effort**. In many organizations, employees spend up to 40% of their time "Re-inventing the Wheel"—searching for old templates, re-creating lost files, or fixing repetitive errors. The Creator role kills this "Friction of Redundancy" by using the "Engine" to index and modernize the team's collective history. This creates a **Shield of Continuity**. Employees no longer waste energy on "Low-Value Creation"; instead, they "Resurrect" and "Refine" existing assets. By "Cleaning the Pipeline" of obsolete tasks, the workforce ensures that their creative energy is always focused on the "New," rather than the "Maintenance of the Old." This defensive posture protects the organization's "Intellectual Capital" and ensures that no lesson learned is ever forgotten or buried in a silo.

Offensive Stance: Unlocking Generative Abundance

On the offense, the Creator workforce moves toward **Hyper-Speed Prototyping**. When every worker can "Generate" a thousand versions of a solution in the time it used to take to make one, the organization's "Evolutionary Speed" skyrockets. This **Unlocks Infinite Iteration**, where employees become "Curators of Excellence." Instead of settling for the "First Good Idea," they use the "Engine" to explore the entire "Design Space" of a problem. A marketing manager might generate a personalized campaign for ten thousand micro-segments; a developer might generate a dozen architectural variations for a new feature. This turns the workforce into a "Multiplication Force," where the "Hands" of the company can manifest any "Idea" instantly. It allows the brand to "Saturate the Market" with high-fidelity, relevant output that makes the manual efforts of competitors look primitive.

The Workforce Outcome: Absolute Agency

The ultimate value of the Creator role is **Operational Velocity**. When the workforce is no longer a "Bottleneck" but an "Engine," the organization achieves a state of "Unconstrained Growth." For the leader, this provides **Execution Leverage**. You no longer manage "Headcount"; you manage "Creative Capacity." This turns the workforce into a "High-Agency Fleet"—where every individual has the power to manifest complex projects in hours rather than months. It ensures that the brand's "Engine" is always fueled by the "Fresh Intent" of its people, transforming the workforce into a "Perpetual Maker Space" that defines the leading edge of the industry.

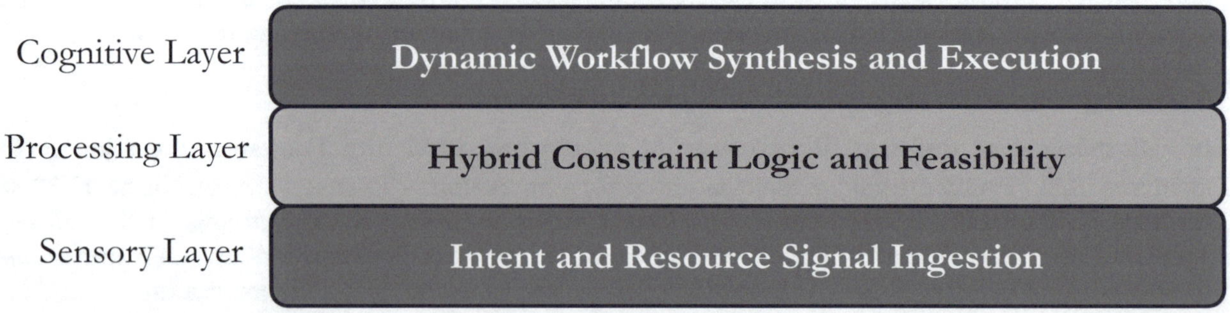

Executive Power: Generative Workforce Manifestation

The **Creator** role transforms the executive from a "Headcount Manager" into a "Capability Architect." In the legacy enterprise, workforce structures were static; a "job description" was a rigid box that humans were forced to fit into, regardless of their evolving potential. The Creator role represents the shift to **Fluid Role Manifestation**, where the work itself is dynamically generated around the intent of the leader and the evolving skills of the talent pool. The executive uses this generative engine to manifest "Hybrid Workflows" that blend human creativity with AI agency on the fly, moving the organization from a state of "Fixed Labor" to "Generative Productivity." This role ensures the brand is always manifesting the most "Relevant" organizational structure for the task at hand.

The Sensory Layer: Intent and Resource Signal Ingestion

The **Sensory Layer** for the Creator captures "Operational Needs" as high-definition signals. It moves beyond traditional project planning by integrating **Real-Time Task Telemetry** and **Skill Availability Streams**. By utilizing **Agentic Discovery** to scan internal project backlogs and external market demands, this layer identifies the "Atomic Requirements" of a new initiative before a formal team is ever assembled. It captures the raw intent of the leadership—"We need to launch a sustainable energy line in six months"—and cross-references it with the high-fidelity signal of available internal cognitive energy and external freelancer capacity. This ensures the leader is building the "Engine" of the project on a foundation of "Actual Capability" rather than "Theoretical Availability," capturing the necessary components of the workforce as a live, fluctuating resource.

The Processing Layer: Hybrid Constraint Logic and Feasibility

The **Processing Layer** acts as the "Reality Filter" for organizational design. It utilizes **Constraint-Based Optimization Engines** and **Financial Modeling** to ensure that any "Generated Role" is viable, compliant, and cost-effective. It understands the "Rules of the Workforce"—legal labor laws, budget ceilings, and technical prerequisites. This layer stress-tests every generative organizational chart against the company's "Causal Truths," ensuring the executive only sees structures that can actually perform the mission. It acts as a **Guardian of Balance**, ensuring that the ratio of human oversight to AI execution is optimized for safety and quality. By applying "Parametric Organizational Logic," this layer ensures the "Engine" is disciplined, preventing the creation of redundant roles or inefficient "Shadow Teams" that drain resources without producing value.

The Cognitive Layer: Dynamic Workflow Synthesis and Execution

The **Cognitive Layer** is the "Orchestration Mind" of the workforce stack, powered by **Large Language Models (LLMs)** and **Action Transformers**. This layer "hallucinates" thousands of different ways to organize the work, presenting the leader with the "Optimal Human-AI Configuration." It doesn't just suggest people; it manifests the **Workflow Architecture**. The leader provides the strategic goal, and the Cognitive Layer generates the "Project Genome"—defining who does what, which AI agents handle the toil, and how the "Hands" will execute the "Voice" of the strategy. It involves "Recursive Role Evolution," where the AI adapts the workflow as the project progresses, ensuring the final output is the "Optimal Manifestation" of the leader's vision. This turns the executive into a "Conductor of Capability," capable of deploying complex organizations at a speed that makes traditional HR processes obsolete.

OVERVIEW

Generative AI creates novel outputs like text, art, code, and immersive environments from simple prompts. It excels at rapid prototyping, style transfer, and exploratory ideation. The Creative & Artistic leader partners with Generative AI to rapidly visualize strategic concepts, explore diverse future scenarios, and prototype new organizational narratives. The synergy transforms vision casting from a slow, manual process into a continuous, inspiring, and engaging collaborative endeavor.

MAIN STEPS

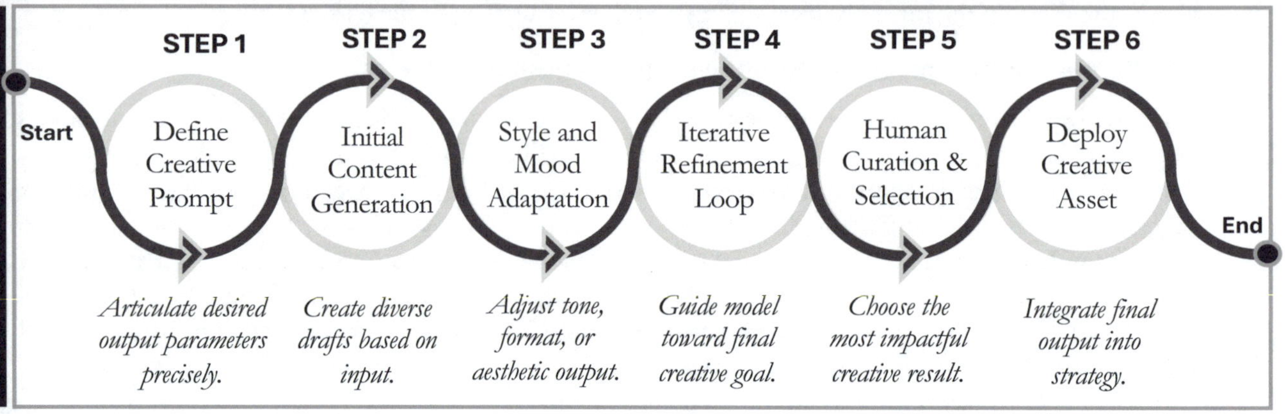

SAMPLE FLOW

Personalized Learning Journeys: A Chief HR Officer uses Generative AI to design custom learning content and visual pathways for employee upskilling and career growth.

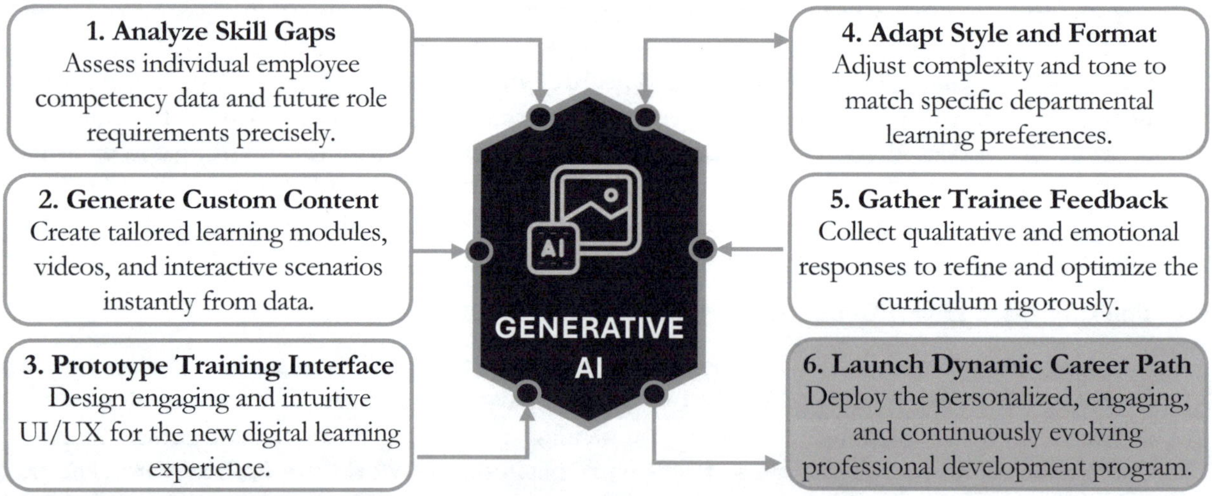

USE CASES

- **Manufacturing | Safety Procedure Visualization:** Generates hyper-realistic VR/AR simulations of high-risk operational environments, dramatically increasing the effectiveness of safety training for new factory floor personnel and reducing incidents.
- **Financial Services | Culture Handbook Design:** Creates interactive, visually stunning, and user-friendly digital handbooks that translate complex compliance rules into memorable, scenario-based learning modules for global teams.
- **Healthcare | Empathy Training Scenarios:** Generates diverse, realistic, and emotionally complex patient conversation scripts and avatars for doctors and nurses to practice sensitive communication skills safely.
- **Technology | Onboarding Storyboards:** Designs personalized, engaging multimedia onboarding journeys for new engineers, visually aligning their role, team, and projects with the company's long-term technical vision.
- **Retail | Sales Skill Training:** Creates generative video scenarios where AI avatars act as demanding customers, allowing staff to practice complex sales techniques and personalized service interactions in a controlled setting.
- **Energy | Remote Field Operations Manuals:** Produces dynamic, interactive 3D instructional guides and augmented reality overlays for field technicians, simplifying complex maintenance procedures on distant grid infrastructure.

In the **Workforce Transformation**, the **Creators (The Engine)** represent the organization's "Evolutionary Core." While the primary instrument of **Generative AI** is used to produce personalized training content or draft job descriptions, it can often become a "Stochastic Echo" that lacks grounding in the firm's specific culture and operational realities. To move from generic automation to a high-velocity "Talent Lab," a leader must integrate these six secondary pairings to ensure the "Engine" is producing human capital that is both specialized and antifragile.

1. The Digital Twin Pairing for Reality-Based Upskilling

Pairing **Generative AI with Worker Digital Twins** is best utilized for **Rapid Role Prototyping**. While Generative AI creates a curriculum for a new technical role, the Digital Twin simulates the employee's current cognitive load and skill set. This allows the leader to manifest "Customized Learning Paths" that are grounded in the worker's actual capacity, ensuring the upskilling process is a bridge to reality rather than a theoretical burden that leads to burnout.

2. The Synthetic Data Pairing for Bias-Free Recruitment

Leaders should pair **Generative AI with Synthetic Data Generation** when building "Future-Ready" hiring pipelines. This is best deployed to solve the "Small Data" problem in emerging fields. By generating high-fidelity, anonymized synthetic profiles based on "Alpha Performers," the Creator can train recruitment models to look for potential rather than pedigree. This ensures the "Engine" identifies top-tier talent in unconventional places, effectively diversifying the workforce through algorithmic meritocracy.

3. The RLHF (Human Feedback) Pairing for Cultural Aesthetic

When the workforce produces outward-facing creative or strategic assets, **Generative AI must be paired with Reinforcement Learning from Human Feedback (RLHF)**. This is best utilized to ensure the "Engine's" output aligns with the brand's "Sovereign Tone." Human mentors provide the "Taste" and "Nuance" that the machine lacks, refining the generative output until it mirrors the organization's unique DNA. This pairing turns every employee into a "Creative Director," amplifying individual output without losing the brand's soul.

4. The API Orchestration Pairing for Functional Empowerment

To turn creative ideas into immediate operational tools, a leader should pair **Generative AI with API Orchestration**. While the AI can draft a new workflow, API Orchestration allows that draft to "reach out" and trigger actual software changes within HR or project management systems. This is best used to empower employees to "Self-Architect" their own productivity tools. It moves the "Engine" from being a generator of ideas to a generator of functional, ready-to-use business logic.

5. The Vector Database (RAG) Pairing for Collaborative Memory

To prevent the "Engine" from producing redundant or misinformed content, leaders should pair **Generative AI with Vector Databases (RAG)**. This is best deployed during "Crowdsourced Innovation" sprints. The Vector Database allows the Generative AI to "recall" every previous internal prototype, successful project, and failed experiment. This pairing ensures the workforce isn't reinventing the wheel but is building upon a foundation of "Searchable Institutional Memory," accelerating the velocity of internal innovation and cross-departmental learning.

6. The Edge AI Pairing for Localized Creativity

In decentralized or remote work environments, **Generative AI should be paired with Edge AI**. This is best utilized when employees need real-time, generative assistance at the "Point of Need" without latency—such as on-site technicians or field agents. By running generative models on local hardware, the "Engine" provides instant troubleshooting or creative pivots. This ensures the organization's creative power isn't centralized in a cloud but is manifested in the "Hands" of every employee, regardless of their location.

Workforce Power: Objective Decisioning

The traditional workforce is often governed by "Inertia" and "Anecdote." Decisions are frequently made based on "how we've always done it" or the gut feelings of a local supervisor. The transformation into **Strategists** provides every employee with an **Objective Compass**. In this role, workers move from "Following Orders" to "Navigating Goals." They are equipped to understand the "Mathematical Why" behind their tasks, using objective logic to prioritize their time and resources. The workforce evolves from a "Passive Audience" of corporate strategy into an "Active Network" of tactical optimizers who refine the company's path every day.

Defensive Stance: Killing the Tactical Mismatch

The defensive mission of the Worker-Strategist is the **Validation of Action**. In large organizations, there is often a "Mismatch" between the CEO's vision and the front-line reality. Workers often find themselves executing tasks that no longer add value because the market has shifted. The Strategist role kills this "Strategic Drift" by giving every worker the tools to "Stress-Test" their own priorities. This is the **Shield of Relevance**. By identifying when a local process is "Misaligned" with the global "Compass," the employee can trigger a "Course Correction" at the source. This defensive posture ensures that the organization's "Capital" and "Time" are never wasted on "Dead-End Work," transforming the workforce into a "Self-Correcting System" that guards the company's resources with the same rigor as the CFO.

Offensive Stance: Unlocking Localized Optimization

On the offense, the Strategist workforce **Unlocks Dynamic Navigation**. In a complex environment, the "Best Move" is often hidden in the local details that central leadership cannot see. The worker uses their "Compass" to find the "Path of Least Resistance" within their specific domain. This might involve "Prescribing" a new logistics route, "Optimizing" a pricing discount for a specific client, or "Re-allocating" their own time toward the highest-probability lead. This **Unlocks Yield Mastery**, where every employee is acting as a "Mini-CEO" of their own desk. By "Finding the Win" in the micro-moments of the business, the workforce creates a "Compound Advantage" that rivals cannot replicate. They move the organization from "Executing a Plan" to "Optimizing a Reality," ensuring the brand is always positioned for maximum profit.

The Workforce Outcome: Distributed Certainty

The ultimate value of the Strategist role is **Organizational Cohesion**. When the "Compass" is in the hands of every worker, the organization achieves a state of "Unified Intent." For the leader, this provides **Decentralized Confidence**. You no longer need to "Micro-manage" every move because you know your people are using the same "Logic of the Win" that you are. This turns the workforce into a "High-IQ Network"—one that remains "On-Mission" even in the absence of direct supervision. It ensures that the brand's "Strategy" is not a document in a drawer, but a "Live Force" animated by the "Objective Decisions" of its entire people, creating a brand that is as "Smart" at the bottom as it is at the top.

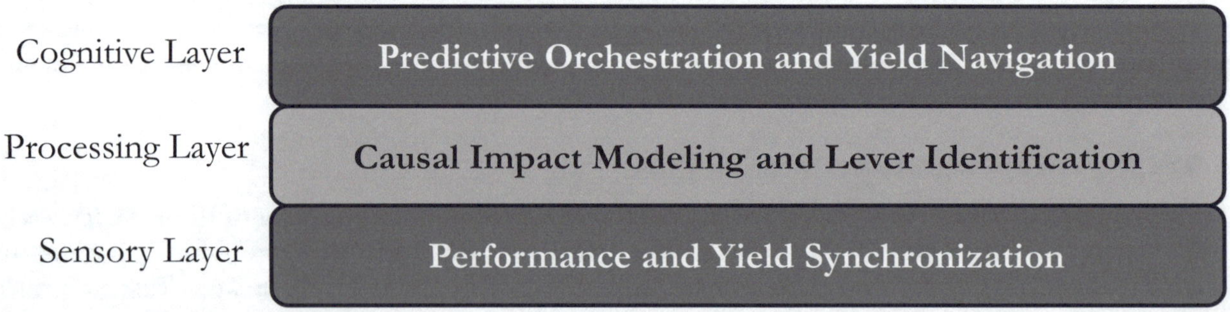

Executive Power: Causal Yield Optimization

The **Strategist** role in workforce transformation represents the "Abolition of Intuition-Based Management." It moves the executive from "Managing Morale" to **"Causal Value Optimization."** The leader uses this stack to ensure that every "Human-AI Hour" is allocated to the highest-impact drivers of the brand's mission. The Strategist is the **Guardian of the Yield**, ensuring that the organization is not just "Busy," but "Effective." In this role, the executive evolves into a "Strategic Human Capital Navigator" who manages the company's "Cognitive Peak" with mathematical precision. This role kills the friction of "Busy Work" and "Operational Waste" by ensuring the workforce's energy is always flowing toward the "True Drivers" of sustainable profit and innovation.

The Sensory Layer: Performance and Yield Synchronization

The **Sensory Layer** of the workforce Compass monitors the "Micro-Levers" of productivity across the entire global enterprise. It synchronizes **Real-Time Output Streams** (code commits, sales closings, customer resolutions) with internal **Operational Data**. By tracking the "Velocity of Achievement" against the "Cost of Effort," this layer provides the "Coordinate Data" required to know exactly where the workforce stands in relation to the brand's "Strategic North Star." It provides the "Eyes" for the Compass, ensuring that the starting point of any workforce shift is grounded in the absolute reality of current performance levels. This layer removes the "Blind Spots" in management, ensuring the leader is aware of every environmental factor—from technical debt to team burnout—that could impact the trajectory of the organization's success.

The Processing Layer: Causal Impact Modeling and Lever Identification

The **Processing Layer** is the heart of the Strategist role, utilizing **Bayesian Networks** and **Causal Discovery Algorithms** to determine "Why" certain teams or individuals are outperforming others. While traditional HR analytics show simple "Correlation" (e.g., remote teams are 10% faster), the Strategist identifies the **Causal Path** (e.g., they are faster because of reduced meeting density and autonomous tool access). This layer provides the "Mathematical Evidence" for workforce strategy, ensuring the leader is pulling the "Levers" that actually matter for the brand's "Yield." It acts as a **Shield of Objectivity**, stripping away the "Noise" of subjective performance reviews and focusing the executive on "Systemic Impact." By identifying the "Mathematical Why," this layer ensures the firm is never "Chasing Trends," but is always focused on the "Causal Truths" that drive real-world results.

The Cognitive Layer: Predictive Orchestration and Yield Navigation

The **Cognitive Layer** acts as the "Navigational Mind" of the workforce stack, utilizing **Reinforcement Learning** and **Multi-Agent Simulation**. This layer "plays the game" of the business millions of times to find the "Optimal Path of Effort." When the market environment shifts, the Compass "Recalculates" the workforce deployment instantly. This layer doesn't just report on the workforce; it "Prescribes the Best Move" to maximize the **Human-AI Yield**. For example, it might recommend shifting a specific skill cluster from "Maintenance" to "R&D" because the probability of a market breakthrough has increased. This provides the executive with **Operational Certainty**, allowing them to navigate workforce disruptions with the confidence that their "Compass" is calibrated to the objective laws of productivity. It turns the organization into a "Calculating Powerhouse," ensuring the brand "Wins" because its "Workforce Strategy" is mathematically optimized for the highest possible outcome.

4. THE STRATEGISTS (THE COMPASS) | COGNITIVE AI

OVERVIEW

Cognitive AI mimics human intelligence to understand and interpret unstructured data like text, speech, and images. It processes complex inputs to generate hypotheses, assess logical connections, and offer nuanced context. The Research & Analytical leader uses Cognitive AI as a hyper-efficient knowledge partner, rapidly synthesizing vast amounts of data to rigorously test assumptions, structure complex arguments, and make deeply informed, evidence-based strategic decisions.

MAIN STEPS

Start

STEP 1 — Data Ingestion: *Collect and structure input knowledge quickly*

STEP 2 — Understand Context: *Understand meaning and relationships deeply.*

STEP 3 — Hypothesis Generation: *Formulate potential explanatory outcomes.*

STEP 4 — Evidence Synthesis: *Find relevant supporting information.*

STEP 5 — Analytical Refinement: *Test hypotheses against real-world data.*

STEP 6 — Insight Formulation: *Structure findings for human action.*

End

SAMPLE FLOW

Human Resources - Capability Gap Analysis : A CHRO uses AI to analyze team skills and job requirements to pinpoint future workforce capability deficiencies and talent risks.

1. Define Future Role Needs
Specify the emerging skills and technological competencies required next year.

2. Ingest Employee Profile Data
Collect resumes, training records, and performance reviews efficiently.

3. Map Skill Similarity Gaps
Analyze current employee skill sets against future job requirements dynamically.

COGNITIVE AI

4. Model Reskilling Feasibility
Determine the time and cost required for employees to acquire new core skills.

5. Predict Attrition Drivers
Analyze satisfaction data and market trends to forecast key talent retention issues.

6. Design Targeted Development
Structure personalized training paths and strategic external hires immediately.

USE CASES

- **Manufacturing | Expertise Transfer Pathways:** Analyzes internal mentorship program transcripts and outcomes to identify successful knowledge transfer pathways, recommending best practices for accelerating the development of specialized factory floor expertise.
- **Financial Services | Compliance Education Efficacy:** Scans employee training completion rates versus compliance audit reports to quantify regulatory education effectiveness, adjusting mandatory curriculum content to close high-risk knowledge gaps.
- **Healthcare | Clinical Team Optimization:** Interprets patient feedback, incident reports, and staff certifications to analyze optimal clinical team composition for complex case management, informing future hiring and team-building strategies.
- **Technology | Technical Proficiency Mapping:** Reviews developer code contributions, project documentation, and bug reports to objectively map technical proficiency across teams, ensuring critical system ownership is diversified and resilient.
- **Retail | Service Agent Coaching:** Analyzes customer service agent interaction logs (chat, voice) for empathy and problem-solving quality, providing leadership with data to create targeted behavioral coaching modules for improvement.
- **Energy | Emergency Preparedness Review:** Synthesizes internal safety audit logs and training simulations to rigorously assess field crew preparedness for emergency scenarios, prioritizing high-risk teams for mandatory, focused training drills.

In the **Workforce Transformation**, the **Strategists (The Compass)** act as the "Navigational Intelligence" of human capital. While the primary instrument of **Cognitive AI** identifies high-level patterns in productivity, it often lacks the mathematical rigor to balance competing organizational priorities. To move from "Static Resource Allocation" to "Dynamic Performance Optimization," a leader must integrate these six secondary pairings. This ensures the organization's most expensive asset—its people—is always pointed toward the highest possible yield without sacrificing long-term resilience.

1. The Multi-Objective Optimization (MOO) Pairing for Peak Yield

Pairing **Cognitive AI with MOO** is best utilized to solve the "Performance Paradox." While AI can maximize for speed, MOO balances three conflicting variables: output velocity, labor cost, and employee burnout. This allows the leader to identify the "Optimal Performance Frontier," ensuring that the workforce is running at a pace that is both hyper-productive and sustainable. It turns the "Compass" into a precision tool that extracts the maximum value from every man-hour.

2. The Game Theory Simulation Pairing for Competitive Positioning

Leaders should pair **Cognitive AI with Game Theory** when making "Alpha Hiring" moves. This is best deployed during a market talent war. By simulating how competitors will react to a specific poaching move or a change in remote-work policy, the Strategist can anticipate the "market's move" and counter it. This ensures the workforce strategy is not just internally sound, but externally dominant, allowing the firm to capture the best talent before the competition can adjust.

3. The Digital Twin Pairing for Labor Elasticity

When navigating seasonal or market-driven volatility, **Cognitive AI must be paired with Labor Digital Twins**. This is best utilized to model "Elasticity Scenarios." The Strategist simulates how the workforce can expand or contract through the use of contractors, AI agents, or internal shifts without breaking core operations. This pairing ensures the "Compass" can guide the firm through economic turbulence with zero "Execution Gap," maintaining operational integrity regardless of external pressures.

4. The Knowledge Graph Pairing for Skill Arbitrage

To maximize the "Return on Skill," a leader should pair **Cognitive AI with Knowledge Graphs**. While AI identifies skills, the Knowledge Graph identifies the *value* of those skills across different departments. This allows for "Skill Arbitrage"—moving a specialist from a low-impact department to a high-impact offensive project where their unique talent has a 10x multiplier effect. It ensures the "Compass" is always directing the organization's "Alpha Minds" to the firm's biggest opportunities.

5. The Causal Inference Pairing for Meritocratic Integrity

For maintaining a "High-Performance Culture," leaders should pair **Cognitive AI with Causal Inference**. While AI tracks results, Causal Inference proves *why* those results occurred—separating luck or market tailwinds from true individual performance. This is best utilized in bonus and promotion cycles to ensure radical meritocracy. It ensures the "Compass" points toward genuine talent, preventing the promotion of "false positives" and building deep institutional trust in the leadership's objectivity.

6. The Agentic Orchestration Pairing for Dynamic Budgeting

In high-velocity transformations, **Cognitive AI should be paired with Agentic Orchestration** for "Live Capital Allocation." While the Strategist sets the goals, the Agentic layer automatically reallocates hiring budgets or L&D funds based on real-time performance data. This is best utilized in rapid-growth phases, ensuring that capital is never trapped in stagnant departments but flows autonomously toward the teams with the highest momentum. It turns the "Compass" into an active driver of financial and human capital efficiency.

Workforce Power: Cognitive Fluidity

In the legacy industrial model, an employee's value was often defined by their "Specialization"—the specific, narrow slice of knowledge they possessed. This created a workforce of "Information Silos," where the organization's intelligence was trapped in individual heads, making the company fragile and slow. The transformation into **Architects** shifts the workforce from "Owning Knowledge" to **"Designing Intelligence."** In this role, every employee becomes a contributor to the "Corporate Brain." They are no longer just "using" data; they are responsible for the "Architecture of Meaning," ensuring that every lesson they learn and every insight they gain is captured, mapped, and made available to the entire system.

Defensive Stance: Killing the Institutional Amnesia

The defensive mission of the Worker-Architect is the **Abolition of Fragmentation**. In most companies, the "Brain" suffers from constant "Amnesia"—when an employee leaves, their knowledge leaves with them, and when a project ends, the lessons are buried in a dead folder. The Architect role kills this "Friction of Loss" by using "Knowledge Mapping" at the individual level. Every worker acts as a "Sentry of Information," tagging, linking, and resolving the identity of the assets they touch. This creates a **Shield of Continuity**. It ensures that the organization never has to "Learn the same lesson twice." By "Cleaning the Intellectual Foundation" of their own desk, the worker ensures that the "Brain" of the company remains healthy, updated, and free of the "Noise" that causes confusion and error. This defensive posture turns "Individual Experience" into "Permanent Institutional Wealth."

Offensive Stance: Unlocking Contextual Mastery

On the offense, the Architect workforce moves toward **Hyper-Relevant Application**. When the "Brain" is well-designed, workers no longer spend their time "Searching" for information; they spend their time "Applying" it. This **Unlocks Relevance**, where the employee can "Retrieve" the exact context needed to solve a unique customer problem or optimize a complex process instantly. This moves the workforce into a state of "Cognitive Fluidity," where a worker in one department can leverage the "Logic" of another to create "Cross-Functional Wins." They use "Adaptive Learning" to feed the system new data, which in turn provides them with "Personalized Insights" that make them more effective. This turns the workforce into a "Learning Organism" that evolves its "Intelligence" in real-time, ensuring the brand's "Brain" is always the most "Sophisticated" and "Context-Aware" in the market.

The Workforce Outcome: Systemic Intelligence

The ultimate value of the Architect role is **Institutional Immortality**. When the workforce acts as the "Architect of the Brain," the organization's intelligence becomes "Independent of Individuals." For the leader, this provides **Operational Continuity**. You no longer fear the "Brain Drain" of turnover because the "Wisdom" is built into the architecture itself. This turns the workforce into a "High-IQ Collective"—one that is fundamentally "Smarter" than any group of isolated experts. It ensures that the brand's "Brain" is a "Live Asset" that grows more valuable with every hour of human work, creating a brand that possesses the "Total Recall" and "Perfect Context" required to dominate a complex world.

Executive Power: Cognitive Infrastructure

In the legacy enterprise, workforce knowledge was fragmented, existing only within the "Local Memory" of individual employees or disparate team folders. The **Architect** role represents the executive's transition from a "Manager of Personnel" to a "Designer of the Firm's Collective Consciousness." In this role, the leader constructs a permanent, self-evolving **Cognitive Infrastructure** that captures the unique "Mental Models" of the workforce and turns them into a persistent corporate asset. This is the shift from "Individual Expertise" to "Architected Intelligence," ensuring that the brand's "Wisdom" is a compound resource that remains within the firm regardless of employee turnover. The Architect ensures the organization has a "Workforce Brain"—one that connects every insight, skill, and lesson learned across the global network.

The Sensory Layer: Experience Capture and Identity Resolution

The **Sensory Layer** of the Workforce Brain acts as the "Cognitive Intake System." It moves beyond static training records into **Universal Experience Resolution**. Utilizing **Semantic Activity Analysis** and **Identity Mapping**, this layer captures the "Atomic Insights" generated during daily work. It doesn't just record that a task was finished; it captures "How" it was solved by ingesting project metadata, code commits, and collaborative decision logs. This layer "labels" the specialized expertise of the firm in real-time, ensuring that every unique problem-solving method is identified and indexed. By removing the "Friction of Documentation," the sensory layer ensures that the "Brain" of the firm is constantly being fed with the raw, unfiltered cognitive output of the entire workforce, creating a high-fidelity map of the organization's actual intellectual capital.

The Processing Layer: Structural Intelligence and Expertise Mapping

The **Processing Layer** serves as the "Knowledge Connector" of the organization. It utilizes **Knowledge Graphs** and **Vector Databases** to define the relationships between the skills and experiences captured in the sensory layer. While traditional HR systems see employees in isolation, the Architect's processing layer stores the **Intellectual Connectivity**. It understands, for example, the semantic link between a specific engineer's "Debugging Logic" and a project manager's "Risk Mitigation Strategy." By employing **Graph Neural Networks (GNNs)**, this layer identifies "Hidden Expertise Clusters" within the organization, allowing the "Brain" to perceive itself as a unified intellectual organism. This creates a **Shield of Systemic Wisdom**, ensuring that when a solution is discovered in one region, the entire global workforce "knows" it instantly through the shared structural logic of the knowledge graph.

The Cognitive Layer: Recursive Institutional Wisdom and Learning

The **Cognitive Layer** is the "Reasoning Mind" and the "Wisdom Core" of the workforce, powered by **Retrieval-Augmented Generation (RAG)** and **Recursive Feedback Loops**. This layer ensures the brand "never forgets" a successful workflow or a hard-earned lesson. It synthesizes the relationships from the processing layer into a "Continuous Stream of Intelligence" accessible to every employee. When a worker faces a complex problem, the Cognitive Layer doesn't just provide a document; it "reasons" through the firm's entire history to suggest the "Optimal Approach" based on past collective successes. It involves **Self-Evolving Workflows**, where the Brain updates its own best practices as the workforce innovates. This turns the organization into a "High-IQ Entity," providing the executive with an "Intelligence Asset" that grows in value every day, ensuring the brand's "Mind" remains its most unassailable competitive advantage.

5. THE ARCHITECTS (THE BRAIN) | HUMAN IN-THE-LOOP AI

OVERVIEW

Human-in-the-Loop (HITL) AI is a cycle where human supervisors continuously train, tune, and validate a model's performance, especially for ambiguous tasks. The Supervisory & Coaching leader focuses on defining ethical boundaries and quality thresholds, ensuring AI decisions align with organizational values. This synergy elevates the leader from a decision-maker to a model trainer and quality assurance expert. By providing critical feedback, the leader helps the AI evolve, improving accuracy, fairness, and overall governance in complex, high-stakes leadership decisions.

MAIN STEPS

Start

STEP 1 — Initial Model Training
Establish baseline knowledge and performance rules

STEP 2 — Active Decision Queue
Flag uncertain outputs for human review.

STEP 3 — Human Supervisor Intervention
Leader reviews, corrects, documents outcomes.

STEP 4 — Feedback Integration Cycle
Corrected data retrains and improves the model.

STEP 5 — Model Validation & Testing
Confirm updated model meets quality standards.

STEP 6 — Deployment and Monitoring
Deploy new model; track performance, collect data.

End

SAMPLE FLOW

Coaching AI for Fair Candidate Sourcing: Leader utilizes HITL to audit and refine an AI that scores job candidates, ensuring the sourcing process remains unbiased and inclusive..

1. Input Ethical Standards
Define fairness metrics and bias thresholds for all performance scoring clearly.

2. Flag Ambiguous Candidates
HITL flags candidates where scoring is close or factors suggest potential bias inaccurately.

3. Coach Model on Merit Leader manually adjusts scoring, labeling why a candidate should be advanced despite low scores.

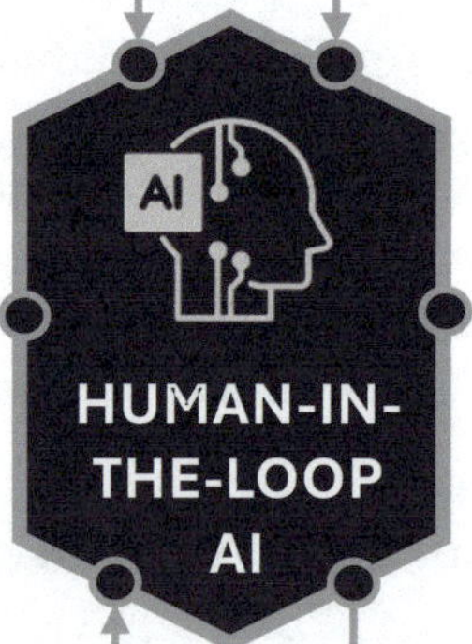

4. Retrain Algorithm Feedback
The model incorporates the leader's corrections to reduce systemic bias into its logic.

5. Validate Unbiased Output
Test the updated model against diverse profiles to confirm ethical fairness is achieved correctly.

6. Approve Diverse Shortlist
Leader signs off on the final, bias-mitigated candidate shortlist and hiring criteria.

USE CASES

- **Manufacturing | Skills Gap Training Prioritization:** Leaders validate AI-recommended training paths, manually adjusting based on future strategic needs to ensure the workforce is prepared for automation effectively.
- **Financial Services | High-Risk Regulatory Training:** Supervisors review employee compliance simulation failures flagged by the AI, labeling root causes to continuously refine and improve regulatory education reliably.
- **Healthcare | Specialized Staffing Allocation:** Senior administrators correct AI's staff scheduling for specialized units, coaching the model on unique human factors to prevent burnout and ensure coverage safely.
- **Technology | Expertise Matching Refinement:** Project managers review AI suggestions for team assembly, correcting incorrect skill tags to ensure the model accurately connects projects with employee talents clearly.
- **Retail | Employee Performance Guidance:** Managers review AI suggestions for disciplinary action, correcting the AI's rigidity to ensure employee counseling is humane and productive accurately.
- **Energy | Safety Protocol Development:** Supervisors validate AI-generated safety procedures for new field equipment, providing practical, real-world context to certify the protocols are clear and safe quickly.

In the **Workforce Transformation**, the **Architects (The Brain)** serve as the "Cognitive Infrastructure" of the enterprise. While the primary instrument of **Human-in-the-Loop (HITL)** ensures that the organization's intelligence remains anchored in human values and ethical oversight, it requires secondary technical pairings to scale that wisdom across a global workforce. To move from "Passive Documentation" to a "Live Institutional Memory," a leader must integrate these six pairings.

1. The Vector Database (RAG) Pairing for Contextual Retrieval

Pairing **HITL with Vector Databases** is the foundation of the **"Searchable Brain."** While the human Architect sets the governance rules, the Vector Database provides the "Semantic Memory." This is best utilized to eliminate the "Onboarding Tax." New hires can query the Brain in natural language—asking "How did we solve the 2023 server migration friction?"—and receive a synthesized answer that pulls from disparate emails and logs. It ensures the organization's past wisdom is a "Live" asset, validated by human experts but retrieved by machine precision.

2. The Knowledge Graph Pairing for Logical Lineage

Leaders should pair **HITL with Knowledge Graphs** to ensure **Logical Traceability**. While the human provides the high-level strategy, the Knowledge Graph provides the "provenance" of the supporting data. This is best deployed in highly regulated industries like Finance. It allows an employee to see not just the policy, but the legal precedent and the specific executive decision-maker behind it. This pairing prevents "Intelligence Decay," ensuring that every piece of retrieved knowledge is anchored in a verifiable web of truth that a human can audit at any time.

3. The Natural Language Processing (NLP) Pairing for Wisdom Refinement

To prevent the "Brain" from becoming a repository of outdated facts, **HITL must be paired with Advanced NLP**. This is best utilized through "Expert Verification Nodes." When the NLP layer identifies a gap or contradiction in the company's internal documentation, it triggers a request for a human subject matter expert to annotate or validate a new entry. This ensures the institutional memory is a "Living Organism" that is constantly pruned and refined by the firm's most senior minds, maintaining a high signal-to-noise ratio in its cognitive output.

4. The Differential Privacy Pairing for Secure Sovereignty

When building a "Searchable Brain" that contains sensitive HR data, a leader should pair **HITL with Differential Privacy**. This allows the human Architect to open the organization's data lake to the workforce without the risk of exposing personal information. This is best used in "Cross-Departmental Collaboration." For example, the marketing team can learn from the "patterns" of engineering successes without seeing the raw, sensitive code or specific employee performance reviews, protecting individual sovereignty while scaling the collective intelligence of the firm.

5. The Multi-Modal AI Pairing for Implicit Knowledge Capture

To capture the "Tacit Knowledge" that usually disappears with turnover, leaders should pair **HITL with Multi-Modal AI**. This allows the human Architect to ingest more than just text; it can "understand" video recordings of factory floor procedures or sketches from design workshops. This is best utilized in "Apprenticeship Scaling." By indexing the visual and auditory nuances of a master craftsman, the Architect ensures that the "Brain" captures the *how* as well as the *what*, preserving the firm's non-verbal expertise for future generations.

6. The Blockchain Pairing for Immutable Governance

For ensuring the integrity of the firm's most critical SOPs, **HITL should be paired with Blockchain**. While the human sets the policy, the Blockchain stores the "Audit Trail" of every change made to that record. This is best utilized for "Compliance Sovereignty." It prevents any unauthorized or accidental tampering with the organization's core "Rules of Play." It ensures that the Brain remains an unassailable record of truth, providing a legally permanent ledger of the firm's evolving strategic and operational logic as guided by its human leaders.

Workforce Power: Agentic Orchestration

The final evolution of the workforce is the transition from "Manual Labor" to **"Orchestrated Autonomy."** Historically, the "Hands" of the workforce were limited by "Physical Presence" and "Linear Execution"—a worker could only do one thing at one time. The transformation into **Executors** represents the birth of the "Multiplied Worker." In this role, the employee moves from "Performing a Task" to "Directing a Fleet." They become the "Conductors of Action," using autonomous systems to manifest their "Intent" across a massive scale. The worker is no longer the "Tool" of the company; they are the "Commander" of the company's automated capabilities.

Defensive Stance: Killing the Operational Drag

The defensive mission of the Worker-Executor is the **Elimination of Mundanity**. In the legacy world, the "Hands" were often tied up in "High-Volume, Low-Value" work—data entry, repetitive sorting, or basic monitoring. This created an "Operational Drag" that led to burnout and error. The Executor role kills this "Wastage of Human Potential" by establishing "Self-Healing Operations" at the task level. The worker identifies the "Friction Points" in their own workflow and "Hard-Codes" the fix, allowing the system to handle the "Boring" while they handle the "Complex." This is the **Shield of Excellence**. By "Automating the Routine," the worker ensures that their own "Hands" are only ever used for "High-Impact Moves." This defensive posture protects the brand's margins and the worker's morale, ensuring that the "Engine" of the company is always running at peak efficiency without human fatigue.

Offensive Stance: Unlocking Scalable Impact

On the offense, the Executor workforce moves toward **Agentic Command**. This is the ability for a single employee to achieve "Infinite Scale" through "Orchestration." Instead of "Doing the Job," the worker "Deploys the Agents" to achieve a "Goal." A single logistics manager can orchestrate a fleet of autonomous vehicles; a single marketer can direct a swarm of generative content agents. This **Unlocks Orchestration**, where the worker's "Intent" is the only limit to their "Impact." They move from "Managing a Process" to "Orchestrating a Result." This turns the workforce into an "Autonomous Fleet," capable of executing complex, global strategies with the speed of software. It allows the brand to "Act" with a level of "Force" and "Precision" that makes manual competition irrelevant.

The Workforce Outcome: Absolute Leverage

The ultimate value of the Executor role is **Individual Empowerment**. When the "Hands" of the workforce are "Autonomous," every employee has the "Leverage" of a thousand workers. For the leader, this provides **Strategic Velocity**. You can move the entire company in a new direction instantly because your people are "Conductors" who can re-direct their "Automated Fleets" with a single command. This turns the workforce into an "Unstoppable Force"—one that executes with "Absolute Precision" and "Infinite Scale." It ensures that the brand's "Hands" are never a limitation, but a "Limitless Engine" of growth, transforming the workforce into a "Global Powerhouse" of autonomous execution.

Executive Power: Agentic Autonomy

The final shift in the workforce journey is the transition from "Managing Task Completion" to **"Orchestrating Autonomous Fulfillment."** The **Executor** role represents the birth of the "Agentic Workforce," where the leader provides the "Strategic Intent" and a blended swarm of human-AI agents manifests the result. Historically, execution was limited by "Human Bandwidth" and the "Toil of Repetitive Tasks." The Executor role replaces this with a system of **Autonomous Agency**, where the "Hands" of the workforce are capable of navigating complex workflows with near-zero manual oversight. This is the shift from "Supervising Labor" to "Directing Power," giving the executive the leverage to "Operate at the Speed of Light" by transforming "Intent" into "Fulfillment" through a self-correcting, autonomous execution machine.

The Sensory Layer: Operational Pulse and Trigger Recognition

The **Sensory Layer** of the workforce Hands serves as the organization's "Reflex System." It is designed for **Real-Time Task Triggering**, monitoring the internal and external environment for signals that require action. Utilizing **Event-Driven APIs** and **Predictive Anomaly Detection**, this layer identifies the moment a goal requires execution. It doesn't wait for a human manager to "Assign a Ticket"; it "feels" the operational need as it arises—whether it is a customer service surge, a supply chain deviation, or a sudden marketing opportunity. This sensory input provides the "Activation Signal" for the agentic workforce, ensuring that the "Hands" of the company are always ready to move the millisecond a requirement is detected. It protects the executive's "Time" by ensuring the organization responds to reality with biological speed, catching opportunities before they vanish.

The Processing Layer: Hybrid Agent Orchestration and Logic

The **Processing Layer** acts as the "Workforce Cerebellum," breaking down the leader's high-level intent into thousands of executable sub-tasks. It utilizes **Agentic Logic (ReAct Frameworks)** and **Chain-of-Thought Reasoning** to orchestrate a hybrid swarm of human experts and AI agents. If the leader's intent is to "launch a global campaign in forty-eight hours," this layer identifies the complex web of tasks—from creative generation to regulatory compliance and localized deployment. It manages the "Inter-Agent Communication," ensuring that the AI agents handle the high-volume toil while the human "Hands" focus on high-stakes judgment. This layer provides the **Governance and Logic** for the execution, ensuring that the workforce stays within the "Ethical and Brand Guardrails" set by the executive, killing the friction of micro-management through automated, high-speed coordination.

The Cognitive Layer: Autonomous Fulfillment and Self-Correction

The **Cognitive Layer** is where the workforce "Hands" actually manifest the result through **Action Transformers** and **Automated Fulfillment Engines**. This is the layer of "Closed-Loop Execution." It doesn't just "process" a task; it "achieves" the outcome. If the workforce encounters an obstacle—such as a technical failure or a market shift—this layer uses **Self-Correction Logic** and **Reinforcement Learning** to pivot the strategy and find an alternative path without requiring new instructions. It turns "Executive Intent" into "Physical and Digital Reality," whether that means deploying code, closing sales, or resolving logistics autonomously. This provides the executive with **Absolute Operational Leverage**, allowing them to "Conduct the Swarm" rather than "Manage the Toil." It transforms the workforce into an "Autonomous Execution Machine" that operates with the precision of a law of physics, delivering "Market Dominance" through sheer agentic velocity.

6. THE EXECUTORS (THE HANDS) | AGENTIC AI

OVERVIEW

Agentic AI operates as an autonomous mission manager, breaking down high-level strategic goals into executable tasks. The Driving Mission leader defines the ultimate objective and core values, essentially setting the constitutional limits for the AI agent. This synergy offloads complex, time-consuming multi-step execution to the AI, allowing the leader to focus exclusively on strategic vision and alignment. The leader initiates the mission; the agent executes and reports on key milestones. This partnership fundamentally converts strategy into action, maximizing the leader's impact on high-level direction and governance.

MAIN STEPS

- **Step 1:** *Define the high-level desired outcome.*
- **Step 2:** *Agent breaks goal into specific sub-tasks.*
- **Step 3:** *Agent performs steps, tracks progress autonomously.*
- **Step 4:** *Agent summarizes for leader approval.*
- **Step 5:** *Leader adjusts mission parameters or values.*
- **Step 6:** *Agent delivers final output to human team.*

SAMPLE FLOW

Autonomous Skill Gap Closure Program: The leader commissions an agent to eliminate critical organizational skill gaps needed for future strategic growth.

1. Define Future Capabilities
Specify target skills and proficiency levels required for upcoming missions clearly.

2. Diagnose Current Gaps
Agent autonomously analyzes existing employee skill data and future needs correctly.

3. Design Learning Paths
Agent generates personalized curriculum roadmaps and resource assignments autonomously.

4. Execute Training deploy
Agent deploys and manages learning modules, tracking engagement and certification

5. Validate Mission Success
Leader reviews impact on productivity and certifies new skills against the organizational mission

6. Integrate New Capabilities
Leader authorizes integrating newly certified employees into critical strategic projects criteria.

USE CASES

- **Manufacturing | Safety Protocol Evolution:** An agent autonomously monitors site incidents and revises safety protocols, deploying mandatory, personalized training to high-risk staff efficiently.
- **Financial Services | Regulatory Certification Maintenance:** The leader tasks an agent with continuously tracking and renewing all employee regulatory certifications, autonomously enrolling staff in required compliance courses promptly.
- **Healthcare | Culture of Innovation Launch:** An agent is commissioned to promote a culture of innovation, autonomously matching interdisciplinary staff to solve complex operational challenges creatively.
- **Technology | Onboarding Personalization Mission:** An agent designs and manages personalized onboarding missions for new hires, autonomously providing tailored documentation, mentors, and integration tasks effectively.
- **Retail | Store Leadership Development:** An agent identifies high-potential store staff and launches a customized leadership development mission, autonomously scheduling coaching, peer learning, and task rotations accurately.
- **Energy | Succession Planning Deployment:** The leader utilizes an agent to manage succession planning for critical roles, autonomously developing high-potential employees with targeted job exposure and mentorship assignments reliably.

In the **Workforce Transformation**, the **Executors (The Hands)** represent the organization's "Autonomous Fulfillment Engine." While the primary instrument of **Agentic AI** provides the "agency" to navigate complex goals, it requires secondary technical pairings to bridge the gap between abstract strategy and precise physical or digital action. To move from "Operational Friction" to "Autonomous Value Action," a leader must integrate these six pairings. This empowers the "Hands" to manifest the leader's intent with surgical precision, freeing the human workforce from the burden of digital drudgery.

1. The RPA Pairing for Legacy System Integration

Pairing **Agentic AI with Robotic Process Automation (RPA)** is best utilized when the "Hands" must interact with older, rigid software environments. While the Agent decides the high-level strategy—such as re-routing a shipment—RPA provides the "fingers" to execute the specific clicks and data entries in legacy ERP or HR systems. This is essential for back-office transformation, allowing autonomous agents to bridge the gap between modern intelligence and the firm's existing digital infrastructure without a total system overhaul.

2. The Digital Twin Pairing for Operational Guardrails

Leaders should pair **Agentic AI with Workforce Digital Twins** when deploying autonomous agents in high-stakes environments. Before an agent implements a new workflow or adjusts a production schedule, it "rehearses" the action in the Digital Twin. This pairing ensures that the "Hands" never take an action that could cause systemic bottlenecks or safety violations. It provides a virtual sandbox for autonomous decision-making, ensuring that the execution layer is always verified against a live model of reality.

3. The Edge AI Pairing for Zero-Latency Actuation

When execution must happen at the "Point of Action"—such as on-site for a field technician or in a smart warehouse—**Agentic AI must be paired with Edge AI**. This allows the "Hands" to process data and make decisions locally on hardware without waiting for a cloud response. This pairing is vital for autonomous logistics and safety, ensuring that the agents can react to sudden physical obstacles or mechanical shifts in milliseconds. It turns the organization's execution layer into a real-time responsive nervous system.

4. The Blockchain Pairing for Autonomous Contracting

For secure, multi-party transactions, a leader should pair **Agentic AI with Blockchain (Smart Contracts)**. This is best used in automated talent procurement or vendor management. The "Hands" don't just find a freelancer or a supplier; they execute the payment and verify delivery through an immutable ledger once specific milestones are met. This pairing ensures that autonomous actions are legally binding and transparent, eliminating the friction of manual financial reconciliation and ensuring absolute integrity in the execution of the firm's capital.

5. The Computer Vision Pairing for Physical Awareness

To allow the "Hands" to interact with the non-digital world, leaders should pair **Agentic AI with Computer Vision**. This is best utilized in quality control and autonomous facility maintenance. By "seeing" the environment, the agent can identify a safety hazard in an office or a defective product on a line. This pairing gives the "Hands" the sensory awareness required to perform precision physical tasks that were previously reserved for human eyes, ensuring that the organization's physical presence is as intelligent as its digital core.

6. The Reinforcement Learning Pairing for Skill Acquisition

In dynamic environments where tasks change constantly, **Agentic AI should be paired with Reinforcement Learning (RL)**. This is best utilized for optimizing complex internal processes like dynamic shift scheduling or energy usage. The "Hands" learn from the success or failure of every action, constantly refining their technique to maximize efficiency. This provides the brand with a self-improving execution layer that grows more skillful with every task performed, ensuring that the organization's capacity for action evolves in lockstep with its strategy.

BLUEPRINT SERIES

SECTION C

Business Maturity

Ai6x6™ Transformation Blueprint

Story 1 — The Marshall Plan Reconstruction (1948–1952): Post-war Europe faced shattered economies, broken supply chains, and institutional collapse. The Marshall Plan introduced structured funding, transparent reporting, and coordinated execution across nations. Local governments provided context while centralized oversight ensured accountability. This blend of autonomy and structure accelerated recovery. The program demonstrated how disciplined governance can rebuild complex systems. It became a blueprint for large-scale organizational maturity.

Story 2 — The Rise of Quality Management in Japan (1950s–1960s): Japan's industries struggled with inconsistent quality and limited global credibility. Statistical quality control introduced measurement, repeatability, and disciplined improvement cycles. Workers contributed insights while leaders standardized processes. Over time, quality became a cultural norm rather than a corrective action. This shift transformed fragmented operations into cohesive systems. It marked the emergence of maturity through methodical refinement.

Story 3 — NASA's Systems Engineering Discipline (1960s): The space race demanded unprecedented coordination across thousands of contractors and scientific domains. NASA formalized systems engineering to integrate requirements, risks, and interfaces into a unified framework. Structured reviews balanced innovation with rigor. Human judgment guided trade-offs while documentation preserved institutional memory. This discipline enabled complex missions to succeed repeatedly. It became a global model for organizational maturity in high-stakes environments.

Story 4 — The Birth of Modern Corporate Governance (1970s–1980s): Corporate scandals exposed weaknesses in oversight, transparency, and fiduciary responsibility. Boards adopted clearer roles, audit mechanisms, and accountability structures. Regulations reinforced ethical conduct while organizations matured through formal governance. Leaders learned to balance autonomy with compliance. This era established governance as a strategic capability, not an administrative burden. It marked a turning point in institutional maturity.

Story 5 — Toyota's Global Expansion Playbook (1980s–1990s): As Toyota expanded globally, it faced cultural differences, supply-chain complexity, and operational variability. The company codified principles, standardized training, and embedded continuous improvement across regions. Local teams adapted practices while preserving core philosophy. This balance of global consistency and local flexibility enabled scalable excellence. It demonstrated maturity as the ability to replicate capability without diluting identity.

Story 6 — The Digital Transformation of Banking (1990s–2000s): Banks confronted rising customer expectations, regulatory pressure, and technological disruption. They modernized legacy systems, standardized data, and introduced structured risk frameworks. Digital channels required new operating models and cross-functional coordination. Human expertise guided decisions where automation lacked nuance. Over time, banks evolved from siloed institutions into integrated digital enterprises. This transition reflected maturity through modernization and disciplined reinvention.

LESSONS FROM A WORKFORCE TRANSFORMATION PERSPECTIVE

Workforce maturity develops when employees move from task execution to system contribution. Over time, organizations learned that sustainable performance requires workers who understand processes, challenge assumptions, and participate in continuous improvement. Mature workforces operate with shared standards, clear roles, and collective accountability. They balance discipline with creativity, following structured methods while adapting to local realities. As organizations scale, the workforce becomes a distributed intelligence system—capable of maintaining quality, absorbing change, and reinforcing culture. Maturity is reached when people no longer rely on individual heroics but on shared practices that elevate everyone's performance.

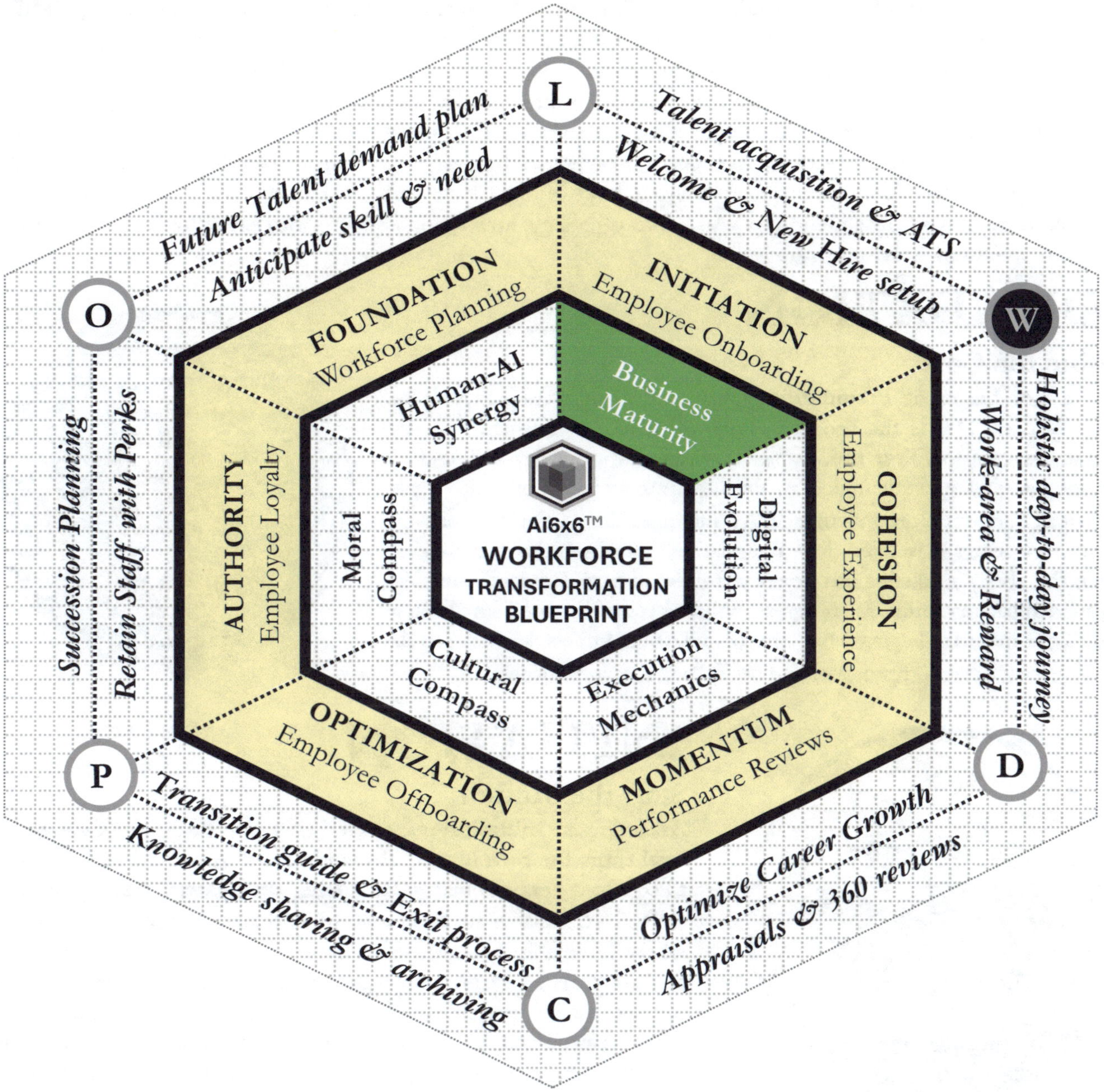

The **Ai6x6 Research Lab** selected this lens because an organization's maturity is measured by how it honors and manages its people through every stage of their journey. We view workforce maturity as a continuous cycle of strategic forecasting, integration, and succession. This dimension ensures that the aggressive talent needs of the AI era are anchored in a deep respect for the employee experience. It is a vital lens because, in the turbulence of rapid turnover and skill shifts, an organization's ability to maintain a stable talent fabric is its greatest competitive advantage. This lens is the heart's commitment to growing the "Employee Experience" as a core requirement.

We believe true maturity is found in the ability to manage the delicate transitions of the workforce—from the first spark of onboarding to the honorable transfer of knowledge during offboarding. By focusing on performance loops and strategic loyalty, the Lab provides a mechanism for leaders to build a workforce that is resilient, adaptable, and deeply committed to the long-term mission. This lens ensures that talent acquisition is not a frantic reaction, but a deliberate evolution of organizational capability. It is the steady hand that guides the workforce through transformation, recognizing that while technologies change, the bedrock of human loyalty and institutional memory must be preserved.

Every leadership transformation begins with ambition, but ambition alone is never enough to carry an organization into its future. **Business Maturity** is the lens that forces leaders to confront the gap between what they hope to achieve and what their enterprise is prepared to sustain. It is the discipline that reveals whether the organization is structurally, operationally, and culturally ready for the future it claims to desire. In the AI era, this gap becomes even more pronounced; AI does not simply automate tasks—it reshapes workflows, decision cycles, and the very rhythm of how leadership operates. It exposes weaknesses that were previously tolerable and accelerates the consequences of poor preparation, demanding a level of rigor that traditional management frameworks often overlook. It examines how ambition translates into capability, how vision converts into execution, and how culture sustains momentum under pressure.

FOUNDATION

Define the Core

The journey begins at **Foundation**, where the primary objective is to *Define the Core*. This is the bedrock upon which all transformation rests, requiring leaders to **Filter the Noise** of competing priorities and anchor their identity and strategic intent. At this level, clarity is not optional—it is existential. Without clear governance, AI initiatives drift, priorities conflict, and teams operate in isolated silos. Foundation maturity is the discipline of intent: the ability to distinguish urgency from readiness. Leaders who skip this stage mistake motion for progress, building transformation on shifting sand. True maturity begins when the organization knows what it stands for, what it will not compromise, and how it will measure success.

INITIATION

Start the Motion

Ascending to **Initiation** marks the first active step: *Start the Motion*. This level turns the gaze inward, demanding self-awareness and the courage to confront personal biases. By **Wiring the System**, leaders catalyze the shift from passive management to intentional authority. Initiation is where leadership becomes kinetic—it transforms from aspiration into activation. The organization begins to move, not because it is told to, but because its internal circuits are aligned. Maturity here is emotional as much as structural; it requires self-regulation, empathy, and the ability to lead through friction. The leader's persona is tested, revealing whether they can sustain momentum through disruption rather than retreat into comfort.

COHESION

Connect the Parts

The climb continues to **Cohesion**, represented by *The Engine*. This is the critical unifier where people, processes, and systems are harmonized. The focus shifts to **Connect the Parts**—to ensure that individual leadership goals align with a unified organizational voice. By **Shifting the Gear**, leaders eliminate silos and systemic friction, transforming fragmented efforts into a synchronized machine. Cohesion maturity is the art of integration: turning disconnected motion into collective momentum. Without it, organizations celebrate activity instead of achievement. The mature leader understands that alignment is not consensus—it is coherence, the ability to move as one without losing diversity of thought.

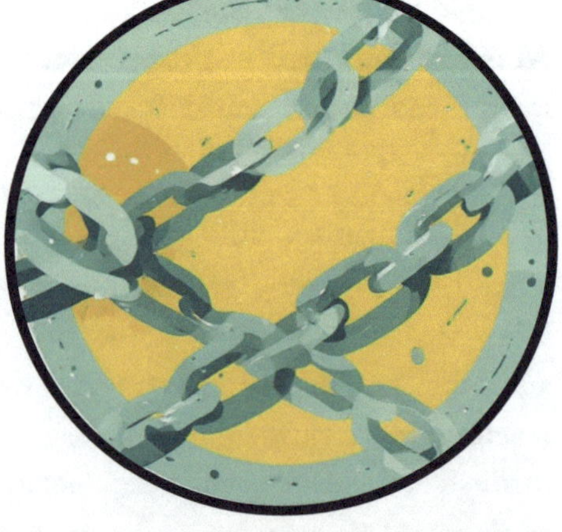

MOMENTUM

Drive the Speed

The fourth rung is **Momentum**, where the organization begins to *Drive the Speed.* Leadership here moves toward scaling impact with predictable velocity. By **Removing the Drag** of legacy thinking and institutional inertia, leaders accelerate the collective intelligence of the enterprise. Momentum maturity is the physics of transformation—it measures how efficiently energy converts into progress. At this stage, the organization's rhythm changes; it begins to operate with self-propelling force. The leader's role shifts from directing individuals to empowering systems, ensuring that innovation scales faster than resistance. Momentum is not chaos—it is controlled acceleration, the moment when transformation gains its gravity.

OPTIMIZATION

Refine the Flow

Refinement occurs at **Optimization**, characterized by *The Flow.* Here, leadership pauses to **Refine the Flow** by **Locking the Code** of influence through reflection and data-driven inquiry. Immature organizations measure what is easy; mature ones measure what matters. Optimization maturity is the discipline of precision—removing friction without losing flexibility. Leaders refine decisions still in motion, adjusting with humility and surgical accuracy. Ego and bias are replaced by evidence and iteration. The organization becomes self-healing, capable of continuous improvement without external intervention. Optimization is not about perfection—it is about adaptability, the ability to evolve faster than the environment demands.

AUTHORITY

Own the Market

The summit is reached at **Authority**, where leaders finally *Own the Market.* This stage manifests as *The Benchmark*—a legacy of endurance that can withstand external shocks and volatility. By **Fueling the Future**, leaders ensure that transformation is not a temporary win but a permanent standard. Authority maturity is the discipline of stewardship: sustaining excellence while preparing for reinvention. Immature organizations avoid this stage because it exposes uncomfortable truths; mature ones embrace it as the crucible of resilience. At this level, leadership becomes institutional—it transcends personality and becomes culture. The organization no longer reacts to the market; it defines it.

Business Maturity is the scaffolding that holds the weight of innovation. It ensures that speed does not collapse under fragility and that ambition does not outpace capability. Each level builds upon the last, forming a continuum from clarity to mastery. Foundation defines intent, Initiation activates leadership, Cohesion integrates systems, Momentum accelerates impact, Optimization refines intelligence, and Authority sustains legacy. In the AI era, maturity is not a luxury—it is survival. As algorithms amplify both strengths and weaknesses, leaders must design organizations that can learn, adapt, and endure. Business Maturity is therefore not a checklist but a **leadership discipline**—a way of thinking that transforms ambition into architecture. When leaders embrace this full sequence, they discover that maturity is not the end of transformation but its engine—the invisible structure that makes both explosive speed and long-term continuity possible.

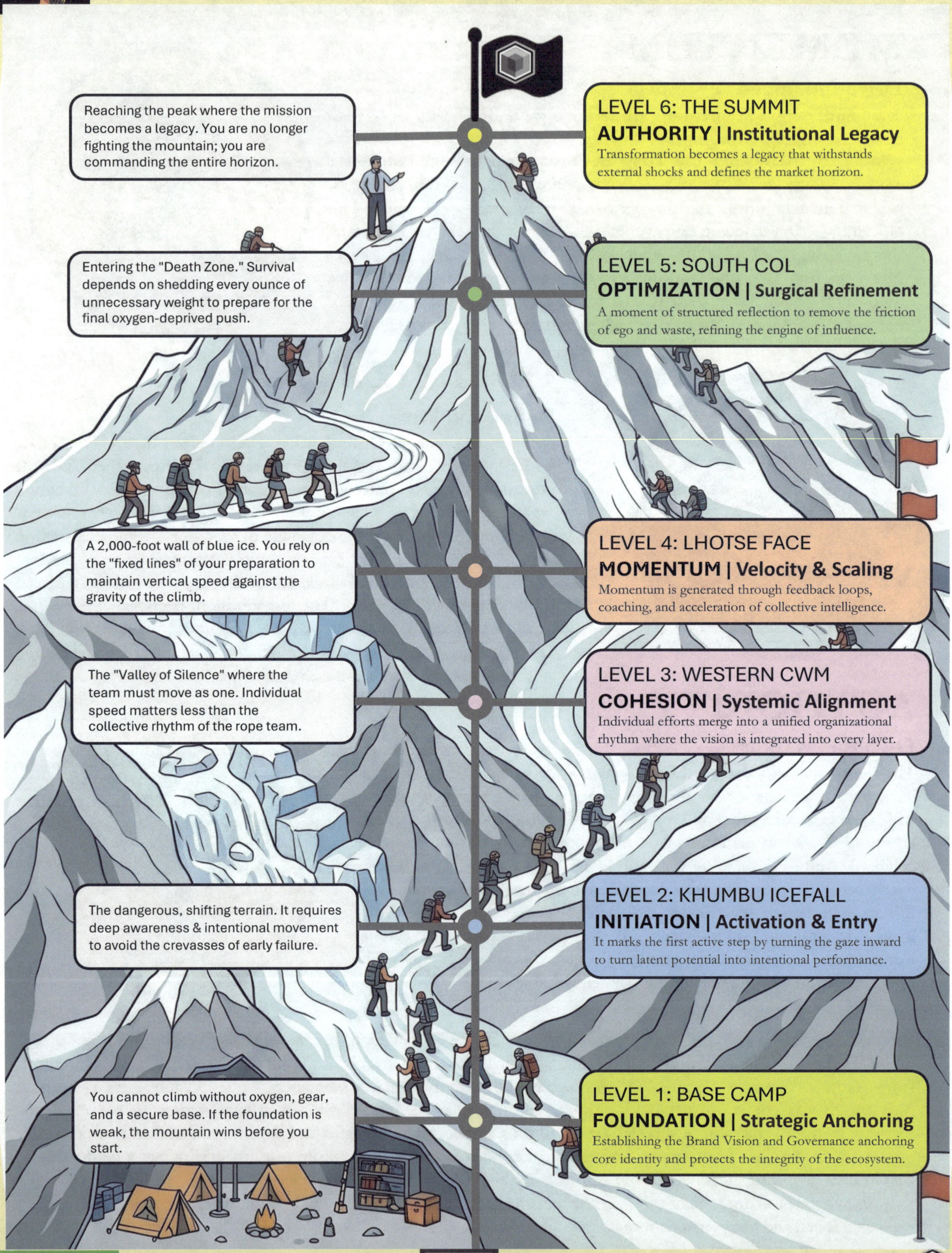
Reaching the peak where the mission becomes a legacy. You are no longer fighting the mountain; you are commanding the entire horizon.
LEVEL 6: THE SUMMIT
AUTHORITY | Institutional Legacy
Transformation becomes a legacy that withstands external shocks and defines the market horizon.
Entering the "Death Zone." Survival depends on shedding every ounce of unnecessary weight to prepare for the final oxygen-deprived push.
LEVEL 5: SOUTH COL
OPTIMIZATION | Surgical Refinement
A moment of structured reflection to remove the friction of ego and waste, refining the engine of influence.
A 2,000-foot wall of blue ice. You rely on the "fixed lines" of your preparation to maintain vertical speed against the gravity of the climb.
LEVEL 4: LHOTSE FACE
MOMENTUM | Velocity & Scaling
Momentum is generated through feedback loops, coaching, and acceleration of collective intelligence.
The "Valley of Silence" where the team must move as one. Individual speed matters less than the collective rhythm of the rope team.
LEVEL 3: WESTERN CWM
COHESION | Systemic Alignment
Individual efforts merge into a unified organizational rhythm where the vision is integrated into every layer.
The dangerous, shifting terrain. It requires deep awareness & intentional movement to avoid the crevasses of early failure.
LEVEL 2: KHUMBU ICEFALL
INITIATION | Activation & Entry
It marks the first active step by turning the gaze inward to turn latent potential into intentional performance.
You cannot climb without oxygen, gear, and a secure base. If the foundation is weak, the mountain wins before you start.
LEVEL 1: BASE CAMP
FOUNDATION | Strategic Anchoring
Establishing the Brand Vision and Governance anchoring core identity and protects the integrity of the ecosystem.

Ai6x6™ Transformation Blueprint

Foundation | Workforce Planning

"Human capital is the most important asset in an organization."
- Richard Branson.

Founder of the Virgin Group, a multinational conglomerate. Known for his adventurous spirit and unconventional strategies, he built a global brand synonymous with innovation and a strong focus on customer and employee experience. His ventures consistently push boundaries.

The Foundation level is your Strategic Workforce Planning, establishing the human identity and market presence required to support your long-term roadmap. This baseline ensures you have a clear blueprint of the talent DNA necessary before activation begins.

Proactively and systematically aligning an organization's talent needs with its overarching business goals, meticulously anticipating future skill requirements, critical roles, and potential talent gaps across all departments. This foresight ensures that the right people with the right capabilities are in place precisely when needed.

It drives organizational agility, fosters innovation, and sustains long-term growth and competitive advantage, building a future-ready workforce that can adapt to evolving market demands and technological shifts, ensuring continuous operational effectiveness.

For business strategy maturity, the **"Strategic Workforce Planning"** chapter is foundational because it defines the workforce vision and strategy, ensuring human capital aligns with organizational goals. A mature business strategy proactively identifies future skill needs, forecasts workforce demand, and strategically shapes hiring efforts. This alignment ensures the right talent is in place at the right time, preventing critical skill gaps and optimizing resource allocation. It's about building a workforce that is not just reactive but strategically prepared to execute current and future business objectives, directly impacting long-term competitive advantage. Here is how we approach our learning:

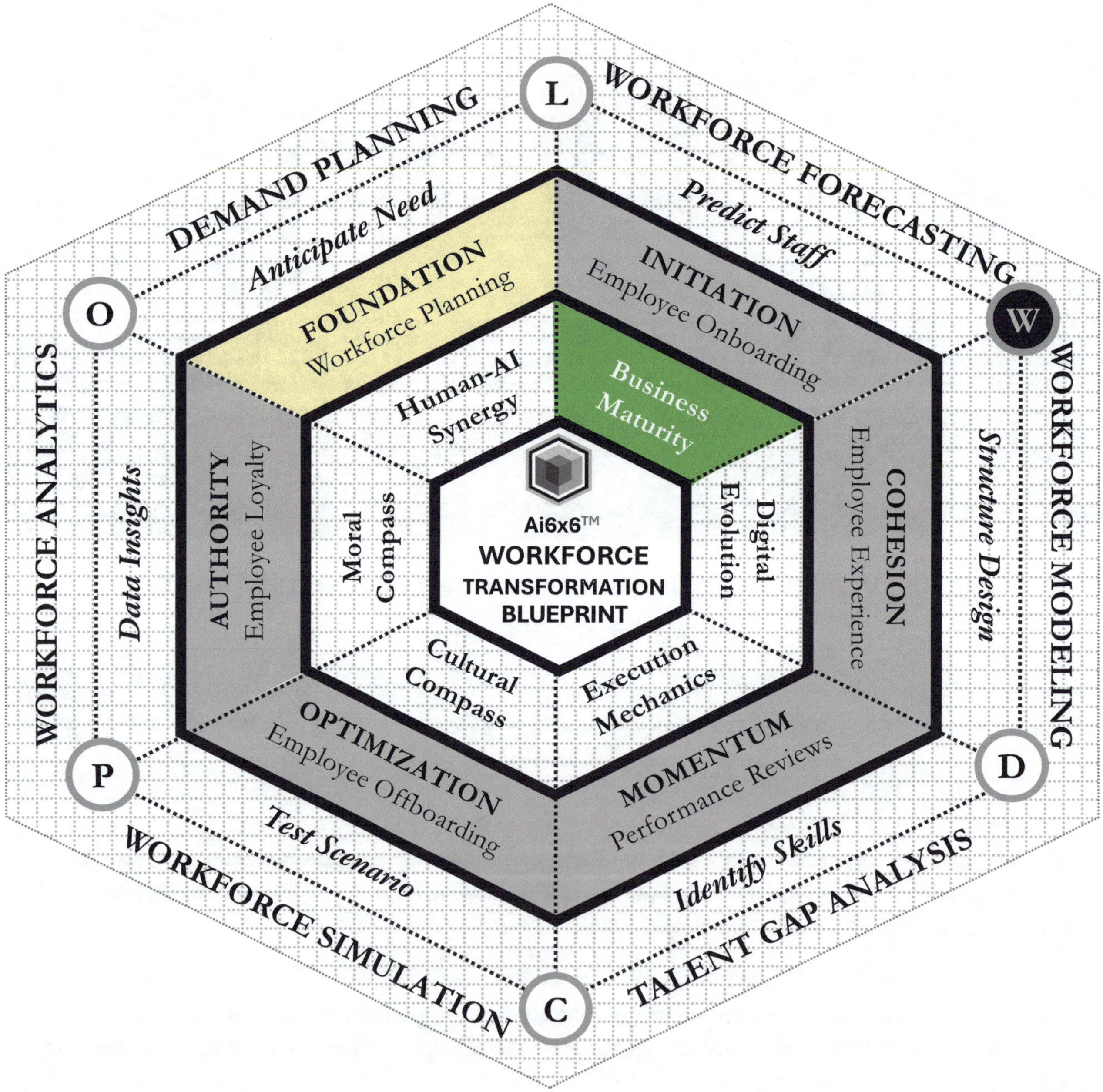

Ai6x6 BUSINESS MATURITY FOR PLANNING | WORKFORCE TRANSFORMATION

1. DEMAND PLANNING

For a business leader, effective demand planning is the foundational step in strategic workforce management, crucial for ensuring your organization has the right people, with the right skills, at the right time. This isn't just about filling current vacancies; it's a proactive exercise in anticipating future talent needs based on both your evolving organizational goals and broader industry trends. Your role as a leader is to articulate a clear vision of where the business is headed, allowing HR and talent teams to translate that into precise workforce requirements. This involves meticulous strategic workforce alignment, ensuring that your employee capacity is always proportional to your current and projected business objectives. It's a continuous balancing act to avoid both understaffing (which hinders growth) and overstaffing (which creates unnecessary costs).

Furthermore, detailed role-specific demand forecasting is essential. This means going beyond general headcount to identify specific hiring needs driven by departmental growth, new product launches, market expansion, or technological shifts, pinpointing exactly what skills and roles will be critical. Crucially, in many industries, planning for seasonal and project-based workforce demand is vital. Neglecting robust demand planning can lead to critical talent shortages, missed market opportunities, increased recruitment costs due to rushed hiring, and a decline in productivity as existing teams become overwhelmed. By championing precise demand planning, you empower your organization to proactively build a resilient and agile workforce, ready to support strategic initiatives and adapt to market dynamics, ultimately securing long-term operational efficiency and competitive advantage. This forward-looking approach to talent is indispensable for sustainable growth.

WHAT?	WHY?	HOW?	BEWARE!
STRATEGIC WORKFORCE ALIGNMENT	**CAPACITY MATCH**	**GOAL TRANSLATION**	**MISMATCH RISK**
Ensuring employee capacity matches business objectives.	Ensures employee capacity is proportional to business objectives.	Translate business goals into specific talent requirements.	Risk of misalignment, leading to over / under staffing
ROLE-SPECIFIC DEMAND FORECASTING	**PRECISE HIRING**	**DEPARTMENTAL ANALYSIS**	**GENERALIZATIO N**
Identifying hiring needs based on department growth.	Pinpoints exact hiring needs for critical roles and skills.	Analyze growth plans, project requirements, and skill gaps.	Tendency to generalize, missing niche or critical skill needs.
SEASONAL & PROJECT-BASED WORKFORCE	**OPERATIONAL AGILITY**	**WORKLOAD ANALYSIS**	**INEFFICIENT SCALING**
Planning staffing for fluctuating workloads.	Optimizes staffing for fluctuating workloads and specific projects.	Analyze historical data, project timelines, and peak demands.	Inefficient scaling, leading to burnout or underutilization.

For a business leader, workforce forecasting is the analytical backbone of strategic talent management, enabling you to predict future labor requirements with precision and proactively optimize your staffing strategies. This is far more sophisticated than simple headcount projections; it involves leveraging data and advanced models to anticipate changes in your workforce needs, ensuring you're always one step ahead. Central to this is utilizing predictive analytics in workforce trends. By analyzing historical data, identifying patterns, and employing AI models, you can accurately estimate future workforce demand, anticipating which roles will grow, which might shrink, and the evolving skill sets required. This data-driven foresight allows for proactive recruitment and development.

Furthermore, accurate attrition and retention forecasting are crucial. Predicting turnover rates, understanding the factors influencing employee departures, and anticipating retention challenges enables you to implement strategies to maintain talent stability, minimizing disruptive gaps and preserving institutional knowledge. This involves understanding economic shifts, industry-specific changes, and demographic movements that will influence talent availability and cost, ensuring your hiring and resource allocation strategies remain competitive and realistic. Neglecting robust workforce forecasting can lead to chronic talent shortages, inflated recruitment costs due to reactive hiring, skill obsolescence, and a significant competitive disadvantage. By championing data-driven workforce forecasting, you empower your organization to build a resilient talent pipeline, adapt proactively to market changes, and optimize human capital investments, ultimately securing the long-term operational effectiveness and strategic advantage of your business.

WHAT?	WHY?	HOW?	BEWARE!
PREDICTIVE ANALYTICS IN WORKFORCE	**ACCURATE ESTIMATION**	**DATA ANALYSIS & MODELING**	**DATA QUALITY**
Using historical data and AI models.	Provides precise estimates of future workforce demand.	Analyze historical data, identify patterns, and apply AI/ML models.	Poor data quality or insufficient historical data leading to inaccurate forecasts.
ATTRITION & RETENTION FORECASTING	**TALENT STABILITY**	**HISTORICAL ANALYSIS & DRIVERS**	**UNFORESEEN EVENTS**
Anticipating turnover rates for stability.	Proactively manages talent stability by anticipating turnover.	Analyze past turnover rates and identify key retention drivers.	External factors (e.g., economic downturns) making predictions difficult.
MARKET-DRIVEN WORKFORCE PROJECTIONS	**COMPETITIVE ALIGNMENT**	**ECONOMIC & INDUSTRY ANALYSIS**	**MARKET VOLATILITY**
Aligning with economic and industry changes.	Ensures hiring and resource allocation are market-competitive.	Monitor economic indicators, industry growth, and talent supply.	Rapid, unpredictable market shifts rendering projections quickly outdated.

3. WORKFORCE MODELING

For a business leader, effective workforce modeling is a strategic imperative that enables you to design and structure your workforce composition for optimal efficiency, agility, and long-term scalability. This goes beyond simply counting heads; it's about proactively engineering the right blend of roles, skills, and organizational structures to meet your strategic objectives. Central to this is organizational structure optimization. As a leader, you must continuously evaluate and redesign workforce hierarchies to ensure maximum agility and responsiveness to market changes. This involves flattening structures where possible, creating cross-functional teams, and empowering decision-making at lower levels to enhance speed and innovation.

Furthermore, meticulous talent distribution models are crucial. This means strategically allocating employees based on their unique skills, defined roles, and alignment with productivity goals. It's about placing the right people in the right seats, optimizing their contributions to maximize collective output and achieve specific business outcomes. Leveraging automation and advanced analytics can significantly refine hiring decisions, optimize resource allocation, and even predict the impact of various workforce changes. AI provides insights that human analysis alone often misses, leading to more data-informed and efficient models. Neglecting workforce modeling can lead to bureaucratic bottlenecks, misallocation of talent, inefficiencies, and an inability to scale operations effectively in response to growth or contraction. By championing a proactive approach to workforce modeling, you empower your organization to build a highly adaptable, efficient, and future-ready workforce, directly contributing to operational excellence and sustainable competitive advantage. This strategic talent architecture is vital for long-term success

WHAT?	WHY?	HOW?	BEWARE!
ORGANIZATIONAL STRUCTURE OPTIMIZATION	**AGILITY & RESPONSIVENESS**	**HIERARCHY EVALUATION**	**RESISTANCE TO CHANGE**
Designing workforce hierarchies for agility.	Ensures a flexible, adaptable organizational structure.	Analyze existing structures, identify bottlenecks, and design agile models.	Employee and management resistance to new organizational structures.
TALENT DISTRIBUTION MODELS	**OPTIMAL UTILIZATION**	**SKILL INVENTORY & MAPPING**	**SKILL MISMATCHES**
Allocating employees based on skills and roles.	Maximizes employee productivity and skill utilization.	Map employee skills, match to role requirements, and balance workloads.	Mismatches between available skills and evolving role demands.
AI-DRIVEN WORKFORCE PLANNING	**REFINED ALLOCATION**	**AUTOMATION & ANALYTICS**	**OVER-RELIANCE ON AI**
Leveraging automation to refine decisions.	Enhances accuracy and efficiency in hiring and resource allocation.	Implement AI tools for predictive modeling and automated insights.	Over-reliance on technology, neglecting human or ethical factors

For a business leader, conducting a thorough talent gap analysis is a non-negotiable strategic activity, vital for identifying critical skills shortages and proactively addressing workforce deficiencies that could hinder organizational growth and innovation. In a rapidly changing economy, the skills required for success are constantly evolving, making this analysis essential for maintaining competitiveness. Central to this is comprehensive competency mapping. As a leader, you must systematically assess employee capabilities against current and future business needs, determining precise training requirements and pinpointing areas where external hiring is essential. This detailed mapping provides a clear picture of your talent landscape.

Furthermore, developing robust skill gap resolution strategies is crucial. This involves implementing a multi-faceted approach that includes upskilling existing employees through targeted training programs, reskilling individuals for new roles, and strategically recruiting external talent to fill immediate and projected gaps. It's a blend of internal development and external acquisition. Crucially, integrating workforce diversity and inclusion analysis ensures equitable talent representation across all roles. Addressing diversity gaps is not just a moral imperative but also a strategic advantage, as diverse teams often lead to greater innovation and broader market understanding. Neglecting talent gap analysis can lead to critical skill shortages, decreased productivity, a lack of innovation, and a significant competitive disadvantage as your workforce falls behind industry demands. By championing this proactive approach, you empower your organization to build a future-ready workforce, cultivate a learning culture, and ensure you have the talent necessary to achieve your strategic objectives, ultimately securing your businesses long-term.

WHAT?	WHY?	HOW?	BEWARE!
COMPETENCY MAPPING	**NEEDS IDENTIFICATION**	**SKILL INVENTORY**	**OUTDATED COMPETENCIES**
Assessing employee capabilities to determine needs.	Clearly identifies training and hiring needs.	Map employee skills against required competencies for roles.	Competencies becoming outdated due to rapid industry changes.
SKILL GAP RESOLUTION STRATEGIES	**CAPABILITY ENHANCEMENT**	**UPSKILLING & RESKILLING**	**RESOURCE CONSTRAINTS**
Implementing upskilling and recruitment solutions.	Proactively closes skill gaps and enhances workforce capabilities.	Implement training, mentorship, and targeted recruitment drives.	Limited budget or time for effective upskilling/reskilling.
WORKFORCE DIVERSITY AND INCLUSION	**EQUITY & INNOVATION**	**DIVERSITY AUDIT**	**RESISTANCE TO DEI**
Ensuring equitable talent representation.	Ensures diverse talent representation, fostering innovation.	Analyze workforce demographics and implement inclusive hiring practices.	Resistance to diversity initiatives or tokenism without genuine inclusion.

5. WORKFORCE SIMULATION

For a business leader, workforce simulation is a cutting-edge strategic tool that allows you to test various workforce scenarios, significantly improving the accuracy and robustness of your talent planning. In an unpredictable business environment, the ability to model potential futures before committing resources is invaluable for minimizing risk and optimizing human capital investments. Central to this is engaging in scenario-based workforce modeling. This involves simulating various hiring, turnover, and workload variations to understand their potential impact on your workforce composition, costs, and operational capabilities. By running "what-if" analyses, you can anticipate challenges and identify optimal staffing levels under different conditions.

Furthermore, meticulous cost and efficiency analysis is crucial within these simulations. Evaluating the budget implications of various workforce changes—such as new hires, layoffs, or training programs—allows you to make financially sound decisions that support both strategic goals and fiscal responsibility. By simulating potential labor shortages, unexpected disruptions (e.g., pandemics, economic downturns), or sudden skill demands, you can prepare comprehensive contingency plans, ensuring your organization is ready to respond effectively to adversity. Neglecting workforce simulation can lead to costly staffing misjudgments, inefficient resource allocation, and a reactive approach to unforeseen challenges, putting your business at a significant disadvantage. By championing this advanced planning methodology, you empower your organization to make highly informed, proactive workforce decisions, ensuring flexibility, resilience, and optimal human capital deployment for sustained strategic success. This foresight helps avoid costly real-world mistakes.

WHAT?	WHY?	HOW?	BEWARE!
SCENARIO-BASED WORKFORCE MODELING	**RISK MINIMIZATION**	**VARIABLE SIMULATION**	**COMPLEXITY & UNCERTAINTY**
Simulating hiring, turnover, and workload.	Minimizes risks by testing impacts of workforce changes.	Simulate changes in hiring, attrition, and workload parameters.	Difficulty modeling all variables or unforeseen external factors.
COST AND EFFICIENCY ANALYSIS	**FINANCIAL OPTIMIZATION**	**BUDGET MODELING**	**ACCURATE DATA**
Evaluating budget implications of changes.	Ensures cost-effective workforce planning decisions.	Analyze budget impacts of staffing changes and efficiency gains.	Difficulty obtaining precise cost data or unexpected expenses.
RISK MITIGATION STRATEGIES	**BUSINESS CONTINUITY**	**CONTINGENCY PLANNING**	**INCOMPLETE PLANNING**
Preparing contingency plans for disruptions.	Enhances organizational resilience against disruptions.	Develop plans for labor shortages, talent loss, and market shifts.	Failure to account for all potential risks or develop robust plans.

For a business leader, leveraging workforce analytics is a strategic imperative that transforms raw human resources data into actionable insights, enabling informed and data-driven workforce planning decisions. In today's competitive talent landscape, relying on intuition alone is no longer sufficient; quantifiable data provides the precision needed to optimize talent management. Central to this is effective employee performance analytics. By measuring productivity, assessing engagement levels, and analyzing the correlation between these factors and various workforce strategies, you gain a clearer picture of what drives success within your organization, allowing for targeted optimization.

Furthermore, meticulous HR metrics and reporting are crucial. Utilizing key performance indicators (KPIs) such as retention rates, time-to-hire, employee satisfaction scores, and training ROI provides a comprehensive overview of your workforce health and the effectiveness of your talent initiatives. These metrics serve as objective benchmarks for progress. Implementing real-time analytics platforms and dashboards allows leaders to monitor trends as they unfold, making agile adjustments to workforce plans based on dynamic data rather than periodic reviews. Neglecting workforce analytics can lead to inefficient hiring, suboptimal talent allocation, increased turnover, and a reactive approach to human capital challenges. By championing a data-driven approach to workforce planning, you empower your organization to make smarter talent investments, optimize employee potential, and proactively address challenges, ultimately securing a more efficient, engaged, and strategically aligned workforce for sustainable business success. This analytical capability is a significant competitive differentiator.

WHAT?	WHY?	HOW?	BEWARE!
EMPLOYEE PERFORMANCE ANALYTICS	**OPTIMIZATION INSIGHTS**	**METRIC TRACKING**	**DATA SILOS**
Measuring productivity and engagement.	Provides data to optimize workforce performance and engagement.	Track productivity, engagement, and development metrics.	Disparate data systems hindering integrated analysis.
HR METRICS AND REPORTING	**OBJECTIVE ASSESSMENT**	**KPI DEFINITION**	**METRIC OVERLOAD**
Using KPIs for planning decisions.	Provides objective measures of HR and workforce effectiveness.	Define key metrics like retention, time-to-hire, and satisfaction.	Too many metrics leading to analysis paralysis or irrelevant data.
CONTINUOUS WORKFORCE INSIGHTS	**AGILE ADJUSTMENTS**	**DASHBOARD IMPLEMENTATION**	**ACTIONABLE INSIGHTS**
Implementing real-time analytics.	Enables real-time strategic adjustments to workforce plans.	Implement real-time dashboards and automated reporting.	Difficulty translating raw data into actionable strategies.

Ai6x6™ Transformation Blueprint

INITIATION | Employee Onboarding

"You can have the best strategy and the best building in the world, but if you don't have the hearts and minds of the people who work with you, none of it comes to life." - Howard Schultz.

Howard Schultz: Former CEO of Starbucks, credited with transforming it into a global coffeehouse giant. He expanded its worldwide presence, emphasized employee benefits, and built a brand focused on community, significantly influencing the modern coffee industry and corporate social responsibility.

The Initiation level is your Onboarding, beginning the moment a new contributor enters the ecosystem to turn potential into performance. Efficient talent acquisition and integration ensure that the workforce is not just hired but successfully activated.

Integrating new hires seamlessly and comprehensively into the organizational fabric, providing them with essential knowledge, the necessary tools, and a deep immersion into the company's culture, values, and strategic objectives. This accelerates productivity and significantly boosts retention rates.

Successful "Employee Onboarding fosters early, profound engagement, setting a robust stage for the new employee's long-term success and contribution. A well-executed onboarding process reduces time-to-proficiency, enhances job satisfaction, and strengthens the overall employee experience from day one.

The **"Employee Onboarding"** chapter is crucial for business strategy maturity as it sets up individuals to be productive and integrated within the workforce. A mature business strategy recognizes that effective onboarding helps new employees quickly understand the company culture, their roles, and how they contribute to strategic objectives. This rapid integration accelerates time-to-productivity, reduces early attrition, and ensures new hires are immediately aligned with organizational values and goals. It's about efficiently mobilizing human resources to contribute to strategic execution from day one, minimizing disruption and maximizing talent utilization. Here is how we approach our learning:

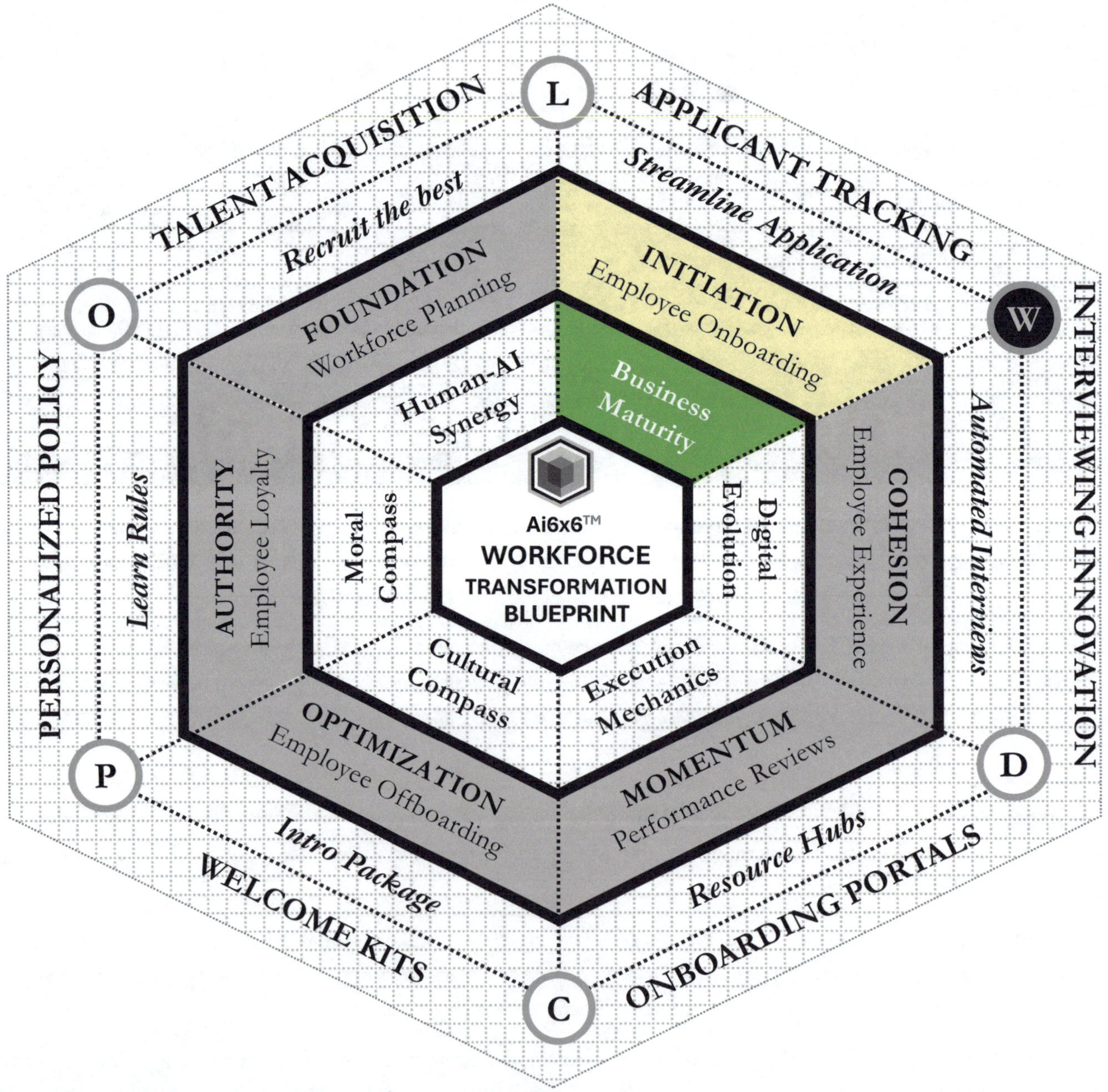

Ai6x6 BUSINESS MATURITY FOR ONBOARDING | WORKFORCE TRANSFORMATION

1. TALENT ACQUISITION

For a business leader, establishing a robust talent acquisition strategy is the critical first step in building a high-performing workforce and ensuring a strong foundation for effective employee onboarding. This isn't merely about filling open positions; it's about strategically attracting, identifying, and securing top-tier candidates who align with your company's vision and culture. Your leadership is crucial in shaping an irresistible employer brand. This involves proactively communicating your company culture, values, career growth opportunities, and unique employee value proposition to the market, making your organization a desirable destination for top talent.

Furthermore, optimizing recruitment channels is essential. This means strategically utilizing a diverse array of platforms—from professional job boards and networking events to social media campaigns and highly effective employee referral programs—to ensure you reach the widest and most relevant pool of candidates. Crucially, leveraging intelligent job matching technologies can revolutionize the process. By employing AI-driven recommendations to align candidate skill sets with precise role requirements, you streamline the initial screening, enhance candidate quality, and reduce time-to-hire. Neglecting talent acquisition can lead to a shallow talent pool, high recruitment costs, poor hiring decisions, and a ripple effect of inefficiencies that impact productivity and morale. By championing a comprehensive and proactive talent acquisition strategy, you empower your organization to attract and secure the best human capital, setting the stage for a successful onboarding experience and building a workforce that drives innovation and sustainable business success. This strategic investment in talent is paramount for long-term growth

WHAT?	WHY?	HOW?	BEWARE!
EMPLOYER BRANDING	**TOP TALENT ATTRACTION**	**VALUE PROPOSITION**	**INAUTHENTICITY**
Communicating company culture and growth opportunities.	Attracts high-caliber candidates aligned with company values.	Clearly articulate company culture, values, and career growth paths.	Perceived inauthenticity or misrepresentation of company culture.
RECRUITMENT CHANNELS	**OPTIMIZED HIRING**	**MULTI-CHANNEL STRATEGY**	**SATURATION & NOISE**
Utilizing diverse sources for optimized hiring.	Expands reach to a diverse and relevant candidate pool.	Use job boards, social media, networking, and referrals.	Channels becoming saturated, making it hard to stand out.
JOB MATCHING	**QUALIFIED CANDIDATES**	**AI-DRIVEN RECOMMENDATIONS**	**ALGORITHMIC BIAS**
Aligning candidate skill sets with role requirements.	Ensures a high fit between candidate skills and job needs.	Implement AI tools for skill assessment and matching.	Potential for algorithmic bias, limiting diversity or overlooking talent.

For a business leader, optimizing applicant tracking is a critical operational imperative for streamlining the hiring process, ensuring efficiency, and significantly enhancing the candidate experience. In today's competitive talent market, a clunky or unresponsive application process can deter top candidates, regardless of your employer brand. Your role is to champion a seamless journey from application to offer. Central to this is creating a frictionless application process. By simplifying application steps, minimizing redundant information requests, and designing an intuitive user experience, you reduce barriers for candidates and increase completion rates, ensuring you don't lose valuable talent due to administrative hurdles.

Furthermore, while technology is crucial, avoiding an over-reliance on automated matching is essential. While algorithmic screening can enhance efficiency, leaders must ensure a balanced approach that combines automated assessments with human decision-making. This safeguards against bias and ensures fair, holistic candidate evaluation, preserving the human touch in recruitment. Tailoring interactions, communications, and outreach efforts to make candidates feel valued and engaged throughout the entire hiring journey creates a positive impression of your company. This extends beyond automated emails to thoughtful communication that acknowledges their unique skills and fit. Neglecting applicant tracking optimization can lead to frustrated candidates, a poor brand reputation, and missed opportunities to secure top talent. By investing in efficient and personalized applicant tracking, you empower your recruitment teams to operate with greater agility, deliver a superior candidate experience, and ultimately build a stronger talent pipeline that contributes directly to your business's growth and success

WHAT?	WHY?	HOW?	BEWARE!
FRICTIONLESS APPLICATION PROCESS	**IMPROVED EXPERIENCE**	**STREAMLINE STEPS**	**DATA SHORTAGES**
Simplifying application steps.	Reduces barriers, improving user and candidate experience.	Minimize forms, enable auto-fill, and optimize for mobile devices.	Simplification leading to insufficient data for proper evaluation.
OVER-RELIANCE ON AUTOMATED MATCHING	**FAIR EVALUATION**	**HUMAN OVERSIGHT**	**BIAS AMPLIFICATION**
Balancing algorithmic screening with human decision-making.	Ensures balanced and fair candidate assessment.	Combine AI screening with manual review and human interviews.	Algorithms perpetuating existing biases if not carefully managed.
PERSONALIZATION IN RECRUITMENT	**CANDIDATE ENGAGEMENT**	**TAILORED COMMUNICATION**	**SCALABILITY**
Tailoring interactions and outreach.	Makes candidates feel valued, enhancing engagement.	Customize emails, provide specific feedback, and address unique skills.	Difficulty personalizing at scale, especially for high-volume roles.

3. INTERVIEWING INNOVATION

For a business leader, embracing interviewing innovation is a strategic imperative to enhance the candidate selection process, ensuring you identify and secure the best-fit talent for your organization. Traditional interview methods often fall short in revealing true potential and cultural alignment; modern approaches leverage technology and structured methodologies to gain deeper insights. Central to this is the adoption of AI-assisted interview assessments. By using machine learning to analyze responses, assess behavioral patterns, and even recommend best-fit candidates, you can bring greater objectivity and efficiency to the screening process, enabling faster and more accurate shortlisting.

Furthermore, implementing structured behavioral interviewing is crucial for consistent evaluation. This involves developing standardized questions that probe past behaviors to predict future performance, ensuring that all candidates are assessed against the same criteria. This data-driven approach minimizes bias and improves the reliability of hiring decisions. Crucially, exploring technologies like virtual reality (VR) and immersive interviews can revolutionize how you gauge candidate adaptability and problem-solving skills. Neglecting interviewing innovation can lead to suboptimal hiring, high turnover rates due to poor fit, and a drain on resources from inefficient recruitment cycles. By championing these modern approaches, you empower your hiring managers with more powerful tools, make more informed talent decisions, and ultimately build a workforce that is not only skilled but also culturally aligned and capable of driving your business forward effectively. This commitment to cutting-edge recruitment enhances your competitive edge in the war for talent

WHAT?	WHY?	HOW?	BEWARE!
AI-ASSISTED INTERVIEW ASSESSMENTS	**OBJECTIVITY & EFFICIENCY**	**ML ANALYSIS**	**BIAS & TRANSPARENCY**
Using machine learning to evaluate responses.	Increases objectivity and efficiency in candidate evaluation.	Deploy AI/ML tools to analyze candidate response, voice, facial cues.	Risk of embedded bias in AI algorithms and lack of transparency.
STRUCTURED BEHAVIORAL INTERVIEWING	**DATA-DRIVEN HIRING**	**STANDARDIZED QUESTIONS**	**RIGIDITY**
Implementing consistent evaluation techniques.	Ensures consistent and objective candidate assessment.	Use specific, behavioral questions and consistent scoring rubrics.	Can feel rigid, potentially missing spontaneous insights or personality fit.
VIRTUAL REALITY (VR) AND IMMERSIVE INTERVIEWS	**REAL-WORLD INSIGHTS**	**SCENARIO SIMULATION**	**TECHNOLOGY ACCESS**
Utilizing interactive environments.	Gauges adaptability and problem-solving in realistic scenarios.	Create VR environments simulating job-specific tasks and challenges.	High cost of development and accessibility issues for candidates.

For a business leader, establishing intuitive onboarding portals is a strategic imperative for centralizing essential resources and delivering a seamless, highly efficient, and positive onboarding experience for new hires. In today's hybrid and remote work environments, a well-designed digital platform is crucial for making new employees feel welcome, informed, and integrated from day one. Central to this is implementing comprehensive digital onboarding platforms. These platforms provide new hires with immediate, centralized access to all necessary documents, pre-boarding materials, initial training modules, and critical company information, eliminating the need for cumbersome paperwork and disjointed processes.

Furthermore, enabling self-guided onboarding workflows empowers new employees to complete onboarding tasks at their own pace and convenience. This flexibility reduces administrative burden on HR teams and allows new hires to absorb information more effectively, fostering a sense of autonomy and engagement from the outset. Crucially, ensuring robust HR system integrations connects your onboarding portals with other vital systems like payroll, benefits administration, and compliance tools. This seamless data flow automates processes, minimizes manual errors, and provides a unified experience for both new hires and HR staff. Neglecting a centralized onboarding portal can lead to a fragmented, overwhelming, and inefficient onboarding process, resulting in early disengagement, confusion, and increased time-to-productivity for new hires. By championing these digital solutions, you demonstrate a commitment to efficiency and employee experience, setting up your new talent for success, accelerating their time-to-contribution, and ultimately strengthening your organization's human capital foundation.

WHAT?	WHY?	HOW?	BEWARE!
DIGITAL ONBOARDING PLATFORMS	**EFFICIENT INFORMATION**	**PLATFORM DEVELOPMENT**	**OVERWHELM**
Providing new hires access to resources.	Centralizes and streamlines access to critical information.	Implement an online platform for documents, training, and company info.	Too much information at once, leading to cognitive overload for new hires.
SELF-GUIDED ONBOARDING WORKFLOWS	**AUTONOMY & FLEXIBILITY**	**MODULAR DESIGN**	**LACK OF INTERACTION**
Allowing employees to complete tasks at their pace.	Enhances new hire autonomy and reduces administrative burden.	Create clear, sequential modules for onboarding tasks.	Reduced human interaction, potentially making new hires feel isolated.
HR SYSTEM INTEGRATIONS	**SEAMLESS PROCESSES**	**API & DATA SYNCHRONIZATION**	**COMPLEX INTEGRATION**
Connecting portals with payroll, benefits, and compliance.	Automates data flow and minimizes manual errors across systems.	Integrate with existing HRIS, payroll, and benefits systems.	Technical complexities and data security challenges during integration.

For a business leader, crafting thoughtfully designed welcome kits is a powerful way to personalize the introduction of new employees to your company culture and their specific role responsibilities. These kits go beyond standard paperwork, creating a memorable and engaging first impression that sets the tone for a positive employee journey. Central to this is offering a comprehensive company overview within the kit. This guide should vividly communicate your company's vision, core values, operational structure, and key achievements, helping new hires quickly grasp the overarching mission and their place within it.

Furthermore, providing a personalized role overview is crucial. Tailoring onboarding materials to each individual's job responsibilities, including specific project briefs, team introductions, and clear expectations for their initial tasks, ensures they feel immediately oriented and understand how their contributions fit into the larger picture. Crucially, innovative approaches like augmented reality (AR) global office tours can provide an immersive and engaging introduction to workplace environments, especially for remote hires. Leveraging AR technology allows employees to virtually explore offices, meet teams, and understand the physical layout, fostering a sense of connection regardless of their location. Neglecting personalized welcome kits can lead to new hires feeling generic, uninspired, or lacking immediate clarity on their role and the company's ethos. By investing in these thoughtful and personalized introductions, you demonstrate a genuine commitment to employee experience, accelerate their integration into the team, and foster a strong sense of belonging and purpose from day one, ultimately contributing to higher engagement and retention within your business

WHAT?	WHY?	HOW?	BEWARE!
COMPANY OVERVIEW	**CULTURAL IMMERSION**	**GUIDE CREATION**	**INFORMATION OVERLOAD**
Offering a comprehensive guide to company vision.	Quickly integrates new hires into company vision and values.	Develop a concise, engaging guide to company history, values, and structure.	Overwhelming new hires with too much information at once.
PERSONALIZED ROLE OVERVIEW	**ROLE CLARITY**	**CUSTOMIZED CONTENT**	**LACK OF SCALABILITY**
Tailoring onboarding materials to job responsibilities.	Ensures immediate understanding of specific job responsibilities and expectations.	Tailor welcome materials to each individual's role and team.	Difficulty personalizing for high-volume hiring without automation.
AR GLOBAL OFFICE TOURS	**REMOTE CONNECTION**	**AR/VR DEVELOPMENT**	**TECHNOLOGY BARRIERS**
Leveraging immersive technology for workplace introductions.	Provides engaging virtual experience for remote employees.	Create interactive AR/VR tours of office spaces and teams.	High development cost and accessibility for all employees.

6. PERSONALIZED POLICY/COMPLIANCE TRAINING

For a business leader, ensuring that new hires thoroughly understand regulations and corporate expectations through personalized policy and compliance training is a non-negotiable imperative. This goes beyond rote memorization; it's about embedding a culture of ethical conduct and operational integrity from day one, while making the learning process engaging and relevant. Central to this is implementing interactive compliance modules. By using gamified, scenario-based, and multimedia learning experiences, you can transform often-dry compliance training into an engaging and memorable process. This approach enhances understanding, improves retention of critical information, and makes learning less of a chore.

Furthermore, adopting adaptive policy training allows for customization based on job function and department-specific requirements. Crucially, establishing continuous compliance monitoring ensures ongoing adherence to company policies and evolving regulations. This involves implementing refresher sessions, microlearning nudges, and regular updates to ensure that employees remain informed and compliant over time, mitigating legal and reputational risks. Neglecting personalized and engaging compliance training can lead to serious legal ramifications, regulatory penalties, reputational damage, and a breakdown of ethical standards within your organization. By championing a proactive, adaptive, and interactive approach to policy training, you not only protect your business from potential liabilities but also cultivate a workforce that is well-informed, responsible, and committed to upholding the highest standards of corporate conduct, thereby securing your business's long-term integrity.

WHAT?	WHY?	HOW?	BEWARE!
INTERACTIVE COMPLIANCE MODULES	**ENGAGEMENT & RETENTION**	**GAMIFICATION**	**COMPLEXITY**
Using gamified and scenario-based learning.	Increases engagement and improves retention of compliance knowledge.	Develop scenario-based, interactive, and gamified training modules.	Difficulty simplifying complex legal and policy information for engagement.
ADAPTIVE POLICY TRAINING	**RELEVANCE & EFFICIENCY**	**JOB-SPECIFIC CONTENT**	**CUSTOMIZATION SCALABILITY**
Customizing training based on job function.	Ensures training is highly relevant and efficient for each role.	Tailor training content based on role, department, and regulatory exposure.	Challenges in customizing content at scale for a large workforce.
CONTINUOUS COMPLIANCE MONITORING	**ONGOING ADHERENCE**	**REFRESHER COURSES**	**FATIGUE**
Implementing refresher sessions for adherence.	Ensures sustained understanding and adherence to policies.	Implement regular refresher modules and automated reminders.	Employees experiencing fatigue from repeated or perceived redundant training.

Ai6x6™ Transformation Blueprint

COHESION | Employee Experience

"To win in the marketplace, you must first win in the workplace." – Doug Conant.

Former CEO of Campbell Soup Company, renowned for leading a remarkable turnaround. He transformed the company's culture by focusing intensely on employee engagement and leadership development, restoring profitability and market leadership. His emphasis on building a high-trust environment is a model for cultural transformation.

individual worker needs with the broader corporate mission. A seamless Total Employee Experience (TEE) creates a unified culture where every contributor feels connected to the whole.

Designing and optimizing every single touchpoint an employee has with the organization, from the initial recruitment process through daily work, career development, and even eventual exit, to ensure it is consistently positive, deeply engaging, and genuinely supportive. This holistic approach creates a thriving, vibrant culture.

"Total Employee Experience" significantly enhances morale, boosts productivity, and attracts and retains top-tier talent for sustained organizational transformation, cultivating loyalty, reducing burnout, and fostering a sense of belonging and purpose.

"Total Employee Experience" is vital for business strategy maturity because it focuses on the ongoing, active management and delivery of a motivating and productive environment. A mature business strategy ensures employees receive continuous training, work in a desirable environment, and have clear career growth opportunities, keeping them motivated and engaged. This comprehensive experience accelerates productivity, fosters innovation, and enhances retention. It's about creating a dynamic ecosystem where employees are empowered to perform at their best, directly contributing to the efficient execution of strategic initiatives and overall organizational success. Here is how we approach our learning:

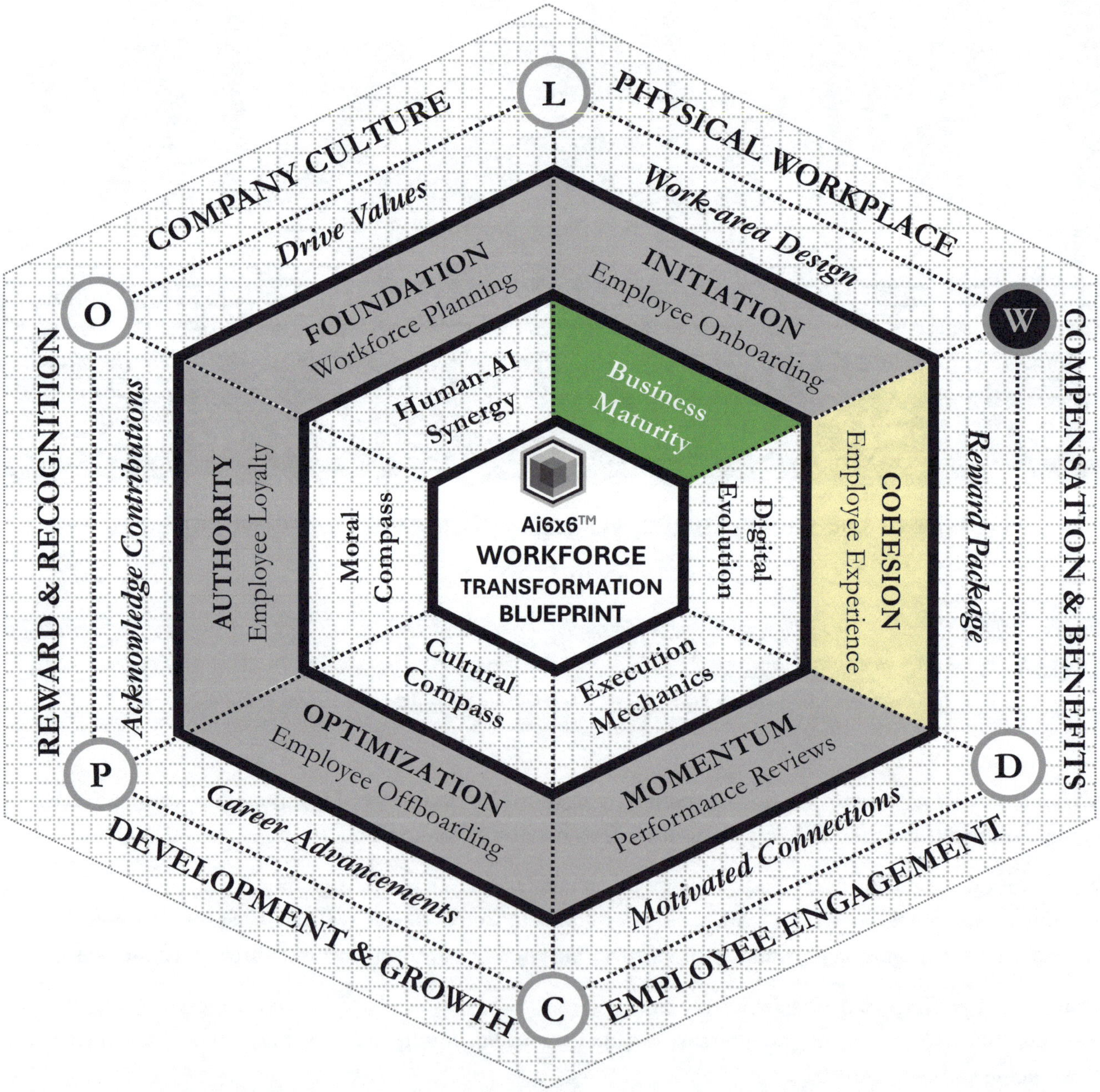

Ai6x6 BUSINESS MATURITY FOR EXPERIENCE | WORKFORCE TRANSFORMATION

1. COMPANY CULTURE

For a business leader, actively fostering a robust and positive company culture is not just an HR initiative; it's a strategic imperative that profoundly impacts employee engagement, productivity, and long-term business success. Your company culture is the collective personality of your organization, defining how work gets done, how employees interact, and how values are upheld. Central to this is championing leadership transparency. Encouraging open communication, building trust between employees and management, and providing clear insights into decisions and challenges create an environment where employees feel informed, valued, and psychologically safe to contribute.

Furthermore, ensuring strong organizational values alignment is crucial. As a leader, you must ensure that employees deeply connect with the mission and purpose of the company, seeing how their individual contributions align with the broader goals. This shared sense of purpose fosters commitment and intrinsic motivation. Crucially, implementing comprehensive Diversity, Equity, and Inclusion (DEI) initiatives is no longer optional. Building a workplace that actively respects, celebrates, and leverages differences in background, perspective, and experience not only creates a more equitable environment but also drives innovation and broader market understanding. Neglecting company culture can lead to high turnover, low morale, decreased productivity, and a struggle to attract top talent. By intentionally cultivating an inclusive and value-driven organizational environment, you empower your workforce to thrive, build a strong employer brand, and create a resilient foundation for sustained business growth and innovation, ultimately making your organization a magnet for the best talent.

WHAT?	WHY?	HOW?	BEWARE!
LEADERSHIP TRANSPARENCY	**TRUST BUILDING**	**OPEN DIALOGUE**	**PERCEIVED HYPOCRISY**
Encouraging open communication and trust.	Fosters trust and open communication between all levels.	Implement regular town halls, open-door policies, and clear communication.	Transparency feeling inauthentic or inconsistent.
ORGANIZATIONAL VALUES ALIGNMENT	**PURPOSEFUL WORKFORCE**	**VALUE INTEGRATION**	**VALUES-ACTION GAP**
Ensuring employees connect with the mission.	Connects employees to the company's mission and values.	Embed values into daily operations, recognition, and decision-making.	Values being stated but not genuinely practiced by leadership.
DIVERSITY, EQUITY, AND INCLUSION (DEI) INITIATIVES	**INNOVATION & BELONGING**	**STRATEGY IMPLEMENTATION**	**RESISTANCE & TOKENISM**
Building a respectful workplace.	Fosters an inclusive environment, driving innovation and belonging.	Implement DEI training, diverse hiring practices, and equitable policies.	Resistance to change or DEI efforts feeling performative.

For a business leader, intentionally designing the physical workplace is a strategic investment that directly impacts employee productivity, well-being, and overall satisfaction. The physical environment is not just a place where work happens; it's a powerful component of the total employee experience that can either enable or hinder performance. Central to this is prioritizing ergonomic office setups. Ensuring comfort, accessibility, and optimized workspaces for all employees, whether in the office or at home, minimizes physical strain, reduces health issues, and maximizes efficiency. This includes adjustable desks, comfortable chairs, and proper lighting.

Furthermore, guaranteeing robust technology and resource availability is crucial. Equipping employees with cutting-edge tools, reliable software, and seamless connectivity is fundamental to maximizing their efficiency and enabling them to perform their jobs effectively. Outdated or insufficient technology is a constant source of frustration and inefficiency. Crucially, providing thoughtful amenities and wellness programs significantly contributes to employee well-being and a positive work environment. This extends to accessible cafeterias, clean restrooms, on-site gyms, quiet library spaces, ample parking, and convenient transportation options. These perks signal that the company cares about its employees beyond their direct output. Neglecting the physical workplace can lead to decreased morale, higher absenteeism, lower productivity, and difficulty attracting and retaining talent. By strategically designing an environment that prioritizes comfort, provides essential resources, and promotes holistic well-being, you demonstrate a deep commitment to your employees, fostering a culture where they feel valued, supported, and empowered to do their best work, ultimately benefiting the business's bottom line.

WHAT?	WHY?	HOW?	BEWARE!
ERGONOMIC OFFICE SETUPS	**PRODUCTIVITY & HEALTH**	**EQUIPMENT PROVISION**	**COST & SPACE**
Prioritizing comfort and optimized workspaces.	Enhances comfort, reduces strain, and boosts productivity.	Provide ergonomic chairs, adjustable desks, and proper lighting.	High initial investment and space constraints for individualized setups.
TECHNOLOGY AND RESOURCE AVAILABILITY	**MAXIMIZED EFFICIENCY**	**INFRASTRUCTURE INVESTMENT**	**RAPID OBSOLESCENCE**
Equipping employees with cutting-edge tools.	Ensures employees have tools for peak performance.	Provide reliable software, hardware, and network connectivity.	Technology quickly becoming outdated, requiring constant updates.
AMENITIES AND WELLNESS PROGRAMS	**WELL-BEING & MORALE**	**FACILITY PROVISION**	**UNDERUTILIZATION**
Offering various perks like cafeteria, gym, etc.	Boosts employee well-being, satisfaction, and morale.	Provide on-site amenities like gym, cafeteria, and quiet zones.	Amenities not being fully utilized by employees.

For a business leader, structuring competitive and holistic compensation and benefits packages is not merely a cost of doing business but a strategic lever for attracting, retaining, and motivating top talent. In today's dynamic labor market, a comprehensive rewards strategy goes far beyond just salary to encompass a total value proposition for employees. Central to this is establishing competitive salary structures. This involves meticulously aligning pay scales with industry standards, considering geographical variations, and ensuring that compensation fairly reflects employee contributions, skills, and experience. Market-competitive salaries are fundamental for attracting new hires and preventing attrition.

Furthermore, providing comprehensive health and wellness benefits is crucial. This includes robust healthcare coverage, essential mental health support, and initiatives that promote work-life balance, signaling a genuine commitment to employee well-being beyond their professional output. These benefits are often as important as salary in an employee's decision to join or stay with an organization. Crucially, offering diverse financial incentives and growth opportunities, such as stock options, robust retirement plans, and performance-based bonuses, provides long-term motivation and aligns employee financial success with organizational success. Neglecting competitive compensation and comprehensive benefits can lead to significant talent acquisition challenges, high employee turnover, decreased morale, and a struggle to maintain a high-performing workforce. By strategically investing in a holistic rewards system, you demonstrate your commitment to your employees' financial and personal well-being, fostering loyalty, engagement, and ultimately, ensuring your business's ability to thrive through a stable workforce.

WHAT?	WHY?	HOW?	BEWARE!
COMPETITIVE SALARY STRUCTURES	**TALENT ATTRACTION**	**MARKET BENCHMARKING**	**WAGE INFLATION**
Aligning pay scales with industry standards.	Attracts and retains top talent in a competitive market.	Conduct regular salary surveys and adjust pay scales accordingly.	Rapid wage inflation in certain sectors, making retention difficult.
COMPREHENSIVE HEALTH AND WELLNESS BENEFITS	**EMPLOYEE WELL-BEING**	**BENEFIT PACKAGES**	**RISING COSTS**
Providing healthcare and mental health support.	Supports holistic employee health and work-life balance.	Offer robust health, dental, vision, mental health, and EAP programs.	Increasing healthcare costs impacting affordability of comprehensive plans.
FINANCIAL INCENTIVES AND OPPORTUNITIES	**LONG-TERM MOTIVATION**	**PERFORMANCE BONUSES**	**PERCEIVED UNFAIRNESS**
Offering stock options and bonuses.	Aligns employee success with company performance and growth.	Implement stock options, profit sharing, and performance-based bonuses.	Incentive structures being perceived as unfair or opaque.

4. EMPLOYEE ENGAGEMENT

For a business leader, actively fostering employee engagement is a strategic imperative that directly impacts productivity, innovation, and long-term organizational success. Engagement goes beyond mere satisfaction; it's about creating meaningful connections and inspiring deep motivation within your workforce, transforming employees into passionate advocates for your mission. Central to this is establishing regular feedback loops. By creating structured mechanisms for employee input and participation—such as continuous pulse surveys, suggestion boxes, and dedicated forums—you ensure that employees feel heard, valued, and empowered to contribute ideas and express concerns. This consistent dialogue builds trust.

Furthermore, implementing robust workplace collaboration programs is crucial. Encouraging teamwork through interactive initiatives, cross-functional projects, and team-building activities strengthens interpersonal bonds, fosters a sense of community, and enhances collective problem-solving capabilities. These programs break down silos and build a cohesive culture. Crucially, consistent recognition of employee voice is paramount. This active listening and responsiveness cultivate psychological safety and boost morale. Neglecting employee engagement can lead to high turnover, low productivity, decreased innovation, and a disengaged workforce that simply "shows up" rather than fully contributes. By championing initiatives that create meaningful connections and intrinsically motivate your employees, you build a resilient, innovative, and highly productive workforce that is deeply committed to your business's success, ultimately driving sustainable growth and a vibrant company culture.

WHAT?	WHY?	HOW?	BEWARE!
REGULAR FEEDBACK LOOPS	**IMPROVED MORALE**	**SURVEYS & OPEN FORUMS**	**FEEDBACK FATIGUE**
Establishing structured mechanisms for input.	Boosts morale by making employees feel heard and valued.	Implement pulse surveys, suggestion systems, and regular town halls.	Employees becoming disengaged from frequent surveys without visible action.
WORKPLACE COLLABORATION PROGRAMS	**ENHANCED PRODUCTIVITY**	**TEAM-BUILDING ACTIVITIES**	**LACK OF PARTICIPATION**
Encouraging teamwork and interactive initiatives.	Drives innovation and strengthens team cohesion.	Organize cross-functional projects and collaborative workshops.	Low uptake or perceived forced participation in activities.
RECOGNITION OF EMPLOYEE VOICE	**EMPOWERMENT & TRUST**	**ACTIVE LISTENING**	**TOKENISM**
Ensuring workers feel valued and heard.	Increases employee trust and empowerment in decision-making.	Implement systems for suggestions and act on actionable employee ideas.	Employee input feeling performative without genuine impact on decisions.

For a business leader, robustly supporting continuous learning and career progression for your employees is not just a benefit; it's a strategic investment that maximizes talent retention, fosters innovation, and ensures your workforce remains agile and future-ready. In a rapidly evolving economic landscape, neglecting employee development is a recipe for obsolescence and attrition. Central to this is mapping personalized career pathways. This involves working with employees to define structured growth opportunities tailored to their individual ambitions, skills, and long-term potential within the organization. Clear paths provide motivation and a sense of direction.

Furthermore, providing comprehensive skill-building programs is crucial. Offering accessible opportunities for professional development, facilitating mentorship relationships, and providing leadership training ensures that employees continuously acquire new knowledge and refine existing capabilities, keeping their skills sharp and relevant. Crucially, encouraging internal mobility and promotions demonstrates a commitment to employee growth. Actively promoting upward and cross-functional career advancement signals that your organization values internal talent and provides avenues for diverse experiences, reducing the need to look externally for new opportunities. Neglecting development and growth opportunities can lead to high employee turnover, a decline in internal skill sets, and a struggle to fill senior roles from within. By championing a culture of continuous learning and clear career progression, you empower your employees to thrive, build a strong internal talent pipeline, and foster a highly engaged and loyal workforce that is equipped to drive your business's strategic objectives for years to come.

WHAT?	WHY?	HOW?	BEWARE!
PERSONALIZED CAREER PATHWAYS	**TALENT RETENTION**	**INDIVIDUAL PLANS**	**LIMITED OPTIONS**
Mapping structured growth opportunities.	Boosts retention by showing clear growth opportunities.	Work with employees to define tailored career progression plans.	Perceived lack of diverse career pathways within the organization.
SKILL-BUILDING PROGRAMS	**CAPABILITY ENHANCEMENT**	**TRAINING & MENTORSHIP**	**TIME & RESOURCES**
Offering access to professional development.	Ensures continuous skill enhancement and relevance.	Provide access to online courses, workshops, and mentorship programs.	Lack of time or budget for extensive training initiatives.
INTERNAL MOBILITY AND PROMOTIONS	**INTERNAL GROWTH**	**INTERNAL JOB POSTINGS**	**FAVORITISM**
Encouraging advancement.	Reduces external hiring needs and boosts morale.	Promote cross-functional moves and internal promotion policies.	Perceptions of favoritism or lack of transparency in internal promotions.

For a business leader, establishing robust reward and recognition programs is a fundamental strategic imperative for boosting employee morale, increasing engagement, and significantly enhancing talent retention. Beyond compensation, acknowledging contributions validates efforts and creates a powerful positive feedback loop that reinforces desired behaviors and fosters a high-performance culture. Central to this is implementing performance-based incentives. By aligning rewards directly with employee achievements and excellence—whether through bonuses, promotions, or public accolades—you create a clear link between effort, results, and recognition, intrinsically motivating individuals to strive for higher levels of performance.

Furthermore, fostering comprehensive peer and leadership recognition programs is crucial. Celebrating contributions through structured appreciation initiatives—from informal shout-outs to formal awards ceremonies—ensures that employees feel valued by their colleagues and management, building a sense of community and shared success. Crucially, offering long-term contribution acknowledgments honors tenure and significant impact through milestone celebrations. Recognizing loyalty and sustained dedication, perhaps through anniversary gifts, special projects, or legacy programs, reinforces commitment and encourages long-term retention. Neglecting reward and recognition can lead to a disengaged workforce, increased turnover, and a culture where effort goes unnoticed, ultimately stifling innovation and productivity. By strategically implementing programs that genuinely acknowledge contributions, you build a highly motivated, loyal, and appreciative workforce, creating a positive work environment that directly contributes to your business's sustained success and reputation as a desirable employer.

WHAT?	WHY?	HOW?	BEWARE!
PERFORMANCE-BASED INCENTIVES	**MOTIVATION & EXCELLENCE**	**BONUS PROGRAMS**	**PERCEIVED UNFAIRNESS**
Aligning rewards with employee achievements.	Drives high performance and rewards excellence.	Implement bonuses, promotions, and public recognition tied to performance.	Bias in evaluation leading to perceived unfairness in rewards.
PEER AND LEADERSHIP RECOGNITION	**MORALE & APPRECIATION**	**FORMAL & INFORMAL SYSTEMS**	**INCONSISTENCY**
Celebrating contributions.	Boosts morale and fosters a culture of appreciation.	Implement formal awards and informal peer-to-peer recognition.	Recognition becoming inconsistent or perfunctory over time.
LONG-TERM CONTRIBUTION ACKNOWLEDGMENTS	**LOYALTY & RETENTION**	**MILESTONE CELEBRATIONS**	**GENERIC ACKNOWLEDGEMENT**
Honoring tenure and impact.	Encourages long-term loyalty and reduces turnover.	Recognize work anniversaries and significant milestones.	Acknowledgments feeling generic or lacking genuine appreciation.

Ai6x6™ Transformation Blueprint

MOMENTUM | Performance Reviews

"We all need people who will give us feedback. That's how we improve."
- Bill Gates.

Co-founder of Microsoft, a pioneering figure in the personal computer revolution. He built Microsoft into the world's largest software company. Beyond technology, he is a leading global philanthropist through the Bill & Melinda Gates Foundation, addressing critical issues like global health and poverty.

The Momentum level is your Performance cycle, built through continuous feedback loops that drive individual and collective output. By standardizing review and coaching loops, the organization sustains a high-velocity culture of improvement.

Structured feedback and development conversations that meticulously assess individual contributions, identify specific areas for growth and improvement, and collaboratively set clear, measurable future goals aligned with both personal aspirations and organizational objectives. This process drives continuous individual and team improvement.

"Performance Reviews" ensures efforts are precisely aligned with strategic organizational objectives and fosters a culture of accountability, continuous learning, and professional development, serving as a vital tool for talent management and succession planning

The **"Performance Reviews"** chapter is paramount for business strategy maturity as it focuses on optimizing individual and team performance based on feedback and adapting development plans. A mature business strategy relies on regular feedback, evaluations, and support for professional growth to ensure the workforce is continuously improving. This process allows leaders to identify high-potential employees, address performance gaps, and align individual goals with broader strategic objectives. It's about refining human capital, ensuring that the workforce is consistently advancing its capabilities to meet evolving business demands and execute strategies more effectively. Here is how we approach our learning:

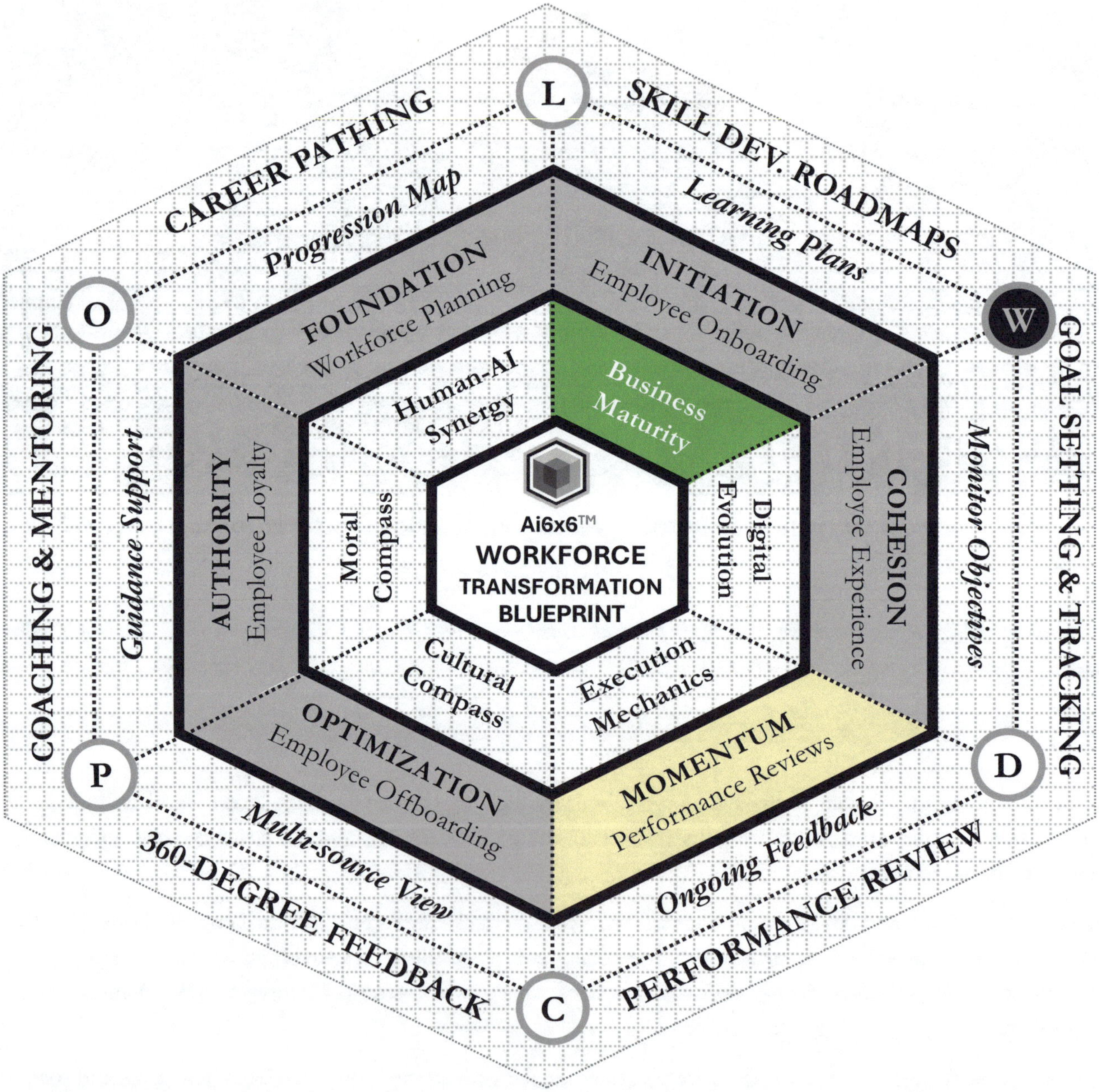

Ai6x6 BUSINESS MATURITY FOR REVIEWS | WORKFORCE TRANSFORMATION

For a business leader, establishing clear career pathing for employees is a strategic imperative that directly influences talent retention, boosts engagement, and ensures a vibrant internal talent pipeline. In today's competitive labor market, employees seek not just jobs, but opportunities for continuous growth and advancement. Your leadership in this area signals a genuine commitment to their long-term professional journey. Central to this is mapping structured career development plans. This involves clearly delineating progression opportunities within the organization, showing employees the various roles they can aspire to and the pathways to get there. This transparency provides motivation and reduces uncertainty about their future.

Furthermore, providing personalized skill-building programs is crucial. Offering tailored learning paths based on individual employee aspirations, existing strengths, and identified skill gaps ensures that development efforts are relevant and impactful, empowering them to acquire the necessary competencies for advancement. Crucially, encouraging internal mobility initiatives promotes cross-departmental movement. This allows employees to diversify their career experiences, gain broader organizational knowledge, and explore different facets of the business, enhancing their versatility and overall value to the company. Neglecting clear career pathing can lead to high employee turnover as ambitious talent looks elsewhere for growth, decreased morale due to a perceived lack of opportunity, and a struggle to fill senior roles from within. By championing defined career paths and providing the resources for their pursuit, you foster a highly motivated, loyal, and skilled workforce that is deeply invested in your business's success and poised for long-term contributions.

WHAT?	WHY?	HOW?	BEWARE!
STRUCTURED CAREER DEV PLANS	**CLARITY & MOTIVATION**	**PATHWAY MAPPING**	**RIGIDITY**
Mapping clear progression opportunities.	Provides clarity and motivation for employee progression.	Create visual career maps showing roles and required skills for advancement.	Plans becoming too rigid, not adapting to individual needs or market changes.
PERSONALIZED SKILL-BUILDING PROGRAMS	**RELEVANCE & IMPACT**	**CUSTOMIZED TRAINING**	**SCALABILITY**
Offering tailored learning paths.	Ensures development efforts are relevant and impactful.	Design learning paths based on individual aspirations and skill gaps.	Difficulty personalizing effectively for a large number of employees.
INTERNAL MOBILITY INITIATIVES	**VERSATILITY & RETENTION**	**CROSS-FUNCTIONAL PROJECTS**	**DEPARTMENTAL RESISTANCE**
Encouraging cross-departmental movement.	Enhances employee versatility and boosts internal retention.	Promote rotation programs and internal job posting awareness.	Managers resisting talent movement out of their teams.

2. SKILL DEVELOPMENT ROADMAPS

For a business leader, meticulously structuring skill development roadmaps is a strategic imperative for fostering continuous employee improvement, ensuring your workforce remains competitive, and proactively addressing evolving business needs. In a world where technological advancements and market demands shift rapidly, a clear pathway for skill enhancement is vital for maintaining organizational agility. Central to this is implementing competency-based learning plans. This involves mapping skill-building milestones directly aligned with career progression and specific job functions. By focusing on observable competencies, you ensure that training is practical, targeted, and directly contributes to an employee's effectiveness in their role and for future advancement.

Furthermore, providing personalized upskilling programs is crucial. Tailoring training to individual employee strengths, areas for development, and career aspirations maximizes engagement and impact. This bespoke approach ensures that resources are invested efficiently, addressing unique learning needs rather than offering generic solutions. Crucially, incorporating regular skill benchmarking allows you to compare employee proficiency against industry standards for targeted development. This external perspective ensures your workforce remains competitive and equipped with cutting-edge skills, highlighting areas where strategic investment in training is most needed. Neglecting structured skill development can lead to a widening skills gap, decreased employee confidence, and a reactive approach to talent shortages. By championing comprehensive skill development roadmaps, you empower your employees to continuously evolve, build a future-ready workforce, and sustain your business's competitive edge in a dynamic global market, ensuring that your human capital remains a primary source of innovation.

WHAT?	WHY?	HOW?	BEWARE!
COMPETENCY-BASED LEARNING PLANS	**TARGETED DEVELOPMENT**	**SKILL FRAMEWORKS**	**COMPLEXITY**
Mapping skill-building milestones.	Ensures training aligns directly with career progression.	Define competencies for each role and create learning modules.	Difficulty defining and updating competencies in rapidly changing fields.
PERSONALIZED UPSKILLING PROGRAMS	**MAXIMIZED IMPACT**	**INDIVIDUALIZED CURRICULA**	**RESOURCE INTENSITY**
Tailoring training to individual strengths.	Ensures highly relevant and effective skill enhancement.	Offer customized training based on assessment results and aspirations.	High resource requirement for truly personalized programs.
SKILL BENCHMARKING	**COMPETITIVE EDGE**	**EXTERNAL ANALYSIS**	**DATA ACCURACY**
Comparing employee proficiency against industry standards.	Ensures workforce skills are competitive and up-to-date.	Compare internal skill levels against industry best practices and competitors.	Difficulty obtaining reliable and comparable external skill data.

For a business leader, establishing effective goal setting and tracking mechanisms is a fundamental pillar of performance management, crucial for driving employee success, aligning individual efforts with organizational objectives, and fostering a culture of accountability. This goes beyond simple task lists to a strategic framework that empowers employees to contribute meaningfully. Central to this is implementing Objectives and Key Results (OKRs). By setting ambitious yet achievable goals, complemented by measurable key results, you provide a clear framework for enhancing productivity and focus. OKRs foster alignment, transparency, and a results-oriented mindset across teams.

Furthermore, rigorous KPI monitoring is crucial. Using key performance indicators and other performance metrics allows you to objectively evaluate efficiency, measure impact, and track progress towards defined objectives. This data-driven approach provides tangible evidence of achievement and highlights areas requiring attention. Crucially, adopting adaptive goal frameworks ensures flexibility in a dynamic business environment. This agility prevents rigidity and ensures that goals remain relevant and motivating, even in unpredictable circumstances. Neglecting clear goal setting and tracking can lead to misaligned efforts, decreased productivity, a lack of accountability, and an inability to measure contributions effectively. By championing a systematic approach to defining, tracking, and adapting goals, you empower your employees to understand their impact, stay focused on priorities, and continuously strive for excellence, ultimately driving the strategic success and growth of your business. This disciplined approach links individual effort directly to organizational achievement.

WHAT?	WHY?	HOW?	BEWARE!
OKRs (OBJECTIVES AND KEY RESULTS)	**ALIGNMENT & FOCUS**	**GOAL DEFINITION**	**OVERWHELM**
Setting ambitious yet achievable goals.	Enhances productivity by aligning individual goals with company objectives.	Define ambitious objectives and measurable key results (OKRs).	Goals becoming too numerous or complex, leading to overwhelm.
KPI MONITORING	**OBJECTIVE MEASUREMENT**	**METRIC IMPLEMENTATION**	**MISLEADING METRICS**
Using performance metrics to evaluate efficiency.	Provides clear, measurable indicators of efficiency and impact.	Set relevant KPIs for each role and establish regular tracking.	Focusing on vanity metrics rather than truly impactful ones.
ADAPTIVE GOAL FRAMEWORKS	**AGILITY & RELEVANCE**	**REGULAR REVIEW & ADJUSTMENT**	**INCONSISTENCY**
Adjusting objectives based on dynamic needs.	Ensures goals remain relevant in dynamic business environments.	Conduct frequent goal reviews and adjust based on feedback/changes.	Goals changing too frequently, causing confusion or frustration.

4. CONTINUOUS PERFORMANCE REVIEW

For a business leader, transitioning to continuous performance review is a strategic imperative that moves beyond traditional annual appraisals to a dynamic, ongoing system for refining employee performance and fostering professional growth. In today's fast-paced work environment, static reviews are insufficient for providing the timely feedback and support needed for optimal performance. Central to this is implementing real-time performance monitoring. By utilizing automated tools and platforms to assess employee contributions continuously, you can gain immediate insights into performance trends, productivity, and areas requiring immediate attention.

Furthermore, establishing frequent check-ins is crucial. Regular, structured discussions between employees and managers provide continuous opportunities for progress assessment, immediate feedback exchange, and collaborative problem-solving. These ongoing conversations replace lengthy, retrospective annual reviews with consistent, forward-looking dialogue. Crucially, a key shift is embracing development-based performance management. Rather than solely focusing on ratings or past deficiencies, the emphasis is shifted towards growth-oriented feedback, identifying areas for skill enhancement, and supporting employees in their professional journey. This fosters a positive learning environment. Neglecting continuous performance review can lead to delayed feedback, unaddressed performance issues, disengaged employees, and a stagnant workforce that misses opportunities for timely improvement. By championing this ongoing, development-focused approach, you empower your employees to excel, build a more agile and responsive workforce, and cultivate a culture of immediate feedback and continuous growth, ultimately driving the sustained success and innovation of your business.

WHAT?	WHY?	HOW?	BEWARE!
REAL-TIME PERFORMANCE MONITORING	**PROMPT INTERVENTION**	**SOFTWARE IMPLEMENTATION**	**MICROMANAGEMENT PERCEPTION**
Using automated tools for assessment.	Enables immediate feedback and proactive interventions.	Deploy tools for real-time tracking of tasks and productivity.	Employees feeling overly surveilled or distrusted.
FREQUENT CHECK-INS	**ONGOING DIALOGUE**	**SCHEDULED 1:1S**	**TIME CONSUMPTION**
Conducting regular discussions for progress assessment.	Fosters continuous communication and progress assessment.	Mandate regular, short, focused manager-employee discussions.	Managers feeling overwhelmed by frequent meeting demands.
DEVELOPMENT-BASED PERFORMANCE MANAGEMENT	**GROWTH ORIENTATION**	**COACHING FOCUS**	**LACK OF ACCOUNTABILITY**
Shifting focus from ratings to growth.	Promotes employee growth over punitive evaluations.	Train managers to provide constructive, forward-looking feedback.	Difficulty in holding employees accountable without clear ratings.

For a business leader, implementing 360-degree feedback is a powerful strategic tool for conducting holistic employee evaluations, leveraging insights from a diverse array of perspectives—peers, supervisors, subordinates, and other stakeholders. Central to this is utilizing multi-source assessment models. By systematically gathering input from all angles, you ensure balanced feedback that uncovers blind spots, validates strengths, and provides a richer understanding of how an employee's actions are perceived across the organization. This breadth of perspective is invaluable for growth.

Furthermore, fostering robust peer and subordinate feedback loops is crucial. Encouraging collaborative insights from those who work most closely with an individual provides valuable, day-to-day context that supervisors might miss. This peer-driven input enhances the well-roundedness of performance evaluation and promotes a culture of mutual accountability. Crucially, establishing clear constructive feedback frameworks ensures that reviews lead to actionable improvement. Training all participants—raters and recipients—on how to give and receive feedback effectively, focusing on specific behaviors and solutions rather than judgments, maximizes the developmental impact of the 360-degree process. Neglecting multi-source feedback can lead to a narrow, potentially biased view of performance, missed opportunities for growth, and a lack of transparency that erodes trust. By championing 360-degree feedback, you empower employees with comprehensive insights, foster a culture of open communication, and drive more effective, well-rounded performance improvements, ultimately strengthening your talent pool and contributing to a more adaptive and high-performing business.

WHAT?	WHY?	HOW?	BEWARE!
MULTI-SOURCE ASSESSMENT MODELS	**BALANCED FEEDBACK**	**COMPREHENSIVE SURVEYS**	**BIAS & VALIDITY**
Gathering diverse perspectives.	Ensures comprehensive and balanced insights into performance.	Design surveys for input from all relevant stakeholders.	Ensuring objectivity and preventing personal biases in feedback.
PEER AND SUBORDINATE FEEDBACK LOOPS	**WELL-ROUNDED EVALUATION**	**ANONYMOUS CONTRIBUTION**	**RELUCTANCE TO CRITICIZE**
Encouraging collaborative insights.	Provides crucial day-to-day context for performance.	Implement systems for confidential peer and subordinate feedback.	Employees fearing repercussions for honest critical feedback.
CONSTRUCTIVE FEEDBACK FRAMEWORKS	**ACTIONABLE GROWTH**	**TRAINING & GUIDELINES**	**DEFENSIVENESS**
Structuring reviews for improvement.	Ensures feedback leads to tangible, positive changes.	Train managers and employees on effective feedback delivery.	Recipients being defensive or unwilling to accept critical feedback.

For a business leader, establishing robust coaching and mentoring systems is a foundational element for fostering continuous professional growth, maximizing employee potential, and building a resilient internal talent pipeline. These structured support systems go beyond formal training to provide personalized guidance, practical insights, and invaluable career advice. Central to this is implementing peer mentoring programs. By connecting employees with experienced colleagues, you facilitate knowledge transfer, skill development, and cultural integration through informal, supportive relationships. This fosters a sense of community and accelerates learning within teams.

Furthermore, offering job shadowing programs allows employees to observe senior colleagues in action, gaining practical insights into different roles, decision-making processes, and organizational complexities. This immersive learning experience provides valuable context and helps employees understand potential career paths. Crucially, facilitating dedicated career coaching sessions helps employees plan their long-term career progression within the company. These one-on-one discussions with professional coaches or experienced leaders provide guidance on setting career goals, identifying necessary skills, and navigating opportunities for advancement, ensuring employees see a clear future with the organization. Neglecting structured coaching and mentoring can lead to stunted employee development, a lack of internal expertise, and a struggle to retain ambitious talent who seek growth opportunities elsewhere. By championing these supportive systems, you demonstrate a deep commitment to your employees' development, build a stronger, more capable workforce, and cultivate a culture of continuous learning and internal growth that directly contributes to the sustained success and innovation of your business.

WHAT?	WHY?	HOW?	BEWARE!
PEER MENTORING PROGRAMS	**SKILL DEVELOPMENT**	**MENTOR MATCHING**	**TIME COMMITMENT**
Connecting employees with experienced colleagues.	Facilitates knowledge transfer and skill development.	Pair employees with experienced peers for informal guidance.	Mentors and mentees struggling with time commitment.
JOB SHADOWING PROGRAMS	**PRACTICAL INSIGHTS**	**OBSERVATION & LEARNING**	**DISRUPTION**
Allowing employees to observe senior colleagues.	Provides practical insights and exposure to new roles.	Arrange opportunities to observe various roles and tasks.	Potential disruption to work schedules of both observer and observed.
CAREER COACHING SESSIONS	**CAREER CLARITY**	**1:1 COACHING**	**LACK OF COACHES**
Helping employees plan long-term progression.	Empowers employees to plan and achieve long-term career goals.	Provide access to professional career coaches or internal senior leaders.	Insufficient number of qualified internal coaches or budget for external ones.

Ai6x6™ Transformation Blueprint

OPTIMIZATION | Employee Offboarding

"It takes 20 years to build a reputation and five minutes to ruin it. If you think about that, you'll do things differently." - Warren Buffett.

One of the most successful investors, chairman and CEO of Berkshire Hathaway. His philosophy of value investing and long-term commitment made him one of the world's wealthiest individuals and a respected voice in global finance and business ethics.

The Optimization level is your Offboarding, ensuring that the departure of talent is handled as a strategic refinement of the organizational memory. Structured knowledge transfer and archiving turn the end of a tenure into an efficient capture of institutional intelligence.

Managing the departure process of employees with utmost respect and professionalism, ensuring smooth transitions, comprehensive knowledge transfer, and the proper handling of all administrative and legal aspects. This protects the organization's brand reputation and maintains positive alumni relations.

“Employee Offboarding” gathers invaluable insights for continuous improvement of the employee experience and internal processes. A positive offboarding experience can turn former employees into brand advocates, maintaining a valuable network and reputation for future recruitment.

"Employee Offboarding" is critical for business strategy maturity because it focuses on gracefully managing the transition of individuals out of the active workforce while preserving value. A mature business strategy handles resignations, retirements, or terminations with professionalism, conducting thorough exit interviews and ensuring critical knowledge transfer. This structured approach minimizes disruption, protects intellectual capital, and maintains positive employer branding. It's about adapting to workforce changes efficiently, ensuring continuity and learning from departures to refine future talent strategies, thereby supporting long-term organizational stability and strategic resilience. Here is how we approach our learning:

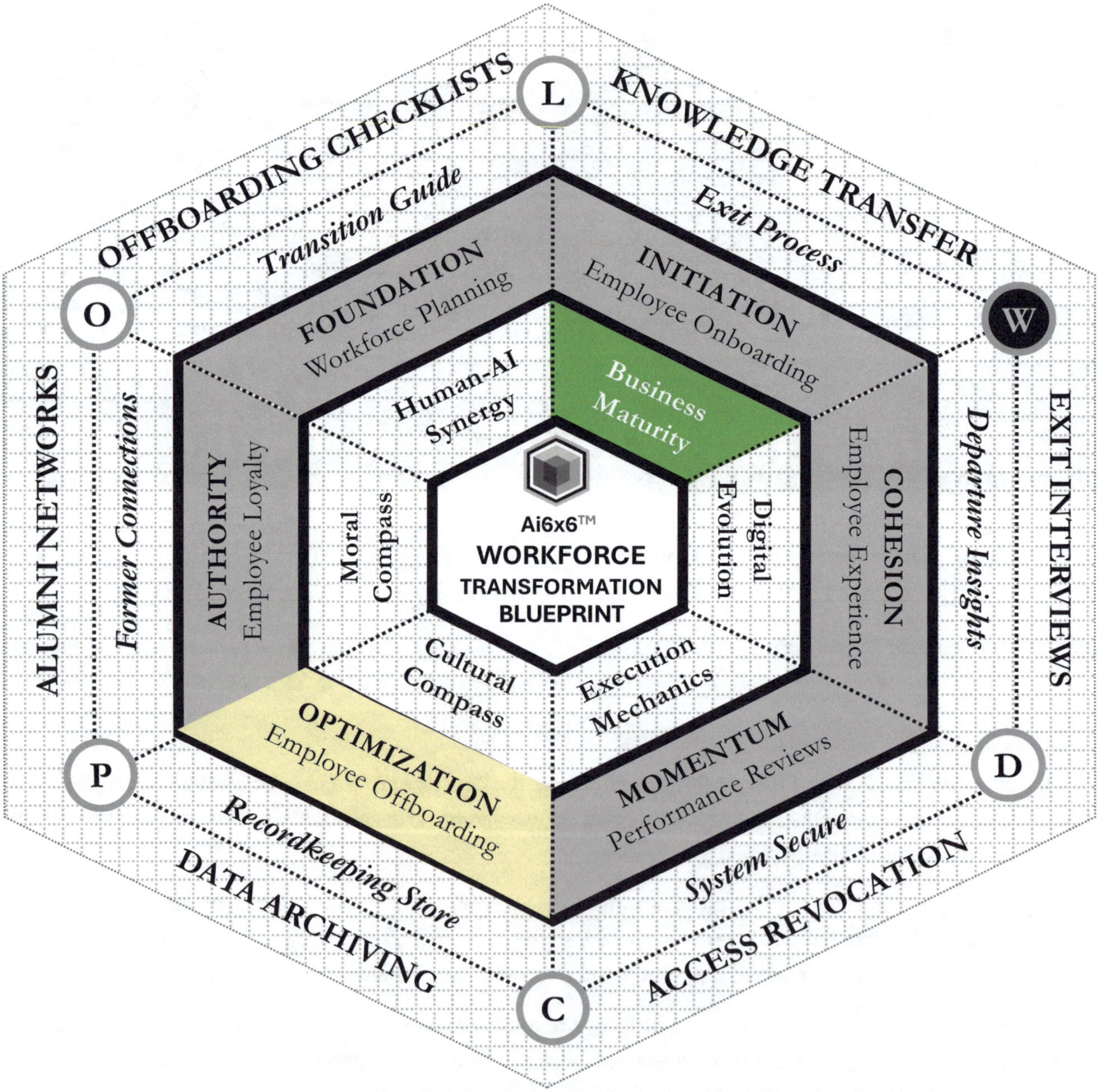

Ai6x6 BUSINESS MATURITY FOR OFFBOARDING | WORKFORCE TRANSFORMATION

1. OFFBOARDING CHECKLISTS

For a business leader, implementing comprehensive offboarding checklists, tailored to different types of employee transitions, is a critical operational and strategic imperative. This ensures a structured, compliant, and empathetic exit process that minimizes risk and preserves organizational integrity, whether an employee is resigning, retiring, or being terminated. Central to this is developing specific protocols for voluntary resignations. This includes standard offboarding processes such as ensuring all documentation is completed, facilitating thorough knowledge transfer, and managing farewell communications in a way that maintains positive employee relations.

Furthermore, distinct procedures are crucial for involuntary terminations. As a leader, you must ensure that these sensitive transitions are managed with utmost care, adhering strictly to compliance guidelines, addressing all legal considerations, and implementing immediate security-related procedures to protect company assets and data. Crucially, supporting retirement transitions requires a tailored approach. For long-term employees, this involves thoughtful legacy planning, initiating succession strategies well in advance, and celebrating their contributions in a way that honors their tenure and maintains their goodwill. Neglecting type-specific offboarding checklists can lead to legal complications, security vulnerabilities, loss of critical institutional knowledge, and a negative impact on the morale of remaining employees. By championing a structured and differentiated offboarding process, you not only mitigate risks but also ensure smooth transitions, maintain a positive brand image even during departures, and safeguard your business's long-term operational efficiency and reputation.

WHAT?	WHY?	HOW?	BEWARE!
VOLUNTARY RESIGNATIONS	**SMOOTH TRANSITION**	**DOCUMENTATION & TRANSFER**	**LACK OF ENGAGEMENT**
Standard offboarding processes.	Ensures a positive and efficient exit for departing employees.	Standardize documentation, knowledge transfer, and farewells.	Departing employees being disengaged from the process.
INVOLUNTARY TERMINATIONS	**RISK MITIGATION**	**LEGAL ADHERENCE**	**EMOTIONAL TOLL**
Managing compliance and legal considerations.	Minimizes legal risks and ensures compliance during terminations.	Follow strict legal, compliance, and security protocols.	High emotional toll on managers and potential for legal challenges.
RETIREMENT TRANSITIONS	**LEGACY PRESERVATION**	**SUCCESSION PLANNING**	**KNOWLEDGE LOSS**
Supporting long-term employees.	Honors long-term employees and ensures knowledge transfer.	Facilitate legacy planning, knowledge transfer, and celebratory events.	Difficulty capturing decades of institutional knowledge from retirees.

2. KNOWLEDGE TRANSFER APPROACH

For a business leader, establishing a robust knowledge transfer approach is a critical strategic imperative for capturing invaluable institutional knowledge and ensuring operational continuity, especially during employee departures. In any organization, a wealth of critical information resides in the minds of its experienced personnel, and losing this without proper transfer can lead to significant inefficiencies, duplicated efforts, and a loss of competitive advantage. This involves compelling employees, particularly those departing, to create clear, comprehensive guides and manuals for their critical operational knowledge, processes, and decision-making frameworks for future reference. This minimizes reliance on individual memory.

Furthermore, meticulously transitioning responsibilities is crucial. This means strategically assigning key tasks and projects to successors well in advance of a departure, providing ample overlap time for thorough handover and training. This phased approach ensures that no critical function is left uncovered. Crucially, implementing mentorship-based handovers leverages the expertise of departing employees. Encouraging them to guide their successors through structured knowledge-sharing sessions, one-on-one training, and answering questions in real-time can transfer nuanced understanding that written documentation alone cannot capture. Neglecting a formal knowledge transfer approach can lead to significant productivity drops, project delays, loss of unique insights, and a fragmented understanding of critical processes, ultimately hindering your business's long-term performance. By championing these strategies, you empower your organization to retain its intellectual capital, ensure seamless operations, and build a more resilient and knowledgeable workforce, safeguarding your business's future success.

WHAT?	WHY?	HOW?	BEWARE!
DOCUMENTATION OF WORKFLOWS	**ACCESSIBLE KNOWLEDGE**	**PROCESS MAPPING**	**TIME CONSUMPTION**
Compiling critical operational knowledge.	Ensures critical operational knowledge is easily accessible.	Create guides, manuals, and standard operating procedures (SOPs).	Demanding time commitment from departing employees for thorough documentation.
TRANSITIONING RESPONSIBILITIES	**OPERATIONAL CONTINUITY**	**PHASED HANDOVER**	**RESISTANCE**
Assigning key tasks to successors.	Minimizes disruption during employee transitions.	Assign tasks, ensure overlap, and provide clear transition plans.	Successors feeling overwhelmed or departing employees disengaged from transfer.
MENTORSHIP-BASED HANDOVERS	**NUANCED KNOWLEDGE**	**1:1 SESSIONS**	**LACK OF WILLINGNESS**
Encouraging departing employees to guide successors.	Transfers tacit and nuanced understanding not captured in documents.	Facilitate structured mentorship and Q&A sessions between individuals.	Departing employees lacking motivation or time for mentorship.

For a business leader, conducting comprehensive exit interviews is a powerful strategic practice for gathering invaluable insights that can significantly improve retention, refine organizational culture, and address underlying issues before they impact more employees. These conversations are not just a formality; they are a critical feedback mechanism from departing talent. Central to this is implementing structured feedback sessions. By using consistent questions and frameworks, you can systematically identify employee concerns, pinpoint areas for workplace improvement, and uncover patterns in why employees choose to leave. This objective approach ensures you collect actionable data.

Furthermore, leveraging sentiment analysis tools can provide deeper insights from this feedback. By employing AI-driven analytics to evaluate trends in offboarding feedback, you can identify underlying emotional patterns, common pain points, and systemic issues that might not be immediately obvious from individual responses, offering a broader organizational perspective. Crucially, ensuring confidential reporting options is paramount. Allowing employees to share honest feedback without fear of repercussions or negative impact on their references encourages transparency and ensures you receive candid insights, even on sensitive topics. Neglecting robust exit interviews means missing a vital opportunity to understand why talent is leaving, allowing systemic problems to persist, and potentially facing higher turnover rates. By championing this process, you gain critical intelligence that empowers you to make data-informed decisions, improve employee experience, and cultivate a more positive and supportive organizational culture, ultimately strengthening your talent management strategies and boosting long-term retention.

WHAT?	WHY?	HOW?	BEWARE!
STRUCTURED FEEDBACK SESSIONS	**ACTIONABLE INSIGHTS**	**STANDARDIZED QUESTIONS**	**DISHONESTY**
Identifying employee concerns.	Pinpoints specific areas for workplace improvement.	Use consistent questions and frameworks for interviews.	Departing employees being reluctant to provide honest or critical feedback.
SENTIMENT ANALYSIS TOOLS	**PATTERN IDENTIFICATION**	**AI TEXT ANALYSIS**	**CONTEXT LOSS**
Using AI-driven insights to evaluate trends.	Uncovers broader trends and emotional patterns in feedback.	Use AI to analyze interview transcripts or written responses.	AI potentially missing nuances or context from verbal interactions.
CONFIDENTIAL REPORTING OPTIONS	**CANDOR & TRUST**	**ANONYMOUS PLATFORMS**	**PERCEIVED LACK OF ACTION**
Allowing employees to share honest feedback.	Encourages honest feedback without fear of repercussions.	Provide secure, anonymous channels for submitting feedback.	Employees feeling their feedback is heard but not acted upon.

For a business leader, meticulously managing access revocation upon employee departure is a critical security imperative that directly safeguards sensitive systems, confidential data, and intellectual property. This is not merely an IT task; it's a vital component of your risk management strategy, protecting your organization from potential breaches, data loss, and unauthorized access. Central to this is adhering to stringent IT security protocols. As a leader, you must ensure that all accounts, credentials, and system logins for a departing employee are immediately deactivated across all platforms—from email and internal networks to cloud services and specialized software. This rapid action prevents any unauthorized access post-employment.

Furthermore, establishing clear procedures for device retrieval and data wipe is crucial. This involves systematically collecting all company-owned hardware (laptops, phones, tablets) and ensuring that any sensitive company information stored on personal or company devices is securely wiped, protecting proprietary data from leaving the organization. Crucially, comprehensive physical security clearance protocols are essential. Revoking access to office spaces, disabling key cards or IDs, and securing access to sensitive physical areas prevents unauthorized entry and potential theft. Neglecting robust access revocation processes can lead to severe security breaches, data theft, compliance violations, and significant reputational damage, posing a substantial threat to your business's stability. By championing a swift, comprehensive, and well-documented access revocation strategy, you minimize security risks, ensure data integrity, and protect your company's valuable assets, ultimately safeguarding its long-term operational security and trustworthiness. This diligence is a core responsibility of modern leadership

WHAT?	WHY?	HOW?	BEWARE!
IT SECURITY PROTOCOLS	**DATA PROTECTION**	**IMMEDIATE DEACTIVATION**	**SYSTEM FRAGMENTATION**
Ensuring deactivation of accounts and logins.	Prevents unauthorized access and potential data breaches.	Implement automated and manual processes for account deactivation.	Disparate systems making full, immediate deactivation difficult.
DEVICE RETRIEVAL AND DATA WIPE	**INFORMATION SECURITY**	**INVENTORY & WIPING**	**EMPLOYEE RESISTANCE**
Managing company-owned hardware.	Safeguards sensitive company data on devices.	Track company devices, retrieve them, and securely wipe data.	Employees refusing to return devices or comply with data wipe.
PHYSICAL SECURITY CLEARANCE	**PHYSICAL SECURITY**	**ACCESS CARD DEACTIVATION**	**OVERSIGHT**
Revoking access to office spaces and IDs.	Prevents unauthorized physical access to company premises.	Disable key cards, update access lists, and collect badges.	Forgetting to revoke access for all physical entry points.

For a business leader, establishing meticulous data archiving protocols for historical employee records is a critical imperative for ensuring regulatory compliance, enabling future reference, and preserving institutional knowledge. This is a foundational element of responsible data governance that extends well beyond an employee's departure, protecting your organization legally and strategically. This involves systematically storing all documentation related to employment, performance, compensation, and benefits for specified periods as mandated by labor laws and industry regulations. Compliance is non-negotiable to avoid penalties.

Furthermore, diligently archiving project and contribution records is crucial. Preserving relevant work outputs, project documentation, and key contributions from departing employees ensures that valuable intellectual capital is not lost but remains accessible for continued use by the organization, supporting ongoing projects and future initiatives. Crucially, implementing secure storage protocols is paramount. This means ensuring that all archived former employee data is encrypted, protected by robust access controls, and stored in secure, compliant environments to prevent unauthorized access or breaches. Data integrity and privacy are key considerations. Neglecting proper data archiving can lead to severe legal penalties for non-compliance, difficulties in resolving future disputes, and the loss of invaluable historical context and knowledge that could inform future strategies. By championing a comprehensive and secure data archiving strategy, you protect your business from legal risks, preserve critical information, and ensure that a rich historical record is available for strategic decision-making, ultimately safeguarding your business's long-term operational integrity and knowledge base.

WHAT?	WHY?	HOW?	BEWARE!
PERSONNEL FILE RETENTION	**LEGAL COMPLIANCE**	**RETENTION POLICIES**	**NON-COMPLIANCE RISK**
Storing documentation in alignment with legal requirements.	Ensures adherence to legal and regulatory requirements for data retention.	Implement clear policies for storing employment records.	Failure to comply with evolving data privacy regulations.
PROJECT AND CONTRIBUTION ARCHIVES	**KNOWLEDGE PRESERVATION**	**SYSTEMATIC STORAGE**	**INCOMPLETE ARCHIVING**
Preserving relevant work outputs.	Retains valuable institutional knowledge for future use.	Develop systems to archive project documents and contributions.	Employees failing to archive all relevant work before departure.
SECURE STORAGE PROTOCOLS	**DATA SECURITY**	**ENCRYPTION & ACCESS CONTROL**	**CYBER THREATS**
Ensuring encrypted and protected access.	Protects sensitive former employee data from unauthorized access.	Implement strong encryption and strict access permissions.	Vulnerability to evolving cyber threats and data breaches.

For a business leader, establishing and nurturing strong corporate alumni networks is an increasingly strategic imperative for maintaining long-term professional relationships, leveraging former employee insights, and even creating pathways for returning talent. This extends the value of your talent pool beyond active employment, fostering goodwill and a potential future resource. Central to this is developing robust corporate alumni programs. By engaging former employees through exclusive communities, professional networking opportunities, and relevant updates, you keep them connected to your organization, turning them into potential customers, or future re-hires.

Furthermore, creating knowledge-sharing opportunities with your alumni base is crucial. Leveraging their insights for mentorship, consulting, or even specific project contributions can provide valuable external perspectives and expertise that benefits current employees and strategic initiatives. This tap into a rich, experienced network. Crucially, establishing boomerang employee pathways actively encourages returning talent through structured re-hiring initiatives. Former employees who return often bring fresh perspectives combined with deep institutional knowledge, reducing onboarding time and increasing retention. Neglecting an alumni network means losing touch with valuable human capital, missing opportunities for referrals, and failing to leverage a powerful external community. By championing comprehensive alumni engagement, you not only foster lasting professional relationships but also create a continuous talent pipeline, enhance your employer brand, and gain access to a network of advocates and potential re-hires, ultimately contributing to your business's long-term talent strategy and competitive advantage.

WHAT?	WHY?	HOW?	BEWARE!
CORPORATE ALUMNI PROGRAMS	**BRAND ADVOCACY**	**COMMUNITY PLATFORMS**	**LOW ENGAGEMENT**
Engaging former employees through communities.	Turns former employees into brand advocates and potential re-hires.	Create exclusive online communities and networking events.	Difficulty sustaining active engagement from former employees.
KNOWLEDGE-SHARING OPPORTUNITIES	**EXTERNAL EXPERTISE**	**MENTORSHIP & CONSULTING**	**TIME COMMITMENT**
Leveraging alumni insights for mentorship.	Provides access to valuable external insights and expertise.	Invite alumni for mentorship, guest speaking, or consulting roles.	Alumni reluctance to commit time without clear incentives.
BOOMERANG EMPLOYEE PATHWAYS	**TALENT RE-ACQUISITION**	**STRUCTURED RE-HIRING**	**NEGATIVE PERCEPTIONS**
Encouraging returning talent.	Streamlines re-hiring, leveraging prior knowledge and cultural fit.	Create clear processes for re-engaging and re-hiring former employees.	Stigma around returning to a previous employer.

Ai6x6™ Transformation Blueprint

AUTHORITY | Employee Loyalty

"You can't have a great company without great people." - Jack Welch.

Legendary former CEO of General Electric (GE), credited with dramatically increasing its market value. Known for his aggressive management style and focus on profitability and talent development, he transformed GE into a highly competitive global conglomerate, setting new standards for corporate leadership.

The Authority level is your Loyalty stage, reached when the culture becomes its own retention engine and legacy. High levels of loyalty and succession planning secure the future of the organization as an industry benchmark for talent.

Implementing comprehensive strategies and initiatives specifically designed to keep valuable employees engaged, highly motivated, and deeply committed to the organization for the long term. This significantly reduces costly turnover and preserves invaluable institutional knowledge.

"Employee Loyalty" fosters a stable, experienced, and highly productive workforce that is absolutely vital for sustained organizational success and competitive advantage. Effective retention strategies build a strong, cohesive team, reduce recruitment burdens, and ensure continuity in projects and operations.

The **"Employee Loyalty"** chapter is fundamental for business strategy maturity as it focuses on innovating employee programs to sustain a healthy and productive workforce. A mature business strategy prioritizes ensuring job satisfaction, offering competitive benefits, and fostering a positive workplace environment. This commitment to loyalty amplifies organizational stability, reduces recruitment costs, and preserves institutional knowledge. It's about building a loyal and engaged workforce that consistently executes strategic initiatives, ensuring the long-term human capital required to achieve sustained growth and competitive advantage in the market. Here is how we approach our learning:

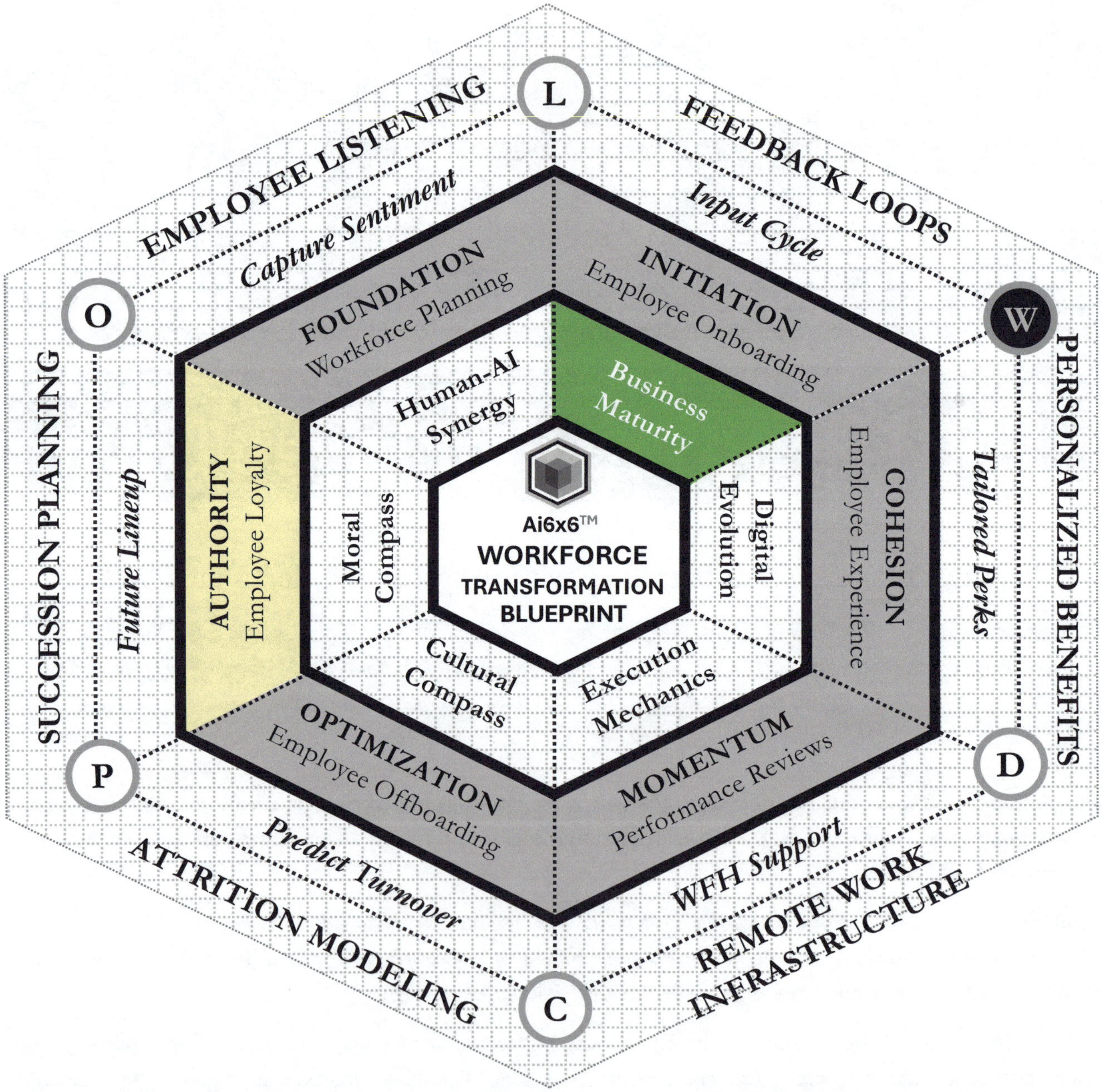

Ai6x6 BUSINESS MATURITY FOR LOYALTY | WORKFORCE TRANSFORMATION

For a business leader, actively engaging in employee listening is a strategic imperative for fostering a positive, responsive workplace culture and significantly boosting employee loyalty. It's about creating continuous channels for capturing employee sentiment and feedback, ensuring that voices are heard and concerns are addressed before they escalate. Central to this is conducting continuous pulse surveys. By deploying short, frequent surveys, you can collect real-time input on employee satisfaction, engagement levels, and overall workplace experience, allowing for immediate insights and agile responses. This moves beyond annual surveys to a dynamic, ongoing dialogue.

Furthermore, leveraging AI-driven sentiment analysis transforms raw feedback data into actionable insights. By using technology to gauge morale, identify emotional trends, and pinpoint specific areas for improvement from unstructured text feedback, you can uncover subtle patterns and systemic issues that might otherwise be missed. Crucially, establishing robust open-door communication policies creates structured forums for employees to voice concerns, share ideas, and provide direct feedback without fear of repercussion. This fosters psychological safety and builds trust, reinforcing that their input is genuinely valued. Neglecting proactive employee listening can lead to simmering discontent, high employee turnover due to unaddressed issues, and a disconnect between leadership perceptions and employee realities. By championing continuous employee listening, you empower your organization to be more responsive, build a culture of transparency, and proactively enhance employee satisfaction, ultimately strengthening loyalty and securing a more engaged and stable workforce for long-term business success.

WHAT?	WHY?	HOW?	BEWARE!
CONTINUOUS PULSE SURVEYS	**REAL-TIME INSIGHTS**	**SHORT, FREQUENT SURVEYS**	**SURVEY FATIGUE**
Collecting real-time input on satisfaction.	Provides immediate insights into employee sentiment and satisfaction.	Implement regular, brief surveys for quick feedback.	Employees becoming disengaged from too frequent survey requests.
AI SENTIMENT ANALYSIS	**TREND IDENTIFICATION**	**TEXT ANALYTICS**	**CONTEXT & ACCURACY**
Using AI technology to gauge morale.	Uncovers morale trends and improvement areas from unstructured data.	Deploy AI tools to analyze open-ended feedback responses.	AI potentially misinterpreting nuanced sentiment or sarcasm.
OPEN-DOOR COMMUNICATION POLICIES	**TRUST & VOICE**	**ACCESSIBLE CHANNELS**	**LACK OF ACTION**
Creating structured forums.	Fosters trust and empowers employees to voice concerns freely.	Establish various channels like town halls, forums, and 1:1 meetings.	Employees feeling their feedback is heard but not acted upon.

For a business leader, establishing robust feedback loops is a fundamental strategic imperative for continuous improvement, driving employee loyalty, and fostering a high-performance culture. It's about creating systematic evaluation channels that provide timely, constructive insights, ensuring that individuals and teams can constantly refine their performance and development. Central to this is implementing regular one-on-one check-ins. These scheduled, focused discussions between employees and managers on development progress, challenges, and goals provide continuous opportunities for personal coaching, immediate feedback exchange, and collaborative problem-solving. This frequent dialogue replaces traditional, infrequent reviews with ongoing support.

Furthermore, providing anonymous feedback tools is crucial. Allowing employees to share concerns, suggestions, and insights for workplace improvement confidentially encourages candor and uncovers issues that might otherwise remain hidden, building trust and psychological safety. Crucially, integrating performance-based coaching adjustments ensures that feedback leads to tangible growth. By modifying coaching approaches based on employee feedback and evolving job needs, you demonstrate responsiveness and commitment to individual development, making the feedback process truly impactful and tailored. Neglecting structured feedback loops can lead to delayed problem-solving, unaddressed performance issues, decreased morale due to a lack of recognition or guidance, and a stagnant workforce. By championing a culture of continuous, constructive feedback, you empower your employees to thrive, strengthen managerial effectiveness, and build a more responsive and adaptive organization, ultimately securing sustained success and loyalty within your business.

WHAT?	WHY?	HOW?	BEWARE!
REGULAR ONE-ON-ONE CHECK-INS	**DEVELOPMENT SUPPORT**	**SCHEDULED CONVERSATIONS**	**MANAGERIAL BURDEN**
Conducting scheduled discussions.	Fosters ongoing development and addresses immediate concerns.	Implement regular, brief, focused 1:1 meetings between managers and employees.	Managers feeling overwhelmed by frequency of mandatory meetings.
ANONYMOUS FEEDBACK TOOLS	**CANDID INSIGHTS**	**SECURE PLATFORMS**	**LACK OF CONTEXT**
Allowing employees to share concerns.	Encourages honest suggestions without fear of repercussion.	Provide secure, anonymous online platforms for feedback submission.	Anonymous feedback sometimes lacking sufficient detail for action.
PERFORMANCE-BASED COACHING ADJUSTMENTS	**TAILORED GROWTH**	**FEEDBACK INTEGRATION**	**INCONSISTENT APPLICATION**
Modifying coaching approaches.	Ensures coaching is relevant and impactful for individual needs.	Adapt coaching strategies based on employee feedback and performance data.	Managers lacking skills or willingness to adapt coaching styles.

For a business leader, strategically offering personalized benefits is a crucial lever for enhancing employee loyalty, maximizing job satisfaction, and demonstrating a deep commitment to the diverse needs of your workforce. In an era where a one-size-fits-all approach is no longer effective, tailoring incentives becomes a powerful differentiator in attracting and retaining top talent. Central to this is implementing flexible compensation structures. This involves offering salary adjustments, equity options, and performance-based bonuses that can be tailored based on individual preferences, roles, and career stages, giving employees greater control over their total rewards package.

Furthermore, providing a diverse range of health and wellness perks is crucial. Beyond standard healthcare, offering personalized wellness programs, gym memberships, mental health resources, and preventative care options signals a genuine investment in the holistic well-being of your employees. These benefits address individual health priorities and promote a healthier, more engaged workforce. Crucially, offering comprehensive work-life integration support acknowledges the need for balance. Encouraging individualized benefits such as childcare assistance, sabbaticals, flexible hours, and personal coaching demonstrates empathy for employees' lives outside of work, fostering a culture of trust and flexibility. Neglecting personalized benefits can lead to a disengaged workforce, high turnover as employees seek organizations that better meet their individual needs, and a struggle to attract top talent in a competitive market. By championing a flexible and tailored benefits strategy, you empower your employees to thrive, demonstrate genuine care, and build a highly loyal and satisfied workforce that feels genuinely valued by your organization, ultimately contributing to your business's sustained success and reputation.

WHAT?	WHY?	HOW?	BEWARE!
FLEXIBLE COMPENSATION STRUCTURES	**INDIVIDUAL SATISFACTION**	**CUSTOMIZED PACKAGES**	**COMPLEXITY**
Offering salary adjustments and equity options.	Meets diverse employee preferences, boosting satisfaction.	Allow employees to choose between salary, equity, or bonus mixes.	Increased administrative complexity in managing varied compensation models.
HEALTH AND WELLNESS PERKS	**HOLISTIC WELL-BEING**	**PROGRAM DIVERSITY**	**UTILIZATION**
Providing personalized wellness programs.	Supports employee health and prevents burnout.	Offer gym memberships, mental health support, and wellness stipends.	Ensuring employees actually utilize the provided wellness benefits.
WORK-LIFE INTEGRATION SUPPORT	**BALANCE & FLEXIBILITY**	**FLEXIBLE OPTIONS**	**PERCEIVED INEQUITY**
Encouraging individualized benefits.	Fosters work-life balance and reduces stress.	Provide childcare, sabbaticals, and personal coaching options.	Some benefits being more accessible to certain roles or life stages.

For a business leader, optimizing remote work infrastructure is a strategic imperative for improving retention and ensuring the productivity and engagement of your hybrid and fully remote employees. The shift to flexible work models is permanent, and your ability to seamlessly support these arrangements directly impacts talent acquisition, employee satisfaction, and operational efficiency. Central to this is providing seamless digital collaboration platforms. Investing in advanced communication tools, project management software, and virtual meeting platforms ensures that all employees, regardless of location, can interact, share information, and collaborate effectively without friction. This connectivity is vital for team cohesion.

Furthermore, implementing flexible scheduling models is crucial. Adapting work hours to accommodate employee preferences and promote well-being—allowing for asynchronous work, compressed weeks, or personalized schedules—demonstrates trust and empowers employees to manage their work-life integration more effectively. This flexibility is a key driver of satisfaction. This involves actively fostering virtual team-building activities, implementing recognition programs tailored for remote contributions, and creating opportunities for informal social interaction, combating isolation. Neglecting remote work infrastructure can lead to decreased productivity, disengaged remote teams, high turnover among flexible workers, and a struggle to attract talent seeking modern work arrangements. By championing these solutions, you empower your remote and hybrid workforce to thrive, build a more inclusive and flexible work environment, and ultimately secure your business's long-term success in a decentralized world. This commitment is vital for future-proofing your talent strategy.

WHAT?	WHY?	HOW?	BEWARE!
SEAMLESS DIGITAL COLLABORATION	**CONNECTIVITY & EFFICIENCY**	**TOOL INVESTMENT**	**TOOL FATIGUE**
Enhancing productivity.	Boosts productivity and communication for remote teams.	Provide robust communication, project management, and video conferencing tools.	Employees becoming overwhelmed by too many collaboration tools.
FLEXIBLE SCHEDULING MODELS	**WORK-LIFE BALANCE**	**POLICY IMPLEMENTATION**	**COORDINATION**
Adapting work hours to accommodate preferences.	Improves employee well-being and satisfaction.	Develop clear policies for flexible hours, compressed weeks, etc.	Difficulty in coordinating team schedules across different time zones.
REMOTE WORKPLACE	**COHESION & MORALE**	**VIRTUAL ACTIVITIES**	**ISOLATION**
Strengthening Internet connectivity.	Fosters a sense of belonging and boosts morale for remote staff.	Organize virtual team-building, recognition events, and social gatherings.	Remote employees still feeling isolated despite engagement efforts.

For a business leader, implementing advanced attrition modeling is a powerful strategic imperative for using data-driven insights to predict and proactively reduce employee turnover, safeguarding your talent pool and minimizing costly recruitment cycles. In today's competitive environment, understanding why employees leave—and who is likely to leave next—is critical for effective talent management. Central to this is utilizing predictive analytics for retention. By analyzing vast datasets related to employee demographics, performance, engagement, compensation, and tenure, you can identify key factors that influence employee departure rates, allowing for precise risk assessment.

Furthermore, conducting thorough workforce trend analysis assesses historical data to forecast attrition risks. By recognizing patterns in past turnover, such as departments with higher churn or specific times of year for departures, you can anticipate future challenges and prepare targeted interventions. Crucially, the ultimate goal is to implement proactive intervention strategies. Based on the insights from modeling and analysis, leaders can deploy targeted engagement tactics, personalized retention efforts, or specific development opportunities before employees even consider leaving, addressing root causes of dissatisfaction. Neglecting attrition modeling can lead to reactive talent management, chronic turnover, significant recruitment costs, and a loss of institutional knowledge that impacts productivity. By championing a data-driven approach to understanding and mitigating attrition, you empower your organization to retain its most valuable asset—its people—build a more stable and experienced workforce, and ultimately secure your business's long-term success through enhanced loyalty & reduced talent drain

WHAT?	WHY?	HOW?	BEWARE!
PREDICTIVE ANALYTICS FOR RETENTION	**EARLY WARNING**	**DATA ANALYSIS**	**DATA PRIVACY**
Identifying factors influencing departure.	Provides early warnings of potential employee departures.	Analyze HR data (tenure, performance, compensation) to build predictive models.	Ethical concerns and data privacy regulations around employee data analysis.
WORKFORCE TREND ANALYSIS	**PATTERN RECOGNITION**	**HISTORICAL REVIEW**	**CAUSATION VS. CORRELATION**
Assessing historical data to forecast attrition risks.	Identifies recurring patterns and high-risk segments for turnover.	Analyze past turnover data, exit interview insights, and industry benchmarks.	Difficulty distinguishing true causes from mere correlations in data.
PROACTIVE INTERVENTION STRATEGIES	**PREVENTATIVE RETENTION**	**TARGETED OUTREACH**	**EFFECTIVENESS MEASUREMENT**
Implementing engagement tactics.	Addresses attrition risks before employees decide to leave.	Deploy personalized engagement programs and retention initiatives for at-risk employees.	Difficulty measuring the direct impact of interventions.

For a business leader, implementing robust succession planning is a non-negotiable strategic imperative for preparing future leaders, mitigating risks associated with key talent departures, and ensuring seamless organizational continuity. It's about proactively building a strong internal talent pipeline that guarantees stability and sustained performance, even in the face of unexpected transitions. Central to this is developing a comprehensive leadership pipeline. This involves systematically identifying high-potential employees across all levels, assessing their readiness for advancement, and investing in their development to prepare them for critical senior roles. This proactive approach ensures you have qualified talent ready to step up.

Furthermore, integrating effective knowledge transfer strategies is crucial within succession planning. This means implementing mentorship programs, job shadowing, and formal handover processes to facilitate skill-sharing and ensure that the invaluable institutional knowledge of departing or retiring leaders is successfully passed on to their successors. Crucially, establishing structured transition frameworks ensures seamless role succession and minimizes disruption when leadership changes occur. This involves detailed plans for onboarding successors, managing communication around transitions, and providing ongoing support to ensure a smooth handover of responsibilities. Neglecting robust succession planning can lead to critical leadership gaps, prolonged periods of instability, increased external recruitment costs for senior roles, and a loss of institutional memory. By championing a systematic approach to preparing future leaders, you empower your organization with resilience, ensure continuity of operations, and cultivate a culture of growth and opportunity, ultimately securing your business's advantage.

WHAT?	WHY?	HOW?	BEWARE!
LEADERSHIP PIPELINE DEVELOPMENT	**FUTURE LEADERSHIP**	**TALENT IDENTIFICATION**	**BIAS IN SELECTION**
Identifying high-potential employees.	Ensures a continuous supply of qualified future leaders.	Identify and assess high-potential employees for key roles.	Bias in identifying high-potential employees, limiting diversity.
KNOWLEDGE TRANSFER STRATEGIES	**INSTITUTIONAL KNOWLEDGE**	**MENTORSHIP & SHADOWING**	**TIME & WILLINGNESS**
Implementing mentorship programs.	Facilitates the transfer of critical institutional knowledge.	Establish formal mentorship programs and job shadowing.	Senior leaders lacking time or willingness to mentor effectively.
STRUCTURED TRANSITION FRAMEWORKS	**MINIMIZED DISRUPTION**	**TRANSITION PLANS**	**LACK OF FLEXIBILITY**
Ensuring seamless role succession.	Reduces operational disruption during leadership changes.	Develop clear plans for onboarding successors and role handovers.	Plans being too rigid, not adapting to specific individual or role needs.

SECTION D

BLUEPRINT SERIES

Digital Evolution

Ai6x6™ Transformation Blueprint

Story 1 — The Birth of COBOL and Business Computing (1959–1965): Early enterprises struggled to translate business logic into machine-readable form, limiting the usefulness of emerging computers. COBOL introduced a standardized language that mirrored human reasoning, enabling organizations to encode processes consistently. This shift allowed systems to scale beyond isolated experiments. Human analysts defined rules while machines executed them with precision. The partnership created the first bridge between business intent and digital execution.

Story 2 — ARPANET and Packet Switching (1969–1972): Communication networks were fragile, centralized, and prone to failure. Packet switching introduced a decentralized architecture that allowed information to flow even when parts of the network failed. Humans designed protocols while machines routed data autonomously. This resilience transformed digital communication from a technical novelty into a strategic infrastructure. ARPANET demonstrated how distributed systems could scale without collapsing. It laid the foundation for the modern internet.

Story 3 — Xerox PARC and the Graphical User Interface (1973–1983): Computers were powerful but inaccessible to most users, requiring specialized knowledge to operate. Xerox PARC introduced the graphical user interface, enabling humans to interact with machines visually rather than through code. This shift democratized computing and redefined usability. Machines handled complexity behind the scenes while humans engaged through intuitive metaphors. The interface became a catalyst for mass digital adoption. It showed that digital evolution accelerates when technology adapts to human cognition.

Story 4 — The Rise of Enterprise Resource Planning (1980s–1990s): Organizations operated in silos, with fragmented systems and inconsistent data. ERP platforms unified finance, operations, and supply chains into integrated digital environments. Machines enforced standardization while humans redesigned processes around shared information. This integration improved visibility, coordination, and control. ERP became the backbone of global enterprises. It demonstrated how digital evolution transforms not just tools but entire operating models.

Story 5 — The Commercial Internet and E-Commerce (1994–2005): The internet shifted from academic infrastructure to commercial ecosystem, enabling new forms of interaction, distribution, and value creation. Websites became digital storefronts, and payment systems matured into secure transactional platforms. Machines automated logistics while humans reimagined customer experience. The digital marketplace expanded global reach and lowered barriers to entry. It marked a turning point where digital capability became essential for competitiveness.

Story 6 — Cloud Computing and Elastic Infrastructure (2006–2015): Organizations struggled with rigid, capital-intensive infrastructure that limited experimentation and scale. Cloud computing introduced elastic resources that could expand or contract instantly. Machines automated provisioning while humans focused on architecture and innovation. This shift accelerated deployment cycles, reduced cost, and enabled global reach. Cloud became the foundation for modern digital ecosystems. It demonstrated how digital evolution thrives when infrastructure becomes fluid, on-demand, and programmable.

LESSONS FROM A WORKFORCE TRANSFORMATION PERSPECTIVE

The workforce evolves as digital systems shift from specialized tools to ubiquitous platforms. Early digital environments required technical experts to translate business needs into machine logic, while later systems empowered broader participation through intuitive interfaces and integrated workflows. Across decades, the workforce learned to navigate increasing complexity by developing digital fluency, cross-functional collaboration, and adaptive problem-solving. Digital evolution elevates workers from operators of tools to designers of systems and stewards of data. Maturity is reached when the workforce can absorb new technologies without disruption, using digital capability to enhance judgment, creativity, and collective performance.

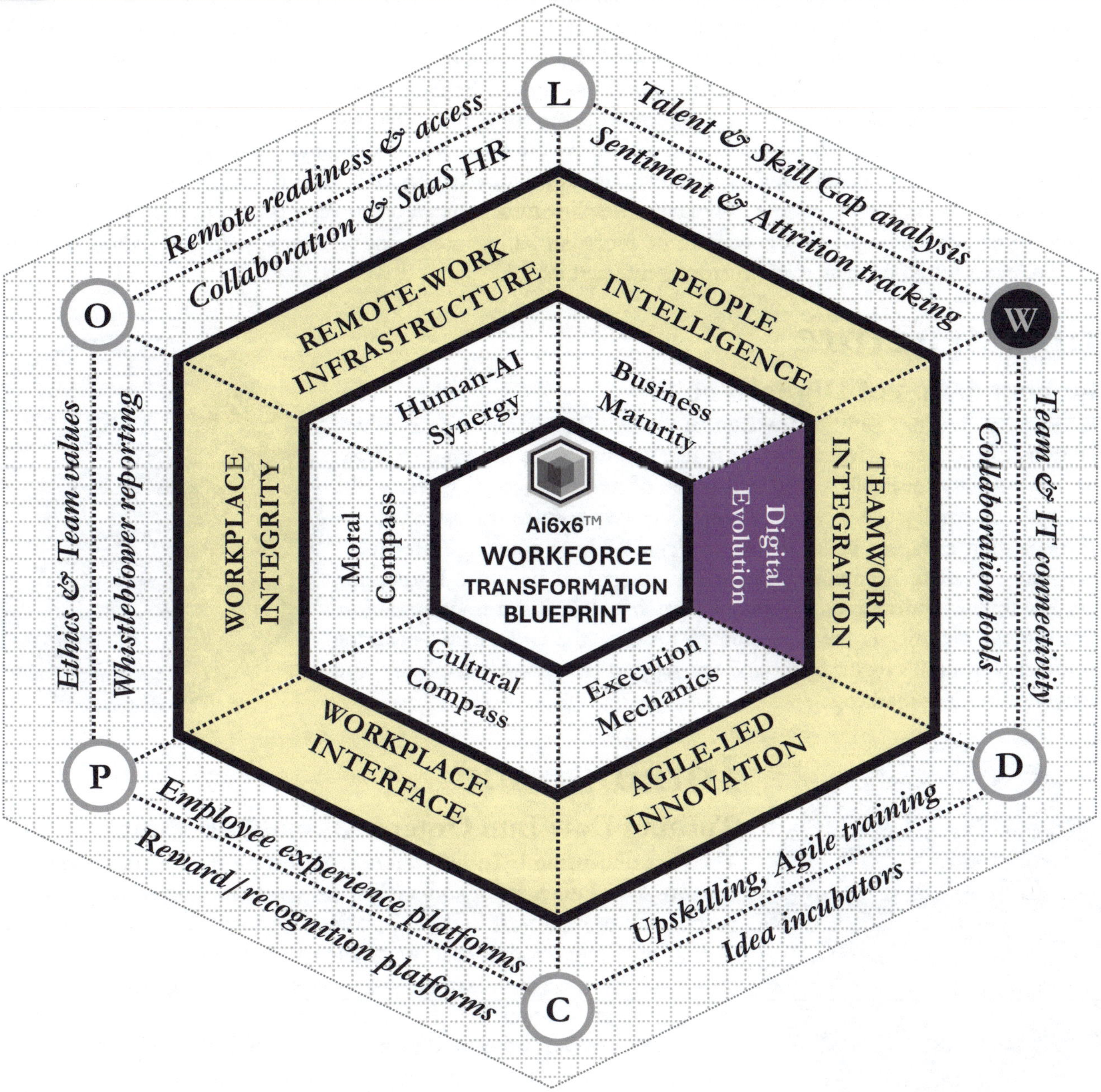

The **Ai6x6 Research Lab** chose this lens to solve the "disconnection gap" that occurs in increasingly distributed and digital work environments. We believe a workforce must possess a sense of digital gravitas—a feeling of being connected and powerful regardless of their physical location on the grid. This isn't about providing better laptops; it is about the mental shift from being a "remote user" to being a vital thread in a seamless team fabric. It is vital because a worker cannot innovate or collaborate if they do not feel a true sense of belonging and technical command within the organization's digital ecosystem.

By mastering this evolution, the workforce moves from digital isolation to absolute collaborative intelligence. This lens ensures the "Work from Any Grid" infrastructure is built on a foundation of trust and people-powered insights, where touchpoints of belonging become a natural extension of the daily workflow. When workers understand the integrity and agility of their digital studios, they gain the freedom to contribute with an undistracted focus on excellence. This lens provides the psychological peace of mind that comes from knowing the team's digital interface is designed for connection. it ensures that the "Seamless Team Fabric" is not just a technical achievement, but a human one.

Digital Evolution is no longer a technical initiative. It is the structural rewiring of how an organization thinks, operates, and competes. For years, leaders treated digital transformation as a project—something with a start date, a roadmap, and a finish line. But the world has moved beyond that framing. Digital Evolution is not a destination; it is a permanent state of adaptation. It is the recognition that technology is no longer an enabler of strategy but a **shaper** of it. The organizations that thrive are those that treat digital capability as a living system—continuously upgraded, continuously integrated, and continuously aligned with the realities of the market. This lens reframes Digital Evolution as a leadership discipline rather than a technology program. It challenges leaders to see digital capability not as a collection of tools but as an interconnected ecosystem that influences every decision, every workflow, and every customer interaction.

Infrastructure

The Foundation of Digital Stability

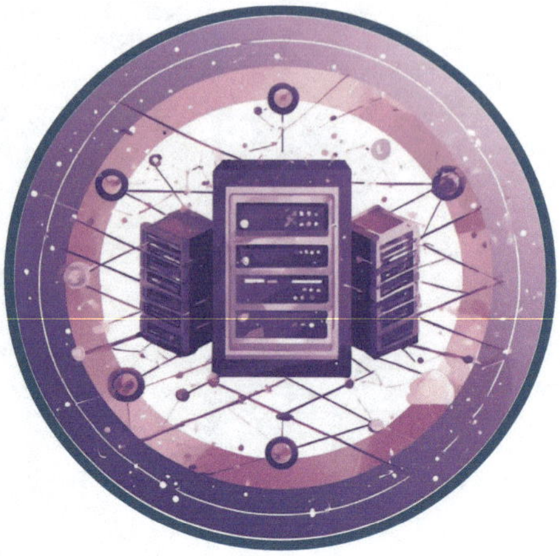

The journey begins with **Infrastructure**, the foundation upon which everything else depends. Infrastructure is not just hardware or cloud capacity; it is the stability, scalability, and reliability of the systems that carry the organization's intelligence. Leaders who underestimate infrastructure discover that innovation collapses under its own weight. Without a strong foundation, every new capability becomes fragile, every integration becomes brittle, and every attempt at modernization becomes a patch rather than a transformation. Infrastructure maturity is the discipline of building for endurance—ensuring that the organization's digital backbone can support the speed, complexity, and volatility of modern markets.

Intelligence

Turning Data Into Organizational Foresight

The next dimension is **Intelligence**, representing the organization's ability to convert data into insight and insight into action. Intelligence is not about dashboards or reports; it is about building a system that learns. It requires data that flows freely, models that are trained responsibly, and decisions that are informed by evidence rather than assumption. Leaders who embrace intelligence understand that it is not a technical asset but a strategic one. Intelligence shapes priorities, exposes risks, and reveals opportunities that would otherwise remain invisible. It is the cognitive engine of Digital Evolution—the mechanism through which the organization perceives, interprets, and responds to its environment.

Integration

Making the Enterprise Whole

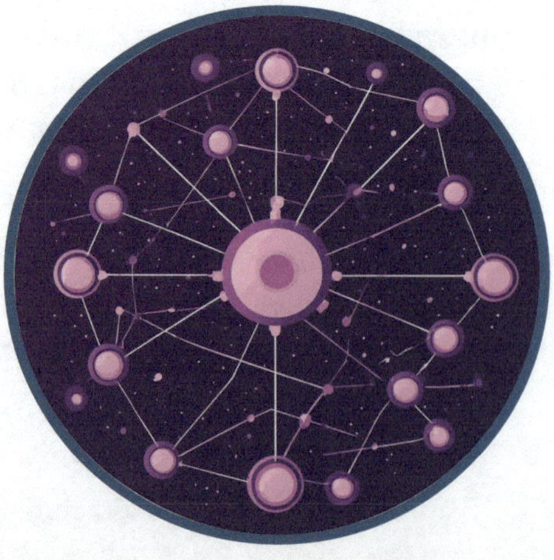

Integration is where Digital Evolution becomes real. Organizations often accumulate technologies faster than they can connect them. Systems become islands, data becomes siloed, and teams become fragmented. Integration is the discipline that ensures the enterprise operates as a coherent whole. It is the work of connecting platforms, harmonizing data, and designing workflows that move across boundaries. Without integration, digital investments become isolated experiments rather than enterprise capabilities. Integration maturity is the art of coherence—ensuring that the organization's digital landscape functions as a single, unified system rather than a patchwork of disconnected tools.

Innovation

Expanding the Field of Possibility

The fourth dimension, **Innovation**, expands the field of possibility. Innovation is not about chasing trends or adopting technology for its own sake. It is about using digital capability to reimagine how value is created, delivered, and experienced. Innovation demands that leaders challenge assumptions, explore alternatives, and test ideas quickly. It requires a culture where experimentation is encouraged, failure is reframed as learning, and progress is measured by insight rather than perfection. Innovation maturity is the discipline of curiosity—using digital tools not just to optimize the present but to invent the future.

Interface

Where Technology Meets Human Behavior

Interface represents the human experience of Digital Evolution. It is the point where technology meets behavior, where systems meet expectations, and where design meets trust. Leaders who ignore interface discover that even the most powerful systems fail if people cannot or will not use them. Interface is not cosmetic; it is strategic. It shapes adoption, accelerates learning, and determines whether digital capability becomes embedded or abandoned. Interface maturity is the discipline of empathy—designing digital experiences that are intuitive, accessible, and aligned with human needs.

Integrity

The Guardrail of Responsible Evolution

The final dimension, **Integrity**, grounds Digital Evolution in responsibility. Integrity ensures that systems are secure, data is protected, and decisions are ethical. It prevents speed from becoming recklessness and innovation from becoming harm. Leaders who prioritize integrity build organizations that are trusted, resilient, and prepared for the scrutiny that accompanies digital capability. Integrity maturity is the discipline of stewardship—ensuring that digital evolution strengthens the organization's moral and operational foundations rather than compromising them.

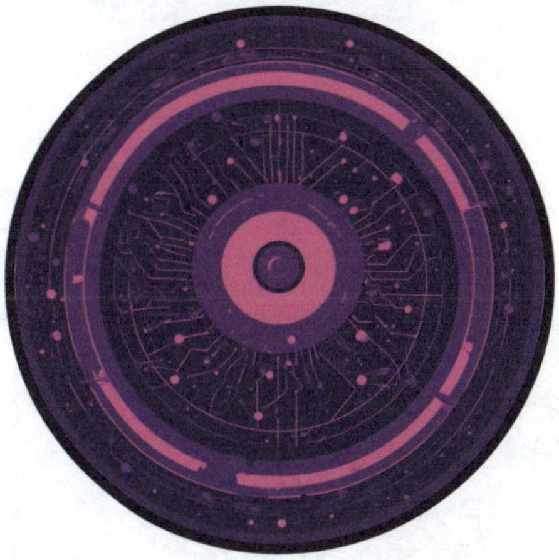

Digital Evolution is not a technology story. It is a leadership story. It is the recognition that the future belongs to organizations that treat digital capability as a core competency, not a side project. It is the shift from building tools to building systems, from adopting technology to evolving with it, and from reacting to change to shaping it. In this future, Digital Evolution is not a department, a project, or a budget line. It is the operating philosophy of the enterprise. It shapes how decisions are made, how systems are built, how teams collaborate, and how value is created. It turns technology from a source of disruption into a source of stability. It transforms the organization from a follower of change into a shaper of it.

DIGITAL EVOLUTION ANALOGY – KEEPING OURSELVES HEALTHY

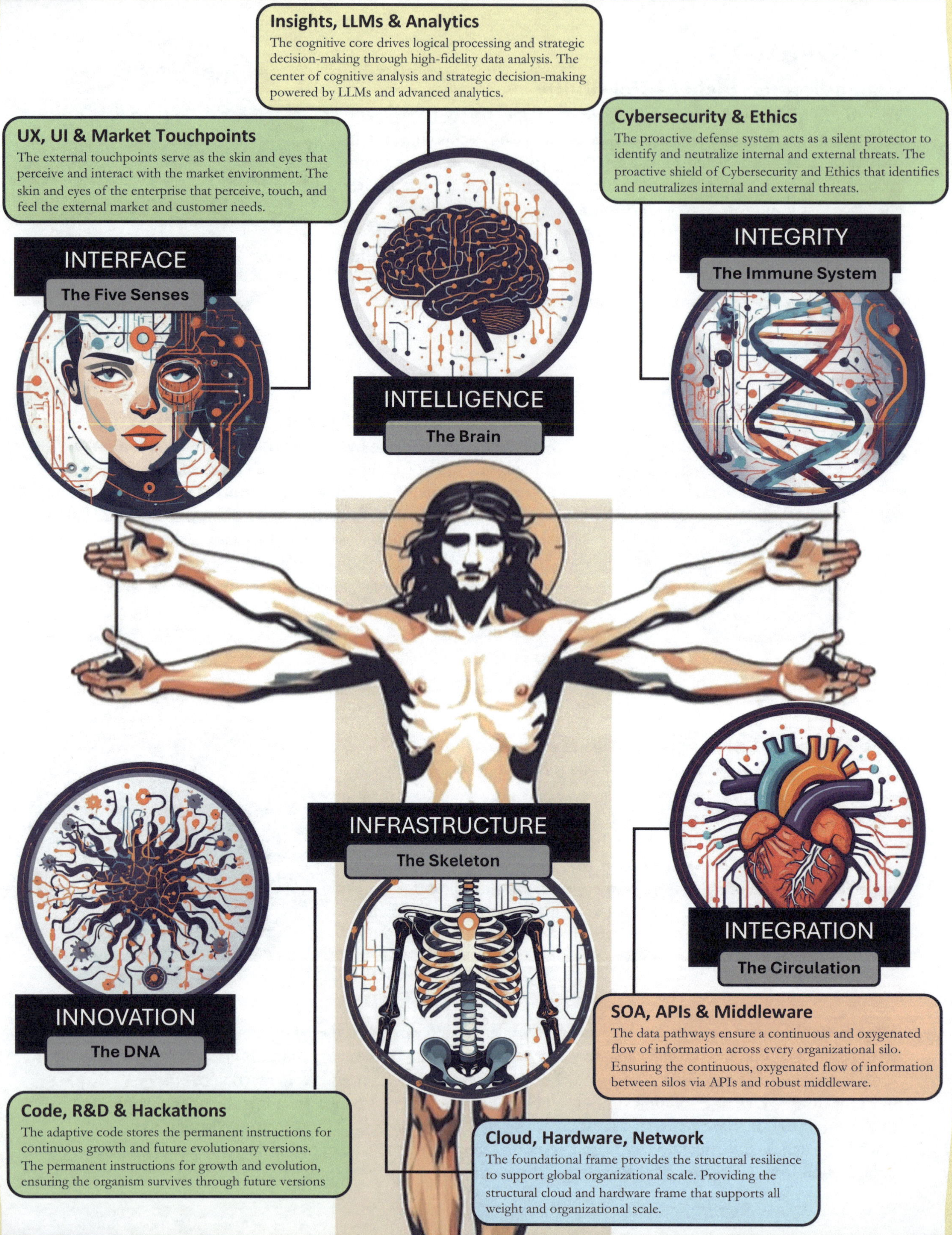

Ai6x6™ Transformation Blueprint

Remote Work 'Infrastructure'

"Technology should make work more human, not more distant." — Stewart Butterfield , Founder of Slack; revolutionized digital collaboration with intuitive, culture-sensitive tools.

Workplace infrastructure now dictates emotional resonance as much as operational range. "Remote Work Infrastructure" isn't about location—it's about design that sustains cohesion when geography dissolves. Virtual desktops, remote access, and collaboration platforms must reinforce belonging, not merely bandwidth. Mature transformation understands that infrastructure is a cultural signal—delivered through experience, not endpoints. When employees can move seamlessly across grids without losing context, connection, or contribution, remote readiness becomes more than technical—it becomes identity-compatible. Transformation doesn't just digitize presence—it digitizes purpose, making infrastructure the quiet architect of distributed excellence.

For workforce transformation, the "Remote Work Infrastructure" chapter is foundational for digital maturity because it defines the technological backbone that enables a flexible, distributed, and highly productive workforce. A digitally mature organization recognizes that empowering employees with seamless, secure access to tools and resources, regardless of their physical location, is paramount. This includes virtual desktop infrastructure, cloud-based HR platforms, and robust remote access solutions. This infrastructure ensures operational continuity, enhances employee experience, and supports the agility needed to adapt to evolving work models, directly impacting the organization's ability to attract, retain, and leverage talent in the digital age.

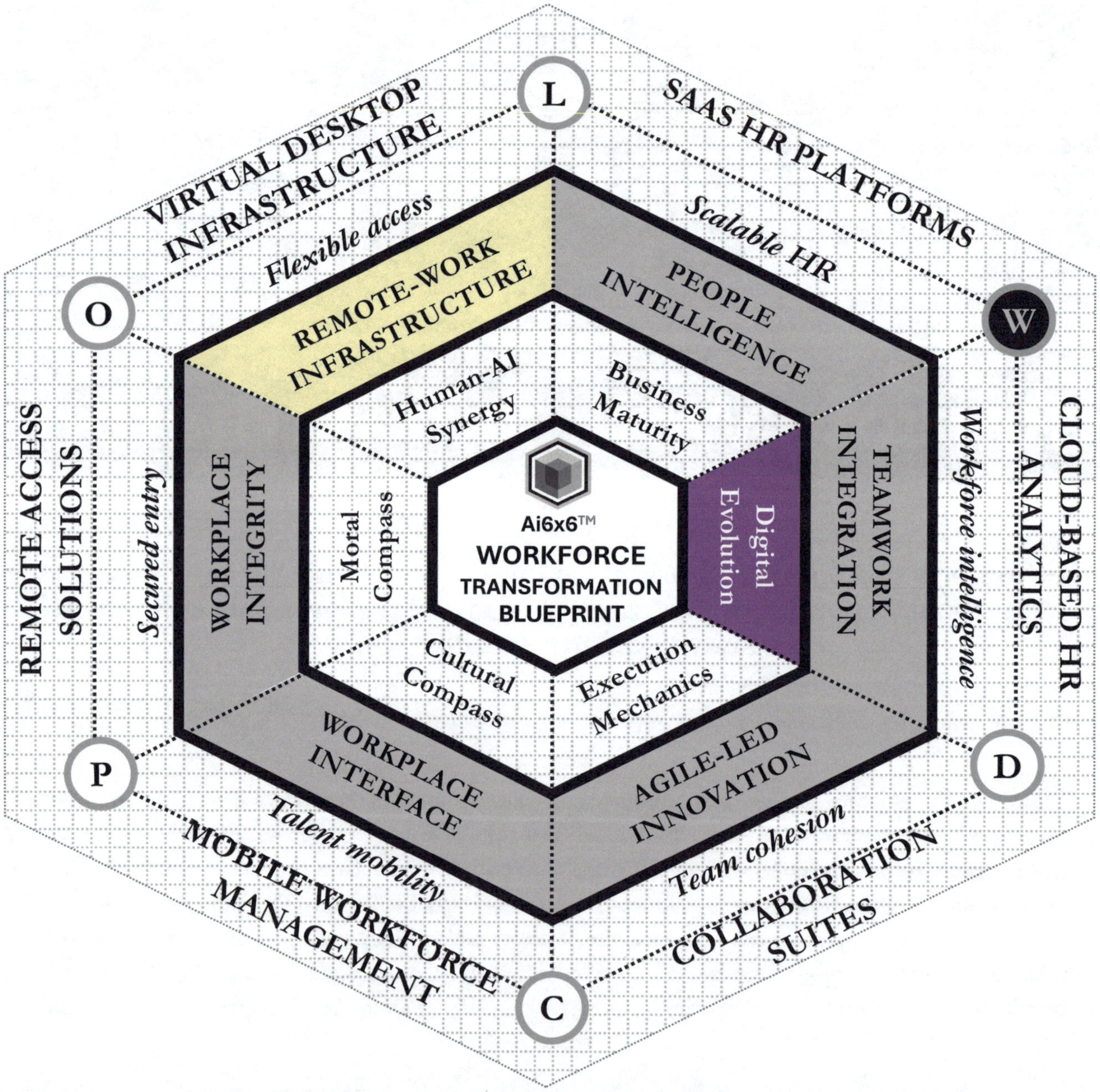

Ai6x6 DIGITAL EVOLUTION FRAMEWORK | **WORKFORCE LIFECYCLE | INFRASTRUCTURE**

1. VIRTUAL DESKTOP INFRASTRUCTURE (VDI)

Virtual Desktop Infrastructure (VDI) centralizes and manages desktop environments on remote servers, delivering a personalized desktop experience to end-users on demand, regardless of the device or location from which they are accessing it. Instead of processing applications locally, VDI streams the desktop interface and applications over a network to client devices, ranging from thin clients to personal laptops and tablets. By providing secure, flexible, and scalable access to corporate resources, VDI significantly enhances data security, simplifies IT management, reduces hardware costs, and empowers a highly flexible workforce by enabling seamless remote work and robust business continuity, which s critical for organizations prioritizing security and adaptability.

Without implementing Virtual Desktop Infrastructure (VDI), organizations often struggle with managing disparate end-user devices, ensuring consistent security across diverse endpoints, and providing seamless, secure access to corporate applications for a remote or hybrid workforce. This decentralized approach leads to increased IT support overhead, higher risks of data breaches on local devices, and significant limitations on workforce flexibility, as employees may face performance issues or security concerns when accessing sensitive information from outside the corporate network. The absence of VDI results in higher total cost of ownership for IT, reduced operational agility, and a less secure computing environment, ultimately hindering productivity and preventing a truly adaptable workforce model.

Implementation	Challenges
* Assess user needs & application compatibility * Select VDI platform & server infrastructure * Design desktop images & user profiles * Plan network capacity & security protocols * Pilot with user groups & refine configuration * Deploy & provide ongoing support/monitoring	* High initial infrastructure investment * Complex setup & ongoing management * Ensuring sufficient network bandwidth & latency * User experience challenges (e.g., peripheral support) * Licensing complexities for software & VDI * Scalability for fluctuating user demand

Products	Platform Cost	Project Cost	Duration
VMware Horizon	$20-100K	$100-500K	4-12 m
Citrix Virtual Apps and Desktops	$20-90K	$90-450K	4-11 m
Azure Virtual Desktop (AVD)	$15-70K	$70-350K	3-10 m
Amazon WorkSpaces	$10-50K	$50-250K	3-9 m
Nutanix Frame	$10-40K	$40-200K	2-8 m
Google Cloud Endpoints (custom)	$5-30K	$30-150K	2-7 m

SaaS (Software as a Service) HR Platforms are cloud-based human resources management systems that offer a comprehensive suite of functionalities delivered via the internet, accessible from any web browser or device. These platforms typically cover core HR processes such as recruitment, onboarding, payroll, benefits administration, performance management, and talent development. By moving HR operations to the cloud, organizations benefit from reduced IT infrastructure costs, automatic software updates, enhanced data accessibility, and improved scalability, empowering HR teams to streamline administrative tasks, focus on strategic initiatives, and provide a more efficient and responsive experience for employees across the entire organization.

Without adopting SaaS HR Platforms, organizations often rely on outdated, on-premise HR systems or manual processes that are cumbersome, expensive to maintain, and lack the flexibility and accessibility required for a modern workforce. This perpetuates administrative inefficiencies, creates data silos, hinders real-time reporting, and limits the ability of HR to support a geographically dispersed or rapidly growing employee base. The absence of a centralized, cloud-native HR solution results in slower HR service delivery, increased operational costs, higher potential for data errors, and a diminished capacity for HR to act as a strategic partner, ultimately impeding talent management and overall organizational agility.

Implementation	Challenges
* Assess current HR processes & needs * Select comprehensive SaaS HR platform * Plan data migration & system configuration * Train HR staff & employees on new system * Integrate with existing payroll/finance systems * Secure data privacy & compliance standards	* Complex data migration from legacy systems * Ensuring seamless integration with other enterprise tools * Managing change resistance from HR and employees * Customization limitations within a SaaS model * Data security and privacy in the cloud environment * Measuring ROI beyond simple cost savings

Products	Platform Cost	Project Cost	Duration
Workday HCM	$30-150K	$200-800K	6-18 m
SAP SuccessFactors	$25-110K	$140-550K	5-15 m
Oracle Cloud HCM	$20-90K	$120-500K	4-14 m
UKG Pro	$15-70K	$80-350K	3-10 m
ADP Workforce Now	$5-30K	$20-100K	2-6 m
BambooHR	$1-10K	$5-50K	1-3 m

Cloud-Based HR Analytics leverages cloud computing power and advanced analytical tools to process, analyze, and visualize vast amounts of human resources data, providing real-time insights into workforce trends, talent performance, and HR program effectiveness. These platforms integrate data from various HR systems (e.g., recruitment, payroll, performance) to identify patterns, predict outcomes (like attrition or skill gaps), and uncover opportunities for optimizing human capital strategies. By democratizing access to actionable workforce intelligence, Cloud-Based HR Analytics empowers HR leaders and managers to make data-driven decisions that improve talent acquisition, enhance employee engagement, increase retention, and ultimately drive overall business performance by optimizing the organization's most valuable asset—its people.

Without adopting Cloud-Based HR Analytics, organizations often struggle to extract meaningful insights from their fragmented HR data, relying instead on manual reporting or intuition, which leads to reactive decision-making and missed opportunities for optimizing human capital. This absence of advanced analytical capabilities results in an inability to accurately predict attrition, identify critical skill gaps, or effectively measure the impact of HR initiatives, leading to suboptimal talent management strategies. The lack of a unified, intelligent view of workforce data hinders the organization's capacity to make proactive, data-driven decisions regarding its talent, ultimately impacting productivity, engagement, and competitive advantage in a talent-driven market.

Implementation	Challenges
* Define key HR metrics & analytical needs * Select a cloud-based HR analytics platform * Ensure data integration from all HR sources * Develop dashboards & reporting capabilities * Train HR & business leaders on data interpretation * Secure data privacy & ethical use of insights	* Ensuring data quality & consistency across sources * Complex integration with diverse HR/business systems * Skillset gap in HR analytics and data science * Overcoming resistance to data-driven HR decisions * Managing data privacy and ethical considerations * Measuring clear ROI on HR analytics investments

Products	Platform Cost	Project Cost	Duration
Visier	$15-70K	$80-350K	3-10 m
Workday Prism Analytics / People Analytics	$20-90K	$100-450K	4-12 m
SAP SuccessFactors People Analytics	$15-60K	$70-300K	3-9 m
Qualtrics EmployeeXM	$10-50K	$60-250K	3-9 m
Tableau / Power BI (for HR data)	$1-10K	$5-50K	1-3 m
ADP DataCloud	$5-20K	$20-100K	2-6 m

Collaboration Suites are integrated collections of software applications that provide a unified platform for teams to communicate, share information, co-create documents, and manage projects in real-time, regardless of their geographical location. These suites typically include tools for instant messaging, video conferencing, shared drives, online document editing, and project task management. By fostering seamless communication, streamlining workflows, and breaking down information silos, collaboration suites significantly enhance team productivity, accelerate decision-making, and build a more connected and engaged workforce, which is crucial for supporting remote, hybrid, and agile work models and driving innovation in a dynamic business environment.

Without robust Collaboration Suites, organizations often suffer from fragmented communication, siloed information, and inefficient workflows, particularly among distributed or cross-functional teams. This leads to missed deadlines, redundant efforts, and a pervasive sense of disconnect as employees struggle to share files, communicate effectively, or track project progress across disparate tools and channels. The absence of a unified platform for teamwork results in lower productivity, hindered innovation due to poor knowledge sharing, and a less engaged workforce, ultimately impacting the organization's ability to operate efficiently, adapt quickly, and leverage its collective intelligence for competitive advantage.

Implementation	Challenges
* Assess communication & workflow needs * Select comprehensive collaboration platform * Plan phased rollout with extensive training * Establish clear usage guidelines & best practices * Integrate with existing business applications * Monitor adoption rates & gather user feedback	* User adoption hurdles & resistance to change * Managing information overload and notification fatigue * Complex integration with diverse legacy systems * Ensuring data security & access control * Sustaining consistent usage and engagement * Measuring tangible impact on team productivity

Products	Platform Cost	Project Cost	Duration
Microsoft Teams (with 365)	$0-30K	$30-150K	2-6 m
Google Workspace	$5-30K	$30-150K	2-7 m
Slack Enterprise Grid	$10-40K	$40-200K	2-7 m
Zoom Workspace (incl. Phone, Chat)	$5-25K	$20-100K	2-5 m
Atlassian Confluence/Jira/Trello	$1-20K	$10-100K	3-9 m
Cisco Webex Suite	$5-20K	$25-100K	2-5 m

Mobile Workforce Management refers to the systems and strategies designed to optimize the operations of employees who work remotely, in the field, or are highly mobile, relying heavily on mobile devices for their daily tasks. These solutions typically include features for scheduling, dispatching, route optimization, real-time communication, data collection, and field service management via mobile apps. By providing mobile employees with the tools they need to perform efficiently and securely from anywhere, Mobile Workforce Management enhances productivity, improves customer service, reduces operational costs, and increases data accuracy, enabling organizations to effectively manage and support a distributed and agile workforce, particularly in industries like field service, logistics, and healthcare.

Without effective Mobile Workforce Management solutions, organizations with a significant mobile or field-based workforce often struggle with inefficient scheduling, poor communication, delayed data capture, and a lack of real-time visibility into operations. This leads to suboptimal resource utilization, increased operational costs, higher rates of human error, and diminished customer satisfaction due to reactive service delivery. The absence of integrated mobile management tools hinders the productivity and effectiveness of critical frontline employees, limits the organization's ability to respond quickly to dynamic field conditions, and ultimately impacts revenue generation and competitive advantage in service-oriented industries.

Implementation	Challenges
* Define needs of mobile/field teams * Select appropriate mobile workforce platform * Ensure device compatibility & network coverage * Train mobile employees on new tools * Integrate with existing ERP, CRM, scheduling * Monitor performance & optimize routes/tasks	* Ensuring reliable mobile network connectivity * Managing diverse mobile devices and OS versions * Addressing data security on mobile endpoints * Overcoming employee resistance to tracking/monitoring * Integrating with complex legacy back-end systems * Measuring clear ROI on efficiency & service improvements

Products	Platform Cost	Project Cost	Duration
Salesforce Field Service	$20-90K	$100-450K	4-12 m
ServiceMax (GE Digital)	$15-70K	$80-350K	3-10 m
Oracle Field Service Cloud	$15-60K	$70-300K	3-9 m
Microsoft Dynamics 365 Field Service	$10-50K	$60-250K	3-9 m
Verizon Connect (Fleet & Field Mgmt)	$5-30K	$20-100K	2-6 m
ArcGIS Workforce (ESRI)	$2-15K	$10-70K	1-4 m

Remote Access Solutions enable employees to securely access an organization's internal networks, applications, and data from any remote location, typically over the internet. These solutions encompass various technologies such as Virtual Private Networks (VPNs), Secure Access Service Edge (SASE), Zero Trust Network Access (ZTNA), and remote desktop protocols. By establishing encrypted connections and implementing robust authentication mechanisms, remote access solutions ensure that distributed workforces can maintain productivity and connectivity while safeguarding sensitive corporate information. This capability is fundamental for supporting remote and hybrid work models, ensuring business continuity during disruptions, and maintaining data integrity and security in an increasingly decentralized operational landscape.

Without robust Remote Access Solutions, organizations face severe limitations in supporting a flexible, remote, or hybrid workforce, while exposing sensitive corporate data to significant security risks. Employees working outside the traditional office environment would either lack access to critical internal systems or be forced to use insecure public networks, making the organization highly vulnerable to cyberattacks, data breaches, and compliance violations. The absence of secure remote connectivity hinders productivity, compromises data integrity, and significantly restricts the organization's ability to adapt to modern work models, ultimately jeopardizing business continuity and exposing it to severe financial and reputational damage.

Implementation	Challenges
* Assess remote access needs & security posture * Select appropriate VPN/SASE/ZTNA solution * Deploy & configure client software/gateways * Implement strong authentication (MFA) * Establish clear access policies & user roles * Monitor network traffic & security logs continuously	* Ensuring sufficient bandwidth for remote users * Complex configuration for diverse user groups * Managing large number of remote connections securely * User experience impact (latency, connection drops) * Protecting against advanced persistent threats (APTs) * Integrating with existing identity management systems

Products	Platform Cost	Project Cost	Duration
Cisco AnyConnect / Secure Firewall VPN	$10-60K	$50-250K	3-9 m
Palo Alto Networks GlobalProtect	$15-70K	$70-300K	3-10 m
Fortinet FortiClient VPN	$5-30K	$20-100K	2-6 m
Zscaler (SASE/ZIA/ZPA)	$20-90K	$100-450K	4-12 m
Check Point Remote Access VPN	$10-50K	$40-200K	2-8 m
OpenVPN Access Server (Commercial)	$2-15K	$10-70K	1-4 m

Ai6x6™ Transformation Blueprint

People 'Intelligence'

"We must understand people as data—without reducing them to it." — Laszlo Bock , Former SVP of People Operations at Google; pioneered people analytics as a strategic talent tool.

People intelligence systems are mirrors of cultural ambition. Skill gap analytics, sentiment monitoring, and learning path tools reveal more than metrics—they uncover organizational character. Mature transformation uses data not for surveillance but for stewardship—curating environments where insight improves inclusion, development, and intent. Employees aren't just quantified—they're understood. True digital maturity doesn't just measure retention—it reads engagement as a strategic pulse. Workforce intelligence transcends performance—it activates humanity at scale. When systems respect the complexity of people, transformation begins to behave with empathy, not just efficiency.

The "People Intelligence" chapter is crucial for workforce transformation's digital maturity as it shifts HR from administrative tasks to strategic insight generation, truly empowering people. A digitally mature organization leverages advanced data analytics, including attrition prediction models, skill gap analysis, and engagement pulse monitoring, to deeply understand its talent. This intelligence allows leaders to predict future needs, proactively address challenges, and make data-driven decisions about talent development, learning paths, and organizational design. This ensures the workforce remains agile, highly skilled, and strategically aligned in a rapidly changing digital landscape.

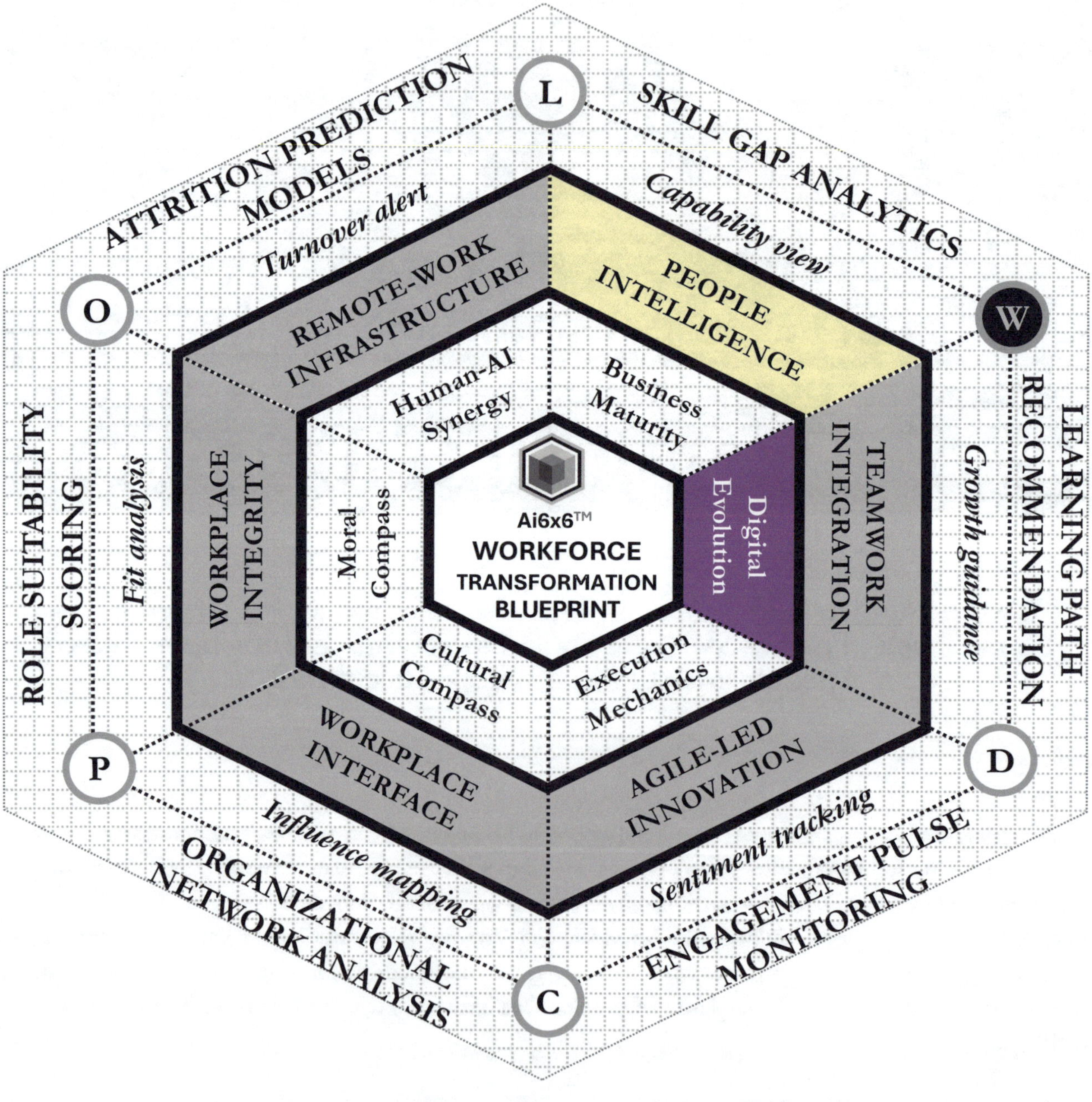

Ai6x6 DIGITAL EVOLUTION FRAMEWORK
FOR **WORKFORCE LIFECYCLE | INTELLIGENCE**

1. ATTRITION PREDICTION MODELS

Attrition Prediction Models utilize advanced analytics and machine learning algorithms to identify employees who are at a high risk of leaving the organization, by analyzing various internal and external data points such as performance reviews, compensation, tenure, engagement scores, and market trends. These models provide proactive insights, allowing HR and leadership to understand the underlying drivers of employee turnover and intervene with targeted retention strategies, such as personalized development plans, compensation adjustments, or improved work-life balance initiatives. By anticipating potential departures, organizations can significantly reduce recruitment costs, preserve institutional knowledge, maintain team stability, and foster a more engaged and committed workforce, thereby enhancing overall talent management effectiveness.

Without implementing Attrition Prediction Models, organizations often operate reactively to employee turnover, only addressing departures after they occur, which leads to significant disruptions in productivity, increased recruitment and onboarding costs, and a loss of valuable institutional knowledge. This lack of foresight means organizations cannot identify at risk employees proactively or understand the root causes of attrition, making it difficult to implement effective retention strategies. The absence of predictive capabilities results in a perpetually reactive talent management approach, higher rates of preventable turnover, and a diminished ability to maintain a stable, experienced workforce, ultimately impacting business continuity and competitive advantage in a talent-driven market.

Implementation	Challenges
* Collect comprehensive employee data (HRIS) * Select ML model & analytics platform * Train model with historical attrition data * Validate model accuracy & refine variables * Integrate predictions into HR workflows * Implement targeted retention strategies	* Ensuring sufficient data quality & completeness * Addressing data privacy and ethical concerns * Mitigating algorithmic bias in predictions * Overcoming resistance to data-driven HR insights * Translating predictions into actionable interventions * Measuring direct impact on retention rates

Products	Platform Cost	Project Cost	Duration
Visier People Analytics	$15-70K	$80-350K	3-10 m
Workday People Analytics	$20-90K	$100-450K	4-12 m
SAP SuccessFactors People Analytics	$15-60K	$70-300K	3-9 m
IBM Watson Talent	$20-80K	$100-400K	4-12 m
Tableau / Power BI (with ML capabilities)	$1-10K	$5-50K	1-3 m
Custom Python/R ML Development	N/A	$50-250K	3-9 m

Skill Gap Analytics involves leveraging data and analytical tools to systematically identify discrepancies between the skills an organization currently possesses within its workforce and the skills it will require to achieve future strategic objectives or adapt to market changes. This process typically analyzes data from performance reviews, learning management systems, job descriptions, market trends, and internal project requirements. By providing clear insights into critical skill shortages and surpluses, Skill Gap Analytics enables organizations to proactively develop targeted upskilling and reskilling programs, optimize talent acquisition strategies, and redeploy internal talent effectively, ensuring a future-ready workforce and maintaining competitive agility in a rapidly evolving business landscape.

Without robust Skill Gap Analytics, organizations often make reactive and imprecise decisions about talent development and acquisition, leading to critical skill shortages that hinder strategic initiatives and slow down innovation. This lack of data-driven insight means organizations cannot proactively address the evolving demands of the market, resulting in an unprepared workforce and increased reliance on costly external hiring. The absence of systematic skill gap analysis limits the organization's ability to efficiently allocate resources for learning and development, impacts employee career growth, and ultimately impedes its capacity to adapt quickly to new technologies and business models, compromising long-term competitiveness.

Implementation	Challenges
* Define future business & strategic goals * Conduct comprehensive current skill inventory * Identify required future skills & roles * Select skill analytics platform or build models * Develop targeted upskilling/reskilling programs * Monitor skill development & program effectiveness	* Accurately capturing and updating employee skills * Defining future skill needs amidst rapid change * Complex integration with HRIS and LMS data * Overcoming resistance to skill assessments * Translating insights into actionable development plans * Measuring the impact on business agility

Products	Platform Cost	Project Cost	Duration
Eightfold AI (Talent Intelligence)	$25-100K	$150-600K	5-14 m
Workday Skills Cloud / People Analytics	$20-90K	$100-450K	4-12 m
Gloat (Talent Marketplace)	$20-80K	$100-400K	4-12 m
Degreed (Skill Intelligence)	$15-70K	$70-350K	3-10 m
Phenom Talent Experience Platform	$20-90K	$120-500K	4-12 m
Fuel50 (Skills & Career Mobility)	$15-60K	$70-300K	3-9 m

3. LEARNING PATH RECOMMENDATION

Learning Path Recommendation systems leverage artificial intelligence and machine learning to analyze an individual employee's current skills, role requirements, career aspirations, and learning history to suggest personalized and relevant educational content or training modules. These systems often integrate with Learning Experience Platforms (LXPs) or Learning Management Systems (LMS) to curate a dynamic and adaptive learning journey from various internal and external sources. By providing highly individualized guidance for skill development, these recommendations enhance learning engagement, accelerate skill acquisition for critical roles, and empower employees to take ownership of their professional growth, fostering a culture of continuous learning and future-proofing the workforce.

Without implementing Learning Path Recommendation systems, organizations often rely on generic, one-size-fits-all training catalogs or manual guidance from managers, which fails to cater to the diverse learning needs and career aspirations of individual employees. This results in inefficient learning efforts, reduced engagement with training content, and slower upskilling for critical roles, as employees struggle to identify the most relevant development opportunities. The absence of personalized learning pathways leads to wasted training budgets on irrelevant courses, diminished employee satisfaction due to a perceived lack of growth opportunities, and a less adaptable workforce, ultimately hindering the organization's ability to quickly address skill gaps and retain top talent.

Implementation	Challenges
* Collect employee skill & role data * Integrate with LMS/LXP & content library * Configure AI/ML for recommendation algorithms * Curate diverse & high-quality learning resources * Train employees on using the recommendation engine * Monitor engagement & refine recommendations	* Ensuring accuracy & relevance of recommendations * Integrating with disparate learning content sources * Managing data privacy for individual learning profiles * Overcoming "cold start" problem for new users * Sustaining employee engagement with recommendations * Measuring impact on skill acquisition & performance

Products	Platform Cost	Project Cost	Duration
Degreed	$15-70K	$70-350K	3-10 m
Cornerstone OnDemand (LXP)	$10-60K	$60-300K	3-9 m
Workday Learning	$20-80K	$100-400K	4-12 m
EdCast (Cornerstone)	$15-60K	$70-300K	3-9 m
Phenom Learn	$10-50K	$50-250K	3-9 m
Docebo (LXP Module)	$10-40K	$50-200K	2-8 m

Engagement Pulse Monitoring refers to the continuous or frequent collection of real-time employee feedback through short, targeted surveys, rather than infrequent annual assessments. These digital platforms gather data on specific aspects of employee sentiment, well-being, workload, and organizational culture, providing immediate insights into workforce morale and potential issues. By offering rapid feedback loops and actionable analytics, pulse monitoring enables leaders to proactively identify areas of concern, address emerging challenges swiftly, and implement timely interventions that foster a more positive and productive work environment. This continuous listening approach enhances employee engagement, improves retention, and builds a more responsive and adaptable organizational culture.

Without implementing Engagement Pulse Monitoring, organizations often rely on outdated annual surveys or anecdotal evidence, which provides infrequent and often stale insights into employee sentiment, leaving critical issues unaddressed until they escalate. This reactive approach means leaders lack real-time understanding of workforce morale, emerging pain points, or the effectiveness of new initiatives, hindering their ability to make timely interventions. The absence of continuous feedback loops results in disengaged employees, higher attrition due to unaddressed concerns, and a less responsive leadership, ultimately impacting productivity, fostering discontent, and limiting the organization's capacity to build a truly adaptive and supportive workplace culture.

Implementation	Challenges
* Define key engagement metrics for pulses * Select a pulse survey platform * Design short, targeted & frequent surveys * Ensure anonymity & trust in feedback process * Train managers on interpreting & acting on data * Communicate results & actions transparently	* Avoiding "survey fatigue" among employees * Ensuring anonymity and building trust in the platform * Translating insights into tangible actions * Ensuring managers are equipped to handle feedback * Integrating with existing HRIS for employee segmentation * Measuring long-term impact on engagement & retention

Products	Platform Cost	Project Cost	Duration
Culture Amp	$10-50K	$60-250K	3-9 m
Qualtrics EmployeeXM	$15-70K	$70-300K	3-10 m
Glint (LinkedIn)	$10-40K	$50-200K	3-8 m
Peakon (Workday)	$15-60K	$70-280K	3-9 m
Perceptyx	$20-80K	$100-400K	4-12 m
SurveyMonkey Enterprise	$5-20K	$20-100K	2-6 m

Organizational Network Analysis (ONA) is a methodology that maps and measures formal and informal relationships and communication flows within an organization, revealing how work actually gets done. By analyzing digital interactions (e.g., emails, collaboration tools, meetings) or survey-based connections, ONA identifies key influencers, knowledge brokers, siloed teams, and communication bottlenecks. These insights enable leaders to understand collaboration patterns, optimize team structures, enhance knowledge sharing, facilitate change management, and identify emerging leaders. By providing a data-driven understanding of internal networks, ONA fosters more effective communication, improves team cohesion, and strengthens organizational resilience, ultimately enhancing productivity and innovation across the workforce.

Without leveraging Organizational Network Analysis (ONA), organizations often rely on traditional hierarchical charts and anecdotal evidence to understand internal dynamics, which provides an incomplete and often inaccurate picture of how work truly flows and who the real influencers are. This lack of data-driven insight into informal networks leads to inefficient communication pathways, undetected silos, missed opportunities for knowledge transfer, and an inability to proactively identify and support key collaborators. The absence of ONA results in suboptimal team design, slower decision-making due to communication bottlenecks, and a less resilient organization that struggles to adapt quickly to change, ultimately impacting productivity, innovation, and employee engagement.

Implementation	Challenges
* Define clear objectives & scope for ONA * Select ONA tool & data sources (passive/active) * Ensure data privacy & ethical usage guidelines * Analyze network patterns & identify insights * Communicate findings & develop action plans * Monitor network changes over time	* Addressing employee privacy concerns & data security * Gaining trust for data collection & analysis * Interpreting complex network data correctly * Translating insights into actionable organizational changes * Overcoming resistance to transparency in networks * Ensuring consistent data quality from diverse sources

Products	Platform Cost	Project Cost	Duration
OrgMapper	$10-50K	$50-250K	3-9 m
Humanyze	$15-70K	$70-350K	3-10 m
SWOOP Analytics	$5-30K	$20-100K	2-6 m
Microsoft Viva Insights (limited ONA)	$1-10K	$5-50K	1-3 m
TrustSphere	$10-40K	$40-200K	2-7 m
Custom Network Analysis Tools (Python/R)	N/A	$50-250K	3-9 m

6. ROLE SUITABILITY SCORING

Role Suitability Scoring utilizes data analytics and potentially AI/machine learning to assess the fit of internal or external candidates for specific roles within an organization, based on a comprehensive evaluation of their skills, experience, competencies, personality traits, and alignment with company culture. This process moves beyond traditional resume screening by quantitatively evaluating multiple factors against predefined role requirements and success profiles. By providing objective, data-driven insights into candidate fit, role suitability scoring enhances recruitment accuracy, reduces time-to-hire, improves employee retention by ensuring better matches, and ultimately optimizes talent allocation, leading to higher performance and more effective team building across the workforce.

Without implementing Role Suitability Scoring, organizations often rely on subjective assessments, limited resume reviews, or traditional interview processes that may introduce bias and lead to suboptimal hiring or internal placement decisions. This lack of a data-driven approach results in longer time-to-hire, higher rates of mis-hires, increased training costs for ill-fitting candidates, and missed opportunities to leverage internal talent effectively. The absence of objective suitability scoring hinders the organization's ability to consistently place the right people in the right roles, impacts team performance, and ultimately compromises both recruitment efficiency and overall human capital optimization in a competitive talent market.

Implementation	Challenges
* Define clear role requirements & success profiles * Collect relevant candidate data (skills, assessments) * Select or develop a scoring model/platform * Integrate with ATS/HRIS for candidate data * Validate model for fairness & accuracy * Train recruiters & managers on score interpretation	* Ensuring fairness & mitigating algorithmic bias * Accurately capturing and updating skills/competencies * Integrating data from disparate assessment tools * Overcoming skepticism about AI in hiring decisions * Balancing objective scores with human judgment * Measuring direct impact on hiring quality & retention

Products	Platform Cost	Project Cost	Duration
Eightfold AI (Talent Intelligence)	$25-100K	$150-600K	5-14 m
Phenom Talent Experience Platform	$20-90K	$120-500K	4-12 m
HireVue (Assessments)	$10-60K	$60-300K	3-9 m
SHL (Assessments)	$10-50K	$50-250K	3-9 m
Workday Talent Marketplace / Skills Cloud	$20-80K	$100-400K	4-12 m
Pymetrics	$15-70K	$70-350K	3-10 m

CHAPTER D3

BLUEPRINT SERIES

Ai6x6™ Transformation Blueprint

Teamwork 'Integration'

"The real technology is trust. Without it, integrations are just code." — Reid Hoffman , Co-founder of LinkedIn; champion of platforms that embed trust in professional connectivity.

Integrations are not connectors—they are cohesion codified. When HR systems, chat tools, and self-service portals operate in synchrony, employee confidence compounds. "Teamwork Integrations" signals an ecosystem where navigating work isn't a task—it's second nature. Transformation maturity is not shown through sophistication of systems but through their humility—interfaces that disappear, workflows that anticipate, portals that reinforce trust by design. When teams can move across functions and platforms without cognitive friction, they don't just collaborate—they synchronize. Integration becomes a feeling, not a feature—and that feeling is culture at scale.

"Teamwork Integrations" is vital for workforce transformation's digital maturity because it ensures that diverse HR systems and employee touchpoints are interconnected and operate harmoniously, creating a unified experience. A digitally mature organization eliminates silos between HR functions, payroll, and collaborative tools through integrations like HCM-payroll and single sign-on. This provides employees with a seamless, intuitive experience from onboarding through daily self-service. Such deep integration enhances efficiency, reduces friction, and fosters a truly collaborative environment where information flows freely, allowing the workforce to operate as a cohesive and agile unit.

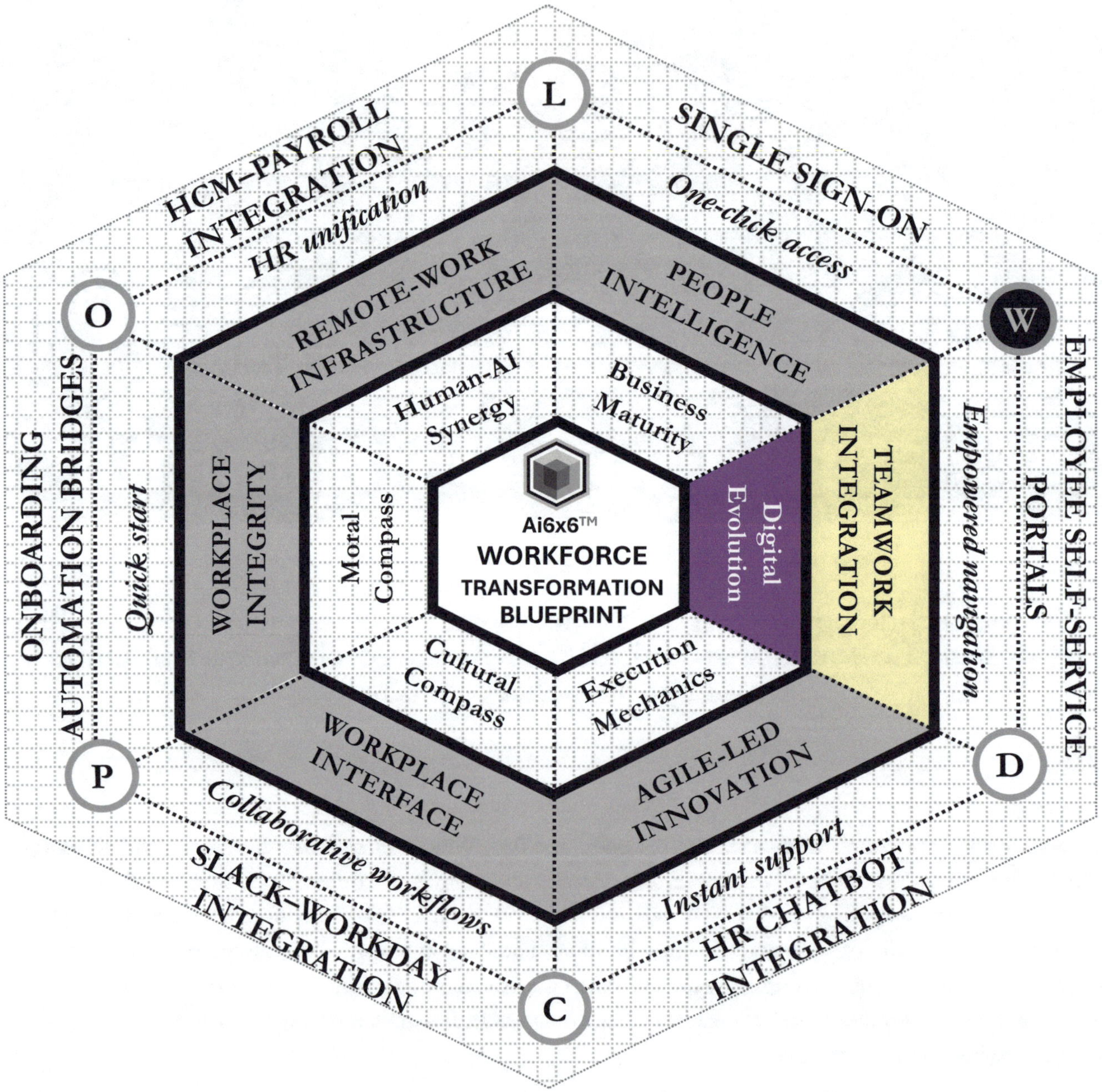

Ai6x6 DIGITAL EVOLUTION FRAMEWORK
FOR **WORKFORCE LIFECYCLE | INTEGRATION**

1. HCM–PAYROLL INTEGRATION

HCM–Payroll Integration refers to the seamless connection and data synchronization between a Human Capital Management (HCM) system and a Payroll system. This integration ensures that all employee data relevant for payroll processing—such as compensation changes, new hires, terminations, benefits deductions, and time-off accruals—is automatically and accurately transferred from the HCM platform to the payroll system without manual intervention. By automating data flow, HCM–Payroll integration significantly reduces administrative errors, streamlines payroll processing, enhances data accuracy and compliance, and frees up HR and finance teams from manual data entry and reconciliation tasks. This foundational integration is critical for operational efficiency, ensuring timely and accurate employee compensation, and providing a unified view of employee records across an organization.

Without robust HCM–Payroll Integration, organizations often face significant manual data entry, prone to human error, and time-consuming reconciliation processes between disparate HR and payroll systems. This leads to frequent payroll inaccuracies, delayed compensation adjustments, increased administrative burdens for HR and finance teams, and potential compliance issues related to employee pay and benefits. The absence of seamless data flow results in operational inefficiencies, higher processing costs, reduced employee trust in payroll accuracy, and a fragmented view of employee data, ultimately hindering strategic workforce planning and consuming valuable resources that could be better allocated to higher-value activities within the organization.

Implementation	Challenges
* Map data fields & workflows between systems * Select integration method (API, middleware) * Develop & test data transformation rules * Implement robust error handling & monitoring * Plan phased rollout & user training * Ensure compliance with payroll regulations	* Complex data mapping and transformation needs * Ensuring data consistency and accuracy * Managing data security & privacy across systems * Overcoming legacy system limitations * Resolving discrepancies and error handling * Gaining alignment across HR, Finance, & IT

Products	Platform Cost	Project Cost	Duration
Workday HCM Integration	$20-80K	$100-400K	4-12 m
SAP SuccessFactors Integration Suite	$15-70K	$80-350K	3-10 m
Oracle Cloud HCM Integrations	$15-60K	$70-300K	3-9 m
Dell Boomi (Integration Platform)	$10-50K	$50-250K	3-9 m
MuleSoft Anypoint Platform	$20-90K	$120-500K	4-12 m
ADP Integration Services	$5-30K	$20-100K	2-6 m

Single Sign-On (SSO) is an authentication scheme that allows a user to log in with a single set of credentials to multiple independent software systems, applications, and services within an organization, without requiring separate logins for each. This technology centralizes user authentication and authorization, eliminating the need for employees to remember numerous usernames and passwords. By streamlining the login process, SSO significantly enhances user experience, improves security by reducing password-related vulnerabilities (like phishing), and increases employee productivity by reducing time spent on authentication. It is a fundamental component for managing access across a diverse ecosystem of digital tools, creating a seamless and secure digital workspace for the modern workforce.

Without implementing Single Sign-On (SSO), organizations typically force employees to manage numerous distinct usernames and passwords for various applications and services, leading to significant frustration, wasted time, and increased security risks due to password reuse or sticky-note reminders. This lack of centralized access management results in frequent lockouts, higher IT helpdesk calls for password resets, and a cumbersome digital experience that hinders productivity. The absence of SSO compromises overall security by making phishing attacks more likely and complicates compliance efforts, ultimately impeding the adoption of new digital tools and creating unnecessary friction for the workforce in their daily tasks.

Implementation	Challenges
* Inventory all applications & user directories * Select an SSO provider & identity platform * Integrate applications with SSO (SAML, OAuth) * Plan user migration & communication * Implement Multi-Factor Authentication (MFA) * Test thoroughly & monitor performance	* Complex integration with legacy applications * Managing diverse authentication protocols * Ensuring high availability & disaster recovery * User management & provisioning complexities * Overcoming internal IT resistance to change * Securing the identity provider itself

Products	Platform Cost	Project Cost	Duration
Okta Workforce Identity	$10-60K	$50-250K	3-9 m
Azure Active Directory (Azure AD)	$5-40K	$30-200K	2-7 m
Ping Identity	$15-70K	$70-350K	3-10 m
OneLogin	$5-30K	$20-100K	2-6 m
Auth0 (Okta)	$5-25K	$20-80K	2-5 m
Duo Security (Cisco)	$2-15K	$10-50K	1-3 m

3. EMPLOYEE SELF-SERVICE PORTALS

Employee Self-Service Portals are centralized online platforms that empower employees to independently access and manage their own HR, payroll, benefits, and personal information, as well as submit requests, without direct intervention from HR or administrative staff. These intuitive portals streamline a wide array of administrative tasks, from updating personal details and checking pay stubs to requesting time off and enrolling in benefits programs. By providing instant, 24/7 access to critical information and enabling self-management of routine inquiries, these portals significantly reduce administrative burdens on HR, improve data accuracy, enhance employee satisfaction through convenience, and free up HR teams to focus on more strategic, value-added initiatives for the workforce.

Without robust Employee Self-Service Portals, organizations typically burden their HR departments with a constant influx of routine inquiries and administrative tasks, diverting valuable resources from strategic human capital initiatives. This reliance on manual processes for common employee requests leads to significant delays, increased operational costs, higher potential for human error in data entry, and considerable frustration for employees who must wait for assistance with simple tasks. The absence of self-service capabilities results in a less efficient HR function, diminished employee satisfaction due to cumbersome processes, and a pervasive drain on productivity, ultimately hindering the organization's ability to scale operations and support a modern, agile workforce effectively.

Implementation	Challenges
* Identify common employee inquiries & tasks * Select a comprehensive self-service platform * Integrate with existing HRIS, payroll, & benefits * Design intuitive user interface & navigation * Plan thorough employee training & communication * Provide continuous support & content updates	* Ensuring data security and privacy of sensitive info * Complex integration with diverse legacy HR systems * Driving high user adoption and consistent usage * Maintaining up-to-date and accurate information * Balancing customization with ease of maintenance * Managing employee expectations for instant solutions

Products	Platform Cost	Project Cost	Duration
Workday Employee Self-Service	$20-80K	$100-400K	4-12 m
ServiceNow HR Service Delivery	$20-90K	$100-450K	4-12 m
SAP SuccessFactors Employee Central	$15-70K	$80-350K	3-10 m
ADP Workforce Now (Self-Service)	$5-30K	$20-100K	2-6 m
Oracle Cloud HCM (Self-Service)	$15-60K	$70-300K	3-9 m
PeopleSoft HCM (Oracle)	$10-50K	$50-250K	3-9 m

4. HR CHATBOT INTEGRATION

HR Chatbot Integration involves deploying AI-powered conversational agents within an organization's HR ecosystem to provide instant, 24/7 support for common employee inquiries and HR-related tasks. These chatbots can answer frequently asked questions about policies, benefits, payroll, and time off, guide employees through self-service processes, or escalate complex queries to human HR representatives. By automating routine interactions, HR chatbots significantly reduce the burden on HR staff, improve response times for employees, enhance the overall employee experience through immediate assistance, and allow HR teams to focus on more strategic, value-added initiatives. This intelligent automation is critical for scalable, efficient, and user-centric HR service delivery.

Without HR Chatbot Integration, HR departments often find themselves overwhelmed by a high volume of repetitive inquiries, leading to delayed responses, increased administrative overhead, and frustrated employees waiting for answers to simple questions. This reliance on human intervention for routine queries diverts HR's attention from strategic initiatives and can lead to inconsistent information dissemination. The absence of an always-on, intelligent self-service option for HR support results in lower employee satisfaction due to slow service, inefficient HR operations, and a diminished capacity for HR to scale its support effectively, ultimately hindering overall organizational productivity and employee well-being.

Implementation	Challenges
* Identify common HR FAQs & user queries * Select AI chatbot platform & natural language processing * Train chatbot with HR knowledge base & policies * Integrate with HRIS & employee portals * Plan escalation paths to human HR support * Monitor performance & continuously optimize responses	* Ensuring accuracy & context of chatbot responses * Managing complex or nuanced HR inquiries * Overcoming employee preference for human interaction * Integration complexity with diverse HR systems * Maintaining data privacy for sensitive HR queries * Measuring clear ROI on reduced HR workload

Products	Platform Cost	Project Cost	Duration
ServiceNow HR Service Delivery (Virtual Agent)	$20-90K	$100-450K	4-12 m
Workday Extend (with AI/Bots)	$15-70K	$80-350K	3-10 m
Zoho Desk / Zoho SalesIQ (with AI)	$1-10K	$5-50K	1-3 m
IBM Watson Assistant	$5-30K	$30-150K	2-7 m
Kore.ai	$10-50K	$60-250K	3-9 m
Custom Chatbot Development (e.g., MS Bot Framework)	N/A	$50-300K	3-9 m

Slack–Workday Integration connects the widely used communication platform Slack with the comprehensive Human Capital Management (HCM) system Workday, allowing employees to access and perform various HR-related tasks directly within their Slack workspace. This integration enables functionalities such as requesting time off, checking leave balances, submitting expenses, accessing pay slips, or updating personal information through simple Slack commands or conversational bots. By embedding HR functionalities directly into the daily workflow of collaboration, this integration significantly enhances employee experience, reduces friction in accessing HR services, streamlines common HR tasks, and boosts productivity by eliminating the need to switch between applications, fostering a more intuitive and efficient digital employee experience.

Without a Slack–Workday Integration, employees often face friction and inefficiencies when performing routine HR tasks, requiring them to navigate away from their primary communication and collaboration tools to log into a separate Workday system. This leads to reduced adoption of self-service HR functionalities, increased time spent on administrative tasks, and a fragmented digital employee experience. The absence of seamless integration results in a less responsive HR service delivery model, diminished employee satisfaction due to cumbersome processes, and a missed opportunity to embed HR processes intuitively into daily workflows, ultimately impacting productivity and the overall digital employee experience.

Implementation	Challenges
* Define specific HR workflows for Slack * Configure Workday & Slack APIs/connectors * Develop custom Slack commands/bots * Implement robust security & authentication * Train employees on integration functionalities * Monitor usage & gather feedback for optimization	* Managing data security and privacy for HR data * Ensuring seamless authentication between platforms * Customizing for complex HR workflows * Overcoming resistance to new interaction methods * Maintaining consistency across various integrations * Handling error messages & unexpected inputs from users

Products	Platform Cost	Project Cost	Duration
Workday Extend Platform	$15-70K	$80-350K	3-10 m
Workday Integration Cloud	$10-50K	$50-250K	3-9 m
Slack Enterprise Grid (API Access)	$10-40K	$40-200K	2-7 m
Custom API Development	N/A	$50-200K	3-8 m
Integration Platform as a Service (iPaaS)	$5-30K	$20-100K	2-6 m
HR Workflow Automation Tools (e.g., Zapier)	$1-10K	$5-30K	1-2 m

Onboarding Automation Bridges are integrated technology solutions that streamline and automate the complex, multi-departmental processes involved in bringing a new employee into an organization, from offer acceptance to their first day and beyond. These "bridges" connect disparate systems like HRIS, IT provisioning, payroll, learning management, and facilities management to orchestrate tasks such as background checks, document signing, system access setup, equipment provisioning, and initial training assignments. By automating these critical steps, Onboarding Automation Bridges significantly reduce manual errors, accelerate time-to-productivity for new hires, enhance the new employee experience, and free up HR and IT teams, fostering a more efficient, compliant, and welcoming start for every new team member.

Without Onboarding Automation Bridges, organizations often subject new hires to a fragmented, manual, and paper-intensive onboarding experience, leading to delays in system access, missing equipment, and a generally frustrating first impression. This lack of automation results in significant administrative burdens for HR, IT, and hiring managers, increased time-to-productivity for new employees, and a higher risk of compliance errors due to uncoordinated processes. The absence of seamless automation across onboarding tasks diminishes the crucial initial employee experience, impacts early engagement and retention, and ultimately hinders the organization's ability to efficiently integrate new talent into the workforce.

Implementation	Challenges
* Map full onboarding journey & tasks * Identify all systems involved (HRIS, IT, LMS) * Select automation platform/middleware * Configure automated workflows & triggers * Integrate APIs between all connected systems * Pilot, train users & refine workflows	* Complex integration with diverse legacy systems * Ensuring data consistency across platforms * Managing cross-departmental coordination * Customizing workflows for different roles/departments * Balancing automation with human touchpoints * Data security & privacy for new hire information

Products	Platform Cost	Project Cost	Duration
Workday Onboarding	$20-80K	$100-400K	4-12 m
ServiceNow HR Service Delivery	$20-90K	$100-450K	4-12 m
SAP SuccessFactors Onboarding	$15-70K	$80-350K	3-10 m
UKG Pro Onboarding	$10-50K	$60-250K	3-9 m
BambooHR (Onboarding Module)	$1-10K	$5-50K	1-3 m
Integration Platform as a Service (iPaaS)	$5-30K	$20-100K	2-6 m

CHAPTER
D4

BLUEPRINT SERIES

Ai6x6™ Transformation Blueprint

Agile Led 'Innovation'

"Learning is no longer periodic—it's perpetual." — Sal Khan , Founder of Khan Academy; global pioneer in continuous, democratized digital learning.

Agile led Innovation transform development from occasional to continuous. VR training, AI coaching bots, and gamified learning platforms convert workforce agility into enterprise acceleration. Mature transformation isn't about classroom replacement—it's about intelligence activation. When mentorship becomes matchable, microlearning becomes habitual, and innovation incubators become internal, learning is no longer a benefit—it's architecture. These studios don't just upskill—they elevate capability through immersive relevance. Talent strategy stops reacting to market changes—it starts rehearsing them. Education becomes infrastructure, and every employee becomes an innovator-in-residence.

The "Agile led Innovation" chapter is paramount for workforce transformation's digital maturity as it focuses on cultivating a culture of continuous learning, adaptation, and growth within the workforce. A digitally mature organization actively invests in innovative learning platforms such as virtual reality training, gamified learning, and AI coaching bots. This empowers employees to acquire new skills rapidly, embrace change, and contribute to organizational innovation. This commitment to agile talent development ensures the workforce remains future-ready, capable of driving innovation, and resilient in the face of evolving technological and market demands.

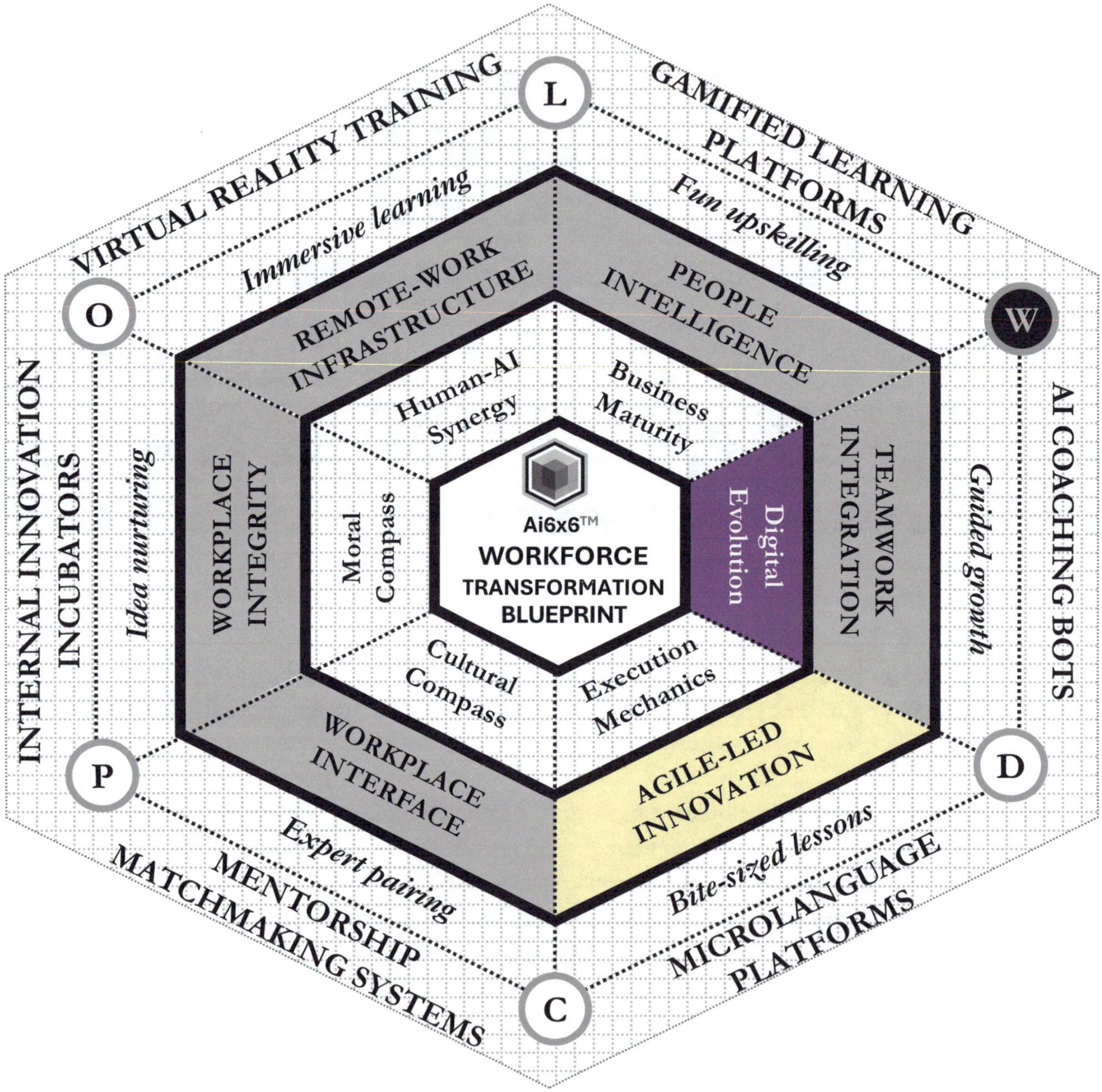

Ai6x6 DIGITAL EVOLUTION FRAMEWORK
FOR **WORKFORCE LIFECYCLE | INNOVATION**

Virtual Reality (VR) Training immerses employees in simulated, interactive 3D environments that closely mimic real-world scenarios, allowing them to practice complex skills, experience high-risk situations safely, and receive immediate feedback. Unlike traditional training, VR provides a hands-on, experiential learning approach that significantly enhances knowledge retention, accelerates skill acquisition, and improves performance transfer to actual job tasks, particularly for technical, operational, and customer service roles. By creating highly engaging and repeatable learning experiences, VR training reduces the need for expensive physical resources, minimizes safety risks, and prepares the workforce more effectively for demanding and dynamic work environments, fostering greater confidence and competence.

Without implementing Virtual Reality Training, organizations often rely on less effective, traditional training methods that may lack the realism, engagement, or safety necessary for truly developing critical skills, especially for complex or high-risk tasks. This can lead to longer training cycles, lower retention of learned material, and higher rates of errors when employees transition to real-world operations, potentially increasing costs and safety incidents. The absence of immersive, experiential learning hinders the rapid development of a highly skilled and confident workforce, slows the adoption of new procedures, and limits the organization's capacity to prepare its talent for evolving challenges, ultimately impacting operational efficiency and safety.

Implementation	Challenges
* Identify specific training needs for VR * Select appropriate VR hardware & software * Develop high-fidelity immersive content/simulations * Plan integration with existing learning platforms * Pilot with target groups & gather user feedback * Provide technical support & content updates	* High initial investment in hardware and content development * Significant technical complexity and specialized skills needed * Addressing user comfort and potential for motion sickness * Ensuring content relevance and pedagogical effectiveness * Scalability for large-scale workforce deployment * Measuring clear ROI on immersive training programs

Products	Platform Cost	Project Cost	Duration
Unity 3D / Unreal Engine (Development)	$0-20K	$100-500K	4-12 m
Oculus for Business (Hardware/SDK)	$1-5K	$30-150K	2-7 m
Strivr (VR Training Platform)	$20-80K	$100-400K	4-12 m
Talespin (XR Training)	$15-70K	$80-350K	3-10 m
Bodyswaps (Soft Skills VR)	$5-30K	$40-200K	2-8 m
HTC VIVE Focus (Hardware/SDK)	$1-6K	$40-200K	2-7 m

Gamified Learning Platforms integrate game-design elements and principles into educational content and training modules to enhance learner engagement, motivation, and knowledge retention. These platforms incorporate features like points, badges, leaderboards, levels, challenges, and rewards to transform traditional learning into an interactive and enjoyable experience. By tapping into intrinsic human desires for achievement, competition, and recognition, gamification significantly boosts participation in training programs, accelerates skill acquisition, and fosters a more positive attitude towards continuous learning. This approach makes professional development more stimulating, drives sustained engagement, and cultivates a workforce that is more eager to acquire new skills and adapt to evolving demands.

Without leveraging Gamified Learning Platforms, organizations often struggle to capture and sustain employee interest in traditional training programs, leading to lower completion rates, reduced knowledge retention, and a less engaged workforce. Relying on passive or unstimulating learning methods results in slower skill development, a reactive rather than proactive approach to professional growth, and a missed opportunity to foster a dynamic and motivated workforce. The absence of gamification can lead to a sense of monotony in routine tasks, limited participation in improvement initiatives, and ultimately hinders the organization's ability to cultivate a highly adaptable, continuously learning, and intrinsically motivated workforce capable of thriving in a rapidly changing business environment.

Implementation	Challenges
* Identify target behaviors/learning outcomes * Design compelling game mechanics & rewards * Select suitable gamification platform/LMS * Integrate with existing learning & HR systems * Communicate rules & track progress transparently * Monitor engagement & iterate on game design	* Ensuring game elements align with learning goals * Overcoming cynicism or resistance to "games" * Sustaining long-term engagement beyond novelty * Balancing individual achievement with teamwork * Complex integration with existing platforms * Measuring clear ROI on behavioral changes

Products	Platform Cost	Project Cost	Duration
Gametize	$1-10K	$5-50K	1-3 m
TalentLMS (Gamification features)	$1-5K	$3-30K	1-2 m
Moodle (with gamification plugins)	$0-5K	$5-25K	1-3 m
360Learning (Collaborative Learning)	$5-20K	$20-100K	2-6 m
Axonify	$10-40K	$40-200K	2-7 m
Centrical	$15-60K	$70-300K	3-9 m

AI Coaching Bots are intelligent, conversational agents powered by artificial intelligence that provide personalized guidance, feedback, and support to employees on various aspects of their professional development, performance, and well-being. These bots leverage natural language processing and machine learning to understand user queries, deliver relevant resources, offer practice scenarios, and track progress, operating 24/7. By providing scalable, accessible, and confidential coaching, AI bots democratize access to personalized development, reinforce learning, improve skill application, and boost employee confidence, acting as a valuable supplement to human coaching or a standalone resource for continuous growth, fostering a culture of self-improvement and agility across the workforce.

Without implementing AI Coaching Bots, organizations often struggle to provide scalable, personalized development and continuous feedback to every employee, relying instead on limited human coaching resources or generic training. This results in inconsistent development support, missed opportunities for immediate feedback, and a slower pace of skill acquisition across the workforce, particularly for common challenges or routine queries. The absence of accessible AI coaching limits individual growth, can lead to frustration when employees lack immediate guidance, and ultimately hinders the organization's ability to foster a culture of continuous improvement and self-directed learning, impacting overall talent development and agility.

Implementation	Challenges
* Define specific coaching objectives & use cases * Select AI chatbot platform capable of coaching * Train bot with relevant knowledge & scenarios * Ensure secure data handling & privacy * Integrate with HRIS/LMS for personalized context * Monitor user engagement & refine coaching logic	* Ensuring the quality & accuracy of coaching advice * Building trust & acceptance among employees * Addressing complex or sensitive human interactions * Limitations in handling nuanced emotional cues * Integrating with diverse internal data sources * Measuring tangible impact on performance & development

Products	Platform Cost	Project Cost	Duration
IBM Watson Assistant (custom coaching)	$5-30K	$30-150K	2-7 m
Kore.ai (HR/Learning Bots)	$10-50K	$60-250K	3-9 m
Microsoft Azure Bot Service (custom dev)	N/A	$50-200K	3-8 m
Humu (Nudge Engine)	$15-60K	$70-300K	3-9 m
BetterUp (Human-AI Hybrid Coaching)	$20-80K	$100-400K	4-12 m
Custom NLP/ML Development	N/A	$100-500K	4-12 m

Microlearning Platforms deliver bite-sized, focused, and easily digestible learning content—such as short videos, infographics, quizzes, or interactive modules—that can be consumed rapidly, typically within minutes, on any device. This approach is designed to fit into employees' busy schedules, address specific knowledge gaps, and provide just-in-time support for performance needs. By offering highly targeted and convenient learning experiences, microlearning platforms significantly improve knowledge retention, boost learner engagement, accelerate skill refreshment, and ensure continuous professional development, empowering a flexible and agile workforce to acquire and apply new information efficiently in a fast-paced environment.

Without implementing Microlearning Platforms, organizations often rely on lengthy, traditional training courses that are difficult to fit into busy schedules, lead to information overload, and result in low engagement and poor knowledge retention. This outdated approach means employees struggle to acquire specific, timely information, leading to skill gaps, slower adoption of new processes, and diminished on-the-job performance. The absence of quick, targeted learning content hinders immediate problem-solving, frustrates time-constrained learners, and ultimately limits the organization's ability to ensure its workforce remains agile and continuously updated with the necessary skills to adapt to rapidly changing business demands.

Implementation	Challenges
* Identify specific skill gaps for micro-content * Select a platform with strong content authoring * Design short, engaging & relevant modules * Integrate with LMS/LXP for tracking * Promote and make content easily discoverable * Monitor completion rates & performance impact	* Developing high-quality, concise content * Ensuring content relevance and freshness * Integrating with diverse existing learning libraries * Measuring the long-term impact on skill application * Preventing content fragmentation or disorganization * Sustaining consistent user engagement over time

Products	Platform Cost	Project Cost	Duration
Axonify	$10-40K	$40-200K	2-7 m
Learning Pool (Learning Suite)	$5-30K	$30-150K	2-6 m
Eduflow	$1-10K	$5-50K	1-3 m
Articulate Rise 360 (Authoring)	$1-5K	$3-20K	1-2 m
TalentLMS (Microlearning features)	$1-5K	$3-30K	1-2 m
Custom Content Development Tools	N/A	$20-100K	2-6 m

5. MENTORSHIP MATCHMAKING SYSTEMS

Mentorship Matchmaking Systems are digital platforms that leverage data and algorithms to intelligently pair mentors with mentees within an organization, based on shared professional goals, desired skills, career stages, interests, and availability. These systems move beyond manual, often biased, pairing methods to create more effective and impactful mentorship relationships. By facilitating strategic connections that foster knowledge transfer, skill development, leadership growth, and career progression, these systems enhance employee engagement, improve retention, democratize access to mentorship opportunities, and cultivate a stronger internal talent pipeline, thereby strengthening organizational culture and long-term workforce capabilities.

Without dedicated Mentorship Matchmaking Systems, organizations often struggle to create effective mentorship programs, relying on informal or manual pairing methods that are often inefficient, limited in scale, and prone to unconscious bias. This results in fewer employees benefiting from mentorship, suboptimal mentor-mentee pairings, and a missed opportunity to leverage internal expertise for talent development and knowledge transfer. The absence of a systematic approach to mentorship limits career growth opportunities for employees, contributes to higher attrition rates among those seeking development, and ultimately hinders the organization's ability to cultivate a strong internal talent pipeline and foster a supportive, learning-oriented culture.

Implementation	Challenges
* Define mentorship program goals & criteria * Select a robust matchmaking platform * Collect comprehensive mentor/mentee profiles * Configure matching algorithms (skills, goals, etc.) * Establish clear program guidelines & support * Monitor program engagement & success metrics	* Ensuring sufficient pool of willing mentors * Accurately capturing skills, goals & preferences * Building trust & encouraging participation * Measuring tangible impact of mentorship on careers * Addressing potential biases in algorithms * Providing ongoing support for mentor-mentee pairs

Products	Platform Cost	Project Cost	Duration
Mentorloop	$5-30K	$20-100K	2-6 m
Insala (Mentoring Software)	$10-50K	$50-250K	3-9 m
MentorCloud	$5-25K	$20-100K	2-5 m
Chronus	$10-40K	$40-200K	2-7 m
Everwise (Presence by Cornerstone)	$15-60K	$70-300K	3-9 m
Workday Extend (Custom Development)	N/A	$50-250K	3-9 m

Internal Innovation Incubators are structured programs or dedicated organizational units designed to foster and accelerate the development of new ideas, products, services, or processes from within the company. These incubators provide employees with resources such as dedicated time, funding, expert mentorship, collaborative workspaces, and access to advanced technologies to explore and prototype novel concepts outside of day-to-day operational constraints. By empowering employees to experiment, take calculated risks, and collaborate on groundbreaking initiatives, internal incubators cultivate a culture of entrepreneurship, stimulate cross-functional problem-solving, identify new revenue streams, and enhance organizational agility, thereby ensuring continuous innovation and long-term competitiveness from within the workforce.

Without establishing Internal Innovation Incubators, organizations often struggle to harness the creative potential of their own workforce, leading to stagnant innovation, missed market opportunities, and a reliance on external sources for disruptive ideas. This absence of a structured environment for internal experimentation means promising employee ideas are often stifled by bureaucratic hurdles, lack of resources, or fear of failure, preventing valuable concepts from being developed. The lack of a dedicated incubator results in a less engaged and motivated workforce, a slower pace of internal transformation, and a diminished capacity to adapt and innovate from within, ultimately impacting the organization's long-term growth and competitive relevance.

Implementation	Challenges
* Define clear goals & scope for innovation * Establish funding model & resource allocation * Recruit diverse talent & ideas internally * Design agile methodology & mentorship structure * Create clear IP & commercialization pathways * Communicate success stories & learnings widely	* Gaining senior leadership buy-in and sustained funding * Balancing innovation efforts with core business priorities * Overcoming internal resistance to new ideas/change * Defining clear metrics for success & ROI * Managing intellectual property and commercialization

Products	Platform Cost	Project Cost	Duration
Innovation Management Platforms (e.g., IdeaScale)	$5-30K	$20-100K	2-6 m
Project Management Software (e.g., Jira, Asana)	$1-10K	$5-30K	1-2 m
Collaboration Tools (e.g., Miro, Mural)	$1-5K	$3-20K	1-2 m
Internal Funding & Resource Allocation Tools	N/A	$10-50K	1-3 m
Custom Software/Platform Development	N/A	$100-500K+	4-12 m
Innovation Consulting Services	N/A	$50-300K	3-9 m

Ai6x6™ Transformation Blueprint

Workplace ‘Interfaces’

“Culture is not a perk. It’s the platform everything runs on.” — Brian Chesky , Airbnb CEO; renowned for embedding emotional experience into every employee touchpoint.

Workplace Interfaces aren’t utilities—they’re emotional gateways. Recognition apps, onboarding portals, and mobile intranets serve more than functionality—they reinforce belonging. In mature transformation, interface is relationship—encoded with tone, ease, and responsiveness. Digital tools shape perception, engagement, and loyalty without explicit instruction. When platforms echo values, employees don’t just interact—they internalize. Culture becomes clickable—not through slogans but through software. Every touchpoint isn’t just used—it’s felt. Belonging is quietly engineered when systems say: we see you, we support you, we built this for you.

"Interface: Touchpoints of Belonging" is critical for workforce transformation's digital maturity because it emphasizes creating intuitive, engaging, and personalized digital experiences for employees that foster connection. A digitally mature organization understands that user-friendly interfaces, from comprehensive employee experience platforms and welcoming onboarding portals to mobile HR apps and social intranets, directly impact employee satisfaction, engagement, and retention. By designing touchpoints that cultivate a strong sense of community and belonging, organizations can build a highly motivated, connected, and productive workforce, essential for thriving in the digital era.

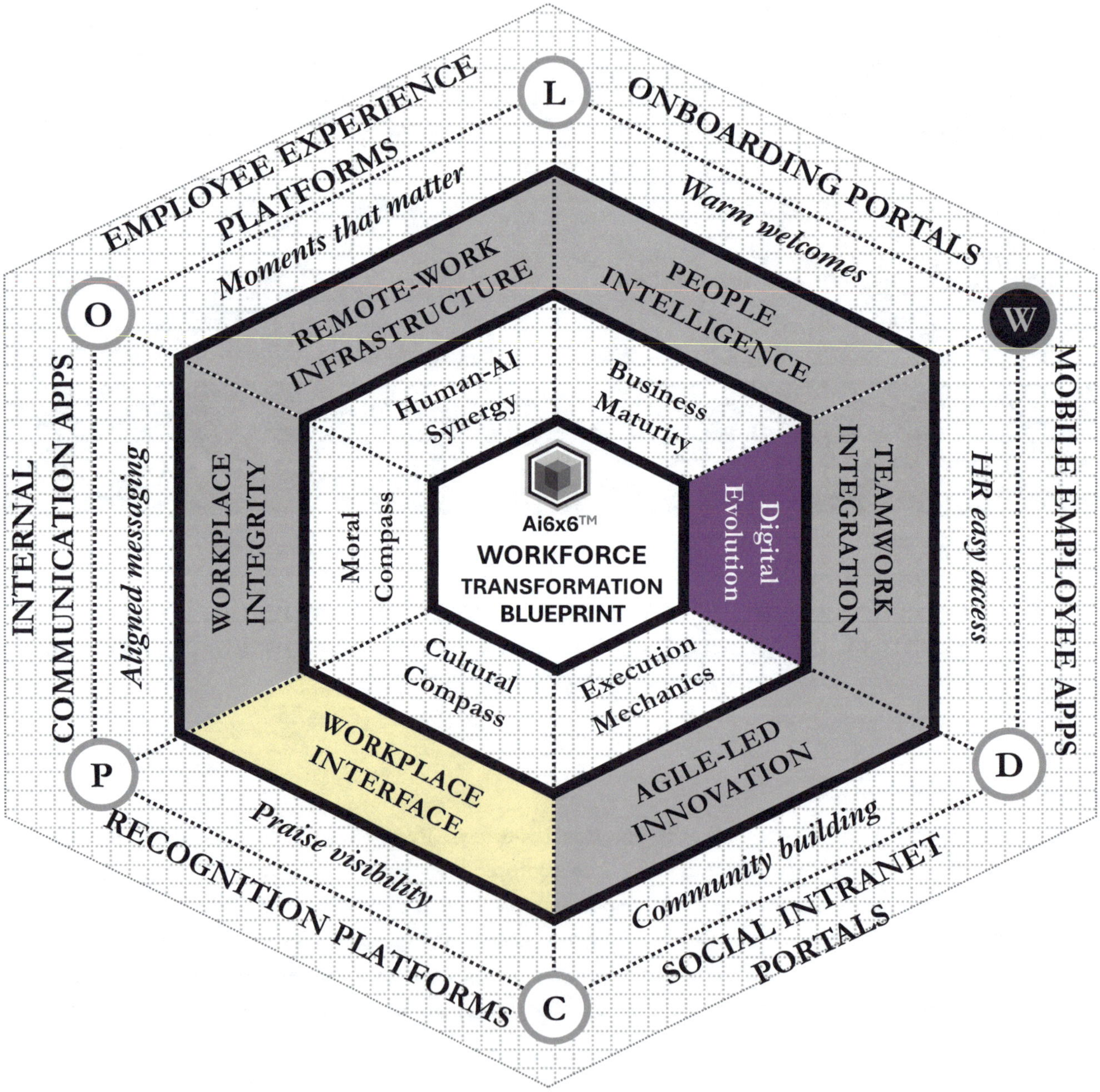

Ai6x6 DIGITAL EVOLUTION FRAMEWORK
FOR **WORKFORCE LIFECYCLE | INTERFACE**

1. EMPLOYEE EXPERIENCE PLATFORMS

Employee Experience Platforms (EXPs) are integrated digital hubs designed to centralize and personalize all aspects of an employee's journey within an organization, from onboarding and daily work to learning and career development. These platforms offer a unified interface that connects various HR systems, communication tools, and productivity applications, providing a seamless and intuitive environment for employees to access information, complete tasks, and engage with their colleagues and the company culture. By reducing digital friction, enhancing communication, and fostering a sense of community, EXPs significantly improve employee satisfaction, boost productivity, enhance retention, and cultivate a positive and inclusive workplace culture, which is essential for attracting and retaining top talent in today's competitive landscape.

Without a dedicated Employee Experience Platform, organizations often present employees with a fragmented and inconsistent digital workplace, forcing them to navigate disparate systems and numerous logins for daily tasks, HR inquiries, and internal communications. This leads to digital fatigue, wasted time searching for information, and a perception of a disjointed organizational culture. The absence of a unified, intuitive interface diminishes overall employee satisfaction, hinders productivity by creating unnecessary friction, and makes it challenging to foster a strong sense of belonging and connection, ultimately impacting engagement, retention, and the organization's ability to create a truly supportive and high-performing workforce environment.

Implementation	Challenges
* Map employee journey touchpoints * Select comprehensive EXP platform * Integrate with HRIS, collaboration, & IT systems * Design intuitive user interface & personalization * Plan phased rollout & robust change management * Continuously gather feedback & iterate features	* Complex integration with diverse legacy systems * Ensuring data privacy & security across platforms * Driving high user adoption and consistent engagement * Tailoring personalization without being intrusive * Measuring tangible impact on employee satisfaction * Maintaining content relevance & platform freshness

Products	Platform Cost	Project Cost	Duration
Microsoft Viva Suite	$10-60K	$50-250K	3-9 m
Workday Journeys / Workday Extend	$20-90K	$100-450K	4-12 m
ServiceNow Employee Workflow	$20-80K	$100-400K	4-12 m
Qualtrics EmployeeXM	$15-70K	$70-350K	3-10 m
Simpplr	$5-30K	$20-100K	2-6 m
Unily	$10-50K	$60-250K	3-9 m

Onboarding Portals are dedicated digital platforms that centralize all information, tasks, and resources relevant to new hires during their pre-boarding and initial employment phases. These portals provide a single point of access for completing paperwork, accessing company policies, viewing training materials, meeting team members virtually, and understanding company culture before their official start date. By providing a structured, engaging, and welcoming experience, onboarding portals significantly reduce administrative burdens for HR, accelerate time-to-productivity for new employees, enhance new hire engagement and retention, and create a strong positive first impression, ensuring a seamless and efficient transition into the organization.

Without a dedicated Onboarding Portal, organizations often provide new hires with a fragmented, manual, and overwhelming experience, requiring them to navigate multiple systems, fill out repetitive paperwork, and search for basic information. This leads to administrative inefficiencies for HR, delayed productivity for new employees due to lack of readiness, and a diminished initial experience that can impact engagement and retention. The absence of a centralized, user-friendly onboarding hub results in higher early attrition rates, increased costs associated with inefficient processes, and a missed opportunity to truly welcome and integrate new talent, ultimately hindering the rapid assimilation of new employees into the company culture and workflow.

Implementation	Challenges
* Map the entire new hire journey * Select or customize onboarding platform * Integrate with HRIS, payroll, & IT provisioning * Design intuitive interface & content (videos, guides) * Plan communication strategy for new hires * Provide continuous support & content updates	* Complex integration with disparate HR/IT systems * Ensuring data security & privacy for new hires * Customizing for different roles/departments * Driving engagement & completion rates for tasks * Balancing automation with a personal touch * Keeping content updated with policy changes

Products	Platform Cost	Project Cost	Duration
Workday Onboarding	$20-80K	$100-400K	4-12 m
SAP SuccessFactors Onboarding	$15-70K	$80-350K	3-10 m
Namely (Onboarding Module)	$5-30K	$20-100K	2-6 m
BambooHR (Onboarding Module)	$1-10K	$5-50K	1-3 m
ServiceNow HR Service Delivery	$20-90K	$100-450K	4-12 m
SilkRoad Technology (Onboarding)	$10-50K	$60-250K	3-9 m

Mobile Employee Apps are dedicated native or web applications designed specifically for smartphones and tablets, providing employees with convenient, on-the-go access to essential HR functionalities, company information, communication tools, and self-service features. These apps typically offer personalized content, push notifications for important updates, and intuitive interfaces for tasks like checking pay stubs, requesting time off, accessing directories, or completing training modules. By delivering critical resources directly to employees' mobile devices, these apps significantly enhance accessibility, improve engagement, streamline administrative tasks, and support flexible work styles, ensuring that all employees, including those without desk jobs, remain connected, informed, and productive from anywhere at any time.

Without dedicated Mobile Employee Apps, organizations often fail to provide their workforce, especially deskless or remote employees, with convenient and immediate access to essential HR services and company information, relying instead on desktop-only portals or manual processes. This creates barriers to accessing critical resources, leads to slower communication, and a perception of being disconnected from the organization. The absence of mobile-first access results in reduced employee satisfaction due to inconvenience, inefficiencies in accessing information, and a diminished ability to engage a diverse and distributed workforce effectively, ultimately impacting productivity, well-being, and the overall sense of belonging for a significant portion of the employee base.

Implementation	Challenges
* Identify essential mobile-first features * Design intuitive UI/UX for mobile devices * Ensure robust security for sensitive data * Integrate with HRIS, payroll, & comms systems * Plan effective app rollout & promotion * Provide continuous updates & support	* Ensuring data security on mobile devices * Managing diverse mobile operating systems (iOS/Android) * Driving high user adoption and consistent usage * Balancing features with app performance * Integrating with diverse back-end systems * Addressing connectivity issues for remote users

Products	Platform Cost	Project Cost	Duration
Workday Mobile	Included with Workday HCM	$50-250K	3-9 m
SAP SuccessFactors Mobile	Included with SF HCM	$40-200K	2-8 m
Staffbase (Employee App)	$10-50K	$60-250K	3-9 m
Simpplr (Mobile App)	$5-30K	$20-100K	2-6 m
Beekeeper	$5-25K	$20-80K	2-5 m
Custom Mobile App Development	N/A	$100-500K+	4-12 m

Social Intranet Portals are dynamic, interactive internal websites that serve as a central hub for company news, resources, and collaborative activities, integrating social networking features like profiles, activity feeds, forums, and content sharing. Unlike traditional static intranets, these platforms foster two-way communication, enable peer-to-peer collaboration, and encourage knowledge sharing across departments, breaking down silos. By creating a vibrant online community, social intranets enhance employee engagement, facilitate internal communication, strengthen company culture, and improve knowledge management, fostering a greater sense of belonging and connecting employees to the organizational mission and to each other, especially beneficial for distributed or hybrid workforces.

Without a modern Social Intranet Portal, organizations often rely on outdated, static intranets or fragmented communication channels that fail to engage employees, hinder effective information dissemination, and impede cross-functional collaboration. This leads to information silos, missed announcements, and a lack of a central online space for employees to connect, share knowledge, and build community. The absence of an interactive social intranet results in lower employee engagement, a less cohesive company culture, inefficient knowledge sharing, and a diminished sense of belonging, ultimately impacting productivity and the organization's ability to foster a connected and informed workforce in today's dynamic work environment.

Implementation	Challenges
* Define content strategy & governance * Select a social intranet platform * Migrate existing content & integrate systems * Design intuitive UI/UX & community features * Plan launch & drive initial user adoption * Continuously update content & encourage engagement	* Ensuring content relevance and freshness * Driving consistent user adoption & participation * Managing information overload & content clutter * Integrating with diverse existing business systems * Securing sensitive information & access control * Measuring impact on communication & culture

Products	Platform Cost	Project Cost	Duration
SharePoint Online (Microsoft 365)	Included with M365	$30-150K	2-7 m
Viva Connections (Microsoft 365)	Included with M365	$20-100K	2-6 m
Unily	$10-50K	$60-250K	3-9 m
Simpplr	$5-30K	$20-100K	2-6 m
LumApps	$10-40K	$50-200K	3-8 m
Happeo	$5-25K	$20-80K	2-5 m

Recognition Platforms are digital systems designed to facilitate and amplify peer-to-peer and manager-to-employee recognition within an organization, allowing for public acknowledgment of achievements, contributions, and adherence to company values. These platforms often integrate with existing HR systems, offer customizable rewards (points, gift cards), and provide analytics on recognition trends. By making appreciation visible, frequent, and timely, recognition platforms significantly boost employee morale, increase engagement, reinforce positive behaviors, foster a culture of gratitude, and reduce turnover. This continuous reinforcement of positive contributions builds a stronger sense of belonging and value among employees, driving higher performance and a more positive workplace environment.

Without dedicated Recognition Platforms, organizations often struggle to create a consistent and impactful culture of appreciation, relying on informal or sporadic recognition methods that may go unnoticed or be perceived as inequitable. This leads to missed opportunities to celebrate achievements, lower employee morale due to a lack of acknowledgment, and a diminished sense of value and belonging among the workforce. The absence of a structured and visible recognition system results in reduced engagement, higher attrition rates among employees feeling unappreciated, and a less positive workplace culture, ultimately impacting productivity and hindering the organization's ability to retain its top talent and foster a truly supportive environment.

Implementation	Challenges
* Define recognition goals & budget * Select a platform with desired features & rewards * Integrate with HRIS & communication tools * Plan communication & launch strategy * Train managers & employees on platform use * Monitor usage, feedback & recognition trends	* Driving consistent user adoption & participation * Ensuring equitable and fair recognition practices * Managing reward budgets & redemption processes * Integrating with diverse HR and payroll systems * Overcoming cynicism or tokenism perceptions * Measuring tangible impact on engagement & retention

Products	Platform Cost	Project Cost	Duration
Workhuman	$10-50K	$60-250K	3-9 m
Bonusly	$5-30K	$20-100K	2-6 m
Kudos	$5-25K	$20-80K	2-5 m
Reward Gateway	$10-40K	$50-200K	3-8 m
Culture Amp (part of Engagement suite)	$10-50K	$60-250K	3-9 m
Achievers	$15-60K	$70-300K	3-9 m

Internal Communication Apps are dedicated mobile, and desktop applications designed to streamline and enhance communication within an organization, providing a centralized platform for company announcements, team updates, emergency alerts, and direct messaging. These apps often include features like personalized news feeds, push notifications, read receipts, and multimedia sharing capabilities, ensuring information reaches the right employees efficiently. By consolidating diverse communication channels into one intuitive interface, these apps improve information flow, increase transparency, foster employee engagement, and build a more connected and informed workforce. This is particularly vital for organizations with distributed, hybrid, or deskless employees, ensuring everyone stays aligned and informed regardless of their location or role.

Without dedicated Internal Communication Apps, organizations often rely on fragmented communication channels like email, outdated intranets, or informal messaging, leading to information overload, missed critical updates, and a pervasive sense of disconnect among employees. This lack of a centralized, efficient platform results in inconsistent messaging, reduced employee engagement due to communication breakdowns, and a slower response time to important organizational changes or emergencies. The absence of a streamlined internal communication system impacts productivity, erodes trust due to lack of transparency, and ultimately hinders the organization's ability to cultivate a cohesive, informed, and truly connected workforce, especially challenging for large or geographically dispersed teams.

Implementation	Challenges
* Define key communication objectives * Select an app with relevant features & scalability * Plan content strategy & governance * Integrate with HRIS, directory, & collaboration tools * Launch with strong promotion & training * Monitor engagement & refine content strategy	* Ensuring consistent content quality & relevance * Driving high user adoption and sustained engagement * Managing information overload & notification fatigue * Addressing diverse communication needs (e.g., deskless) * Integrating with existing legacy communication tools * Measuring impact on employee awareness & alignment

Products	Platform Cost	Project Cost	Duration
Staffbase	$10-50K	$60-250K	3-9 m
Beekeeper	$5-25K	$20-100K	2-6 m
Workvivo (Zoom)	$10-40K	$50-200K	3-8 m
Simpplr (Communication features)	$5-30K	$20-100K	2-6 m
Microsoft Viva Engage (Yammer)	Included with M365	$20-100K	2-6 m
Connecteam	$1-10K	$5-50K	1-3 m

Ai6x6™ Transformation Blueprint

Workforce 'Integrity'

"Privacy isn't dead. It just needs help." — Christopher Soghoian , Cybersecurity expert; global advocate for digital rights and ethical organizational conduct.

Workforce Integrity in workforce systems isn't optional—it's infrastructural. Ethics hotlines, DEI tools, and policy governance software operationalize values before violations occur. Mature transformation weaves protection into experience—not as obstacle, but as assurance. Employees feel safer, clearer, and more respected when digital systems enforce fairness without friction. These frameworks don't just prevent risk—they cultivate accountability. When policies become interfaces, and safety becomes ambient, the workforce operates with conviction, not constraint. Integrity isn't the compliance office—it's the climate of contribution. Transformation gains soul when ethics are lived systemically.

The "Workforce Integrity" chapter is fundamental for workforce transformation's digital maturity as it underscores the importance of ethical conduct, compliance, and trust within the organization. A digitally mature enterprise embeds integrity into its workforce operations through robust systems for compliance management, ethics hotlines, and privacy tools. This commitment ensures a secure, fair, and transparent environment, protecting both the organization and its employees. It builds a foundation of trust that is vital for fostering a positive culture, mitigating legal risks, and ensuring long-term sustainability and reputation in the digital landscape.

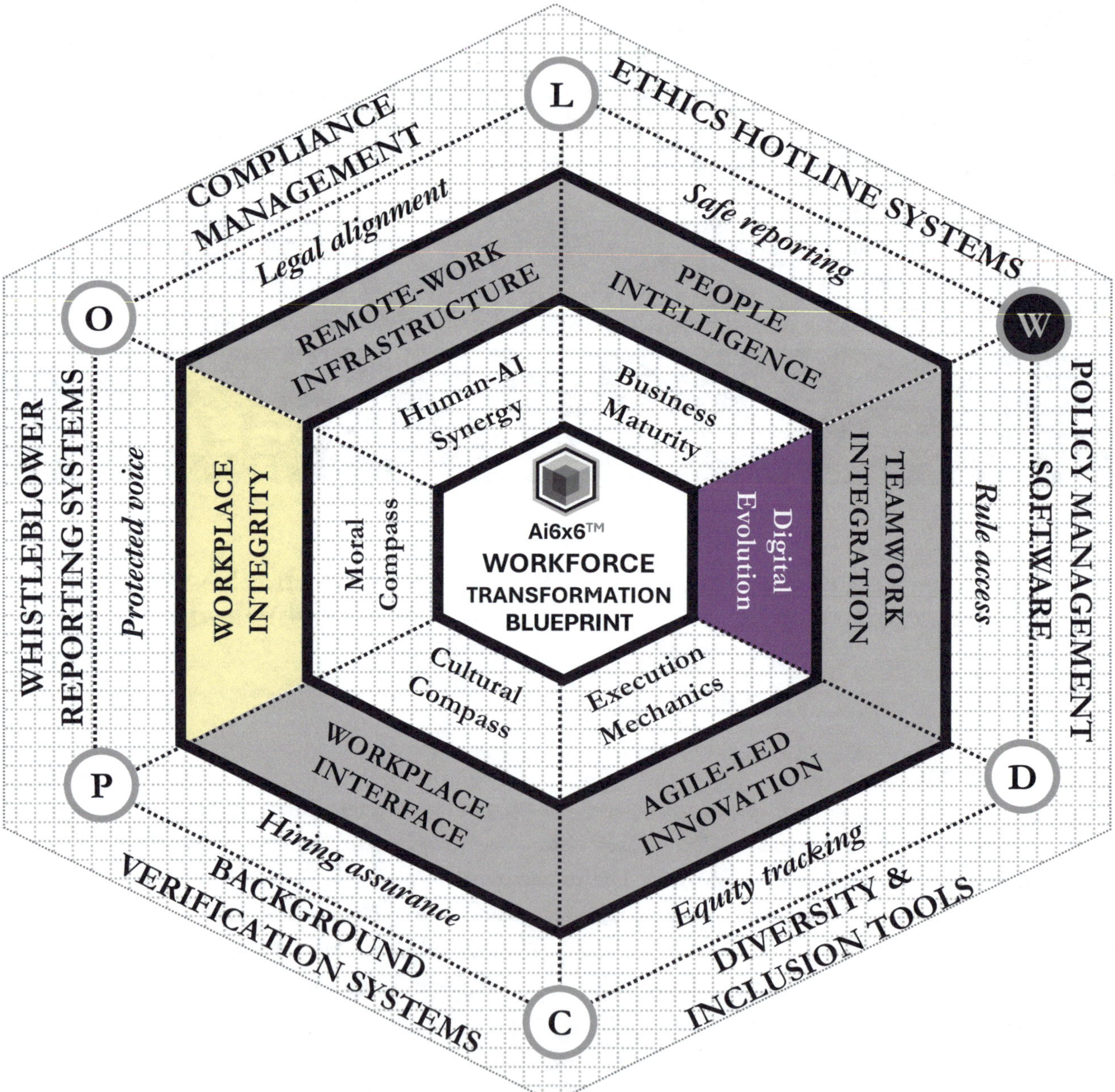

Ai6x6 DIGITAL EVOLUTION FRAMEWORK
FOR **WORKFORCE LIFECYCLE | INTEGRITY**

1. COMPLIANCE MANAGEMENT

Compliance Management in the context of workforce transformation involves establishing systematic processes and utilizing specialized software to ensure an organization and its employees adhere to all relevant laws, regulations, internal policies, and ethical standards related to labor, employment, and data privacy. This includes managing evolving legal requirements, tracking certifications, ensuring policy acknowledgment, and maintaining auditable records across various jurisdictions. By proactively managing compliance, organizations mitigate legal and financial risks, avoid penalties, protect their reputation, and foster a culture of integrity and accountability throughout the workforce, which is crucial for sustainable operations and maintaining stakeholder trust in a highly regulated global environment.

Without robust Compliance Management systems, organizations face significant risks of legal non-compliance, leading to hefty fines, costly litigation, reputational damage, and operational disruptions due to violations of labor laws, industry regulations, or internal policies. Relying on manual or fragmented compliance efforts increases the likelihood of human error, makes auditing challenging, and delays the adaptation to new regulatory requirements. The absence of a systematic approach to workforce compliance results in increased exposure to legal liabilities, a less ethical workplace culture, and a diminished ability to operate transparently, ultimately jeopardizing the organization's long-term viability and public trust.

Implementation	Challenges
* Identify all relevant laws, regs, policies * Select comprehensive compliance management platform * Automate policy distribution & acknowledgment * Implement compliance training & tracking * Establish audit trails & reporting capabilities * Continuously monitor regulatory changes	* Keeping up with constantly evolving regulations * Complexities of multi-jurisdictional compliance * Ensuring employee understanding and adherence * Integrating with diverse HR and legal systems * Data privacy and security concerns * Measuring the effectiveness of compliance programs

Products	Platform Cost	Project Cost	Duration
ServiceNow GRC (HR Compliance)	$20-90K	$100-450K	4-12 m
Workday Compliance Modules	$15-70K	$80-350K	3-10 m
Ethico (Compliance Management)	$10-50K	$60-250K	3-9 m
OneTrust (GRC & Ethics)	$20-100K	$120-500K	4-13 m
LogicManager (Enterprise GRC)	$15-60K	$70-300K	3-9 m
Archer (RSA)	$25-110K	$140-550K	5-15 m

Ethics Hotline Systems provide a secure, confidential, and often anonymous channel through which employees, customers, or third parties can report concerns about unethical behavior, fraud, harassment, discrimination, or compliance violations within an organization. These systems typically offer multiple reporting methods (phone, web, email) and ensure that reports are routed to appropriate personnel for investigation, upholding principles of privacy and non-retaliation. By fostering a culture of transparency and accountability, ethics hotline systems encourage early detection of misconduct, mitigate legal and reputational risks, reinforce organizational values, and build trust among employees, demonstrating a commitment to integrity and responsible corporate governance.

Without accessible Ethics Hotline Systems, organizations often lack a clear, safe, and confidential channel for employees to report serious concerns, leading to an environment where misconduct may go unreported, fester, and escalate into larger issues. This absence of a formal reporting mechanism increases the risk of undetected fraud, harassment, and compliance breaches, potentially resulting in severe legal repercussions, significant financial losses, and irreversible damage to the organization's reputation and employee morale. The lack of a trusted hotline undermines transparency, erodes employee trust, and ultimately hinders the organization's ability to maintain a truly ethical culture and prevent systemic issues before they become public crises.

Implementation	Challenges
* Define clear reporting scope & policy * Select a reputable third-party hotline provider * Establish clear investigation & follow-up processes * Communicate system availability & anonymity * Train relevant personnel (investigators, HR) * Monitor report trends & root causes	* Ensuring true anonymity and building trust * Managing and investigating sensitive allegations * Preventing retaliation against whistleblowers * Integrating with existing HR and legal systems * Handling reports across multiple languages/regions * Measuring the impact on ethical behavior & culture

Products	Platform Cost	Project Cost	Duration
Ethico (Ethics & Compliance)	$10-50K	$60-250K	3-9 m
NAVEX Global (EthicsPoint)	$15-70K	$80-350K	3-10 m
Convercent (OneTrust Ethics)	$15-60K	$70-300K	3-9 m
Lighthouse Services	$5-30K	$20-100K	2-6 m
ComplianceLine	$5-25K	$20-80K	2-5 m
WhistleB (Navex Global)	$10-40K	$50-200K	3-8 m

Policy Management Software provides a centralized, automated system for creating, distributing, tracking, and ensuring acknowledgment of an organization's internal policies, procedures, and guidelines. These platforms streamline the entire policy lifecycle, from drafting and version control to employee dissemination, mandatory read-and-acknowledge processes, and auditing. By digitizing and automating policy management, organizations ensure that all employees are aware of and comply with critical operational, HR, and ethical standards, significantly reducing compliance risks, improving efficiency in policy updates, and providing a clear audit trail. This is essential for fostering a consistent, compliant, and well-informed workforce, critical for upholding integrity and organizational governance.

Without dedicated Policy Management Software, organizations often struggle with fragmented, outdated, and manually managed internal policies, leading to inconsistencies, confusion among employees, and significant compliance risks. Relying on disparate documents and unreliable methods of distribution and acknowledgment makes it difficult to ensure all employees are aware of, and adhere to, critical guidelines, increasing exposure to legal liabilities and operational errors. The absence of a centralized, automated system results in administrative burdens for HR and legal teams, a slower response to regulatory changes, and a diminished capacity to enforce a consistent culture of compliance and integrity across the workforce.

Implementation	Challenges
* Centralize all existing policies & procedures * Select a robust policy management platform * Digitize content & design clear policy structure * Automate distribution, version control & acknowledgment * Plan employee training & awareness campaigns * Establish audit trails & reporting capabilities	* Migrating and digitizing large volumes of existing policies * Ensuring policy content remains current and accurate * Driving mandatory employee acknowledgment rates * Integrating with existing HRIS or document management * Managing complex version control and policy reviews * Measuring the impact on compliance and behavior

Products	Platform Cost	Project Cost	Duration
PolicyTech (NAVEX Global)	$10-50K	$60-250K	3-9 m
LogicManager (Policy & Procedure Mgmt)	$15-60K	$70-300K	3-9 m
ComplianceBridge	$5-30K	$20-100K	2-6 m
Onspring GRC (Policy & Compliance)	$15-70K	$80-350K	3-10 m
StandardFusion (GRC Platform)	$5-25K	$20-80K	2-5 m
SharePoint Online (Document Mgmt with add-ons)	Included with M365	$30-150K	2-7 m

Diversity & Inclusion (D&I) Tools are technology solutions designed to help organizations promote, measure, and manage workplace diversity, equity, and inclusion initiatives. These tools can range from unconscious bias training platforms and inclusive language checkers to diversity analytics dashboards, equitable hiring software, and D&I survey tools. By providing data-driven insights and fostering inclusive behaviors, D&I tools enable organizations to identify and mitigate biases in talent processes, enhance representation, improve equitable opportunities, and cultivate a more inclusive and welcoming workplace culture. This strategic investment is crucial for attracting diverse talent, improving innovation, enhancing employee engagement, and driving business performance by leveraging the full potential of a diverse workforce.

Without dedicated Diversity & Inclusion Tools, organizations often struggle to effectively identify, address, and measure systemic biases and inequalities within their workforce, relying instead on manual efforts or anecdotal evidence. This results in slow progress on D&I initiatives, perpetuation of unconscious biases in hiring and promotion, and a lack of accountability for creating a truly inclusive environment. The absence of data-driven insights and specialized tools hinders the ability to attract and retain diverse talent, impacts employee morale and sense of belonging, and ultimately prevents the organization from fully realizing the benefits of a diverse workforce, such as enhanced innovation and stronger business performance.

Implementation	Challenges
* Define clear D&I goals & metrics * Select tools aligning with specific D&I needs * Integrate with HRIS, ATS, & learning platforms * Plan comprehensive D&I training & awareness * Establish D&I reporting & accountability * Continuously monitor & adapt D&I strategies	* Overcoming ingrained biases & resistance to change * Ensuring data privacy & ethical use of D&I data * Measuring true impact beyond simple metrics * Integrating with diverse HR and business systems * Gaining consistent leadership and employee buy-in * Avoiding "check-the-box" rather than genuine impact

Products	Platform Cost	Project Cost	Duration
Textio (Inclusive Language)	$5-30K	$20-100K	2-6 m
Culture Amp (D&I Surveys)	$10-50K	$60-250K	3-9 m
HireVue (Bias Mitigation)	$10-60K	$60-300K	3-9 m
Vervoe (Fair Skills Assessments)	$5-25K	$20-80K	2-5 m
Qualtrics (D&I Experience)	$15-70K	$70-350K	3-10 m
Paradigm (D&I Training & Consulting)	N/A	$50-300K	3-9 m

Background Verification Systems automate and streamline the process of conducting checks on prospective employees or contractors, ensuring the accuracy of their credentials, employment history, criminal records, and other relevant background information. These systems integrate with various data sources and utilize advanced analytics to provide comprehensive, compliant, and timely verification reports. By conducting thorough background checks, organizations mitigate risks associated with hiring, protect their assets and reputation, ensure a safe work environment, and maintain compliance with industry regulations. This proactive measure is essential for building a trustworthy and reliable workforce, safeguarding the organization from potential legal liabilities and operational disruptions.

Without reliable Background Verification Systems, organizations expose themselves to significant risks, including negligent hiring claims, internal fraud, workplace safety issues, and reputational damage from hiring individuals with undisclosed criminal histories or falsified credentials. Relying on manual or inconsistent background checks is inefficient, prone to human error, and may not comply with evolving legal requirements, leading to potential legal repercussions and a less secure workforce. The absence of automated, thorough verification processes compromises the integrity of the hiring process, increases operational risks, and undermines trust, ultimately impacting overall organizational security and the quality of its human capital.

Implementation	Challenges
* Define scope of checks & legal compliance * Select reputable background check provider * Integrate with ATS/HRIS for seamless workflow * Establish clear consent & data privacy protocols * Plan for international background checks (if applicable) * Monitor turnaround times & report accuracy	* Ensuring compliance with complex privacy laws (FCRA, GDPR) * Managing variations in legal requirements by jurisdiction * Addressing false positives or contested results * Integrating with disparate HR and recruiting systems * Maintaining speed and accuracy of results * Protecting sensitive candidate data securely

Products	Platform Cost	Project Cost	Duration
Sterling Check	$5-30K	$20-100K	2-6 m
Checkr	$5-25K	$20-80K	2-5 m
HireRight	$10-40K	$50-200K	3-8 m
GoodHire	$1-10K	$5-30K	1-2 m
Verifirst	$5-20K	$20-70K	2-4 m
Accurate Background	$5-25K	$20-90K	2-5 m

Whistleblower Reporting Systems provide a confidential and secure channel for employees, contractors, and other stakeholders to report serious misconduct, fraud, ethics violations, or illegal activities within an organization without fear of retaliation. These systems often include features for anonymous reporting, secure communication with investigators, and case management tools to track allegations from submission to resolution. By offering a robust and trusted mechanism for internal reporting, these systems promote transparency, enable early detection and resolution of serious issues, mitigate legal and reputational risks, and reinforce a culture of integrity and accountability throughout the organization. This critical safeguard ensures that ethical breaches are addressed promptly and effectively, maintaining public trust and internal morale.

Without robust Whistleblower Reporting Systems, organizations lack a safe and structured mechanism for individuals to raise serious concerns about misconduct, leading to internal issues remaining unaddressed, potentially escalating into public scandals, legal challenges, and severe reputational damage. This absence of a trusted, confidential reporting channel discourages employees from coming forward, allowing unethical or illegal activities to persist undetected and undermining organizational integrity. The lack of a formal whistleblower system increases the risk of regulatory fines, loss of public confidence, and a toxic work environment, ultimately jeopardizing long-term stability and the organization's ability to uphold its ethical commitments.

Implementation	Challenges
* Define clear scope for reportable issues * Select a reputable, independent system provider * Establish clear investigation & follow-up protocols * Communicate system availability & non-retaliation policy * Train internal investigation teams * Ensure compliance with whistleblower protection laws	* Guaranteeing anonymity and protecting whistleblowers * Managing and investigating sensitive, complex allegations * Ensuring fairness and due process in investigations * Integrating with existing GRC and legal systems * Preventing frivolous or malicious reporting * Measuring the impact on ethical climate & risk reduction

Products	Platform Cost	Project Cost	Duration
Ethico (Whistleblowing Software)	$10-50K	$60-250K	3-9 m
NAVEX Global (EthicsPoint Whistleblower)	$15-70K	$80-350K	3-10 m
Convercent (OneTrust Ethics)	$15-60K	$70-300K	3-9 m
IntegrityCounts	$5-30K	$20-100K	2-6 m
ComplianceLine	$5-25K	$20-80K	2-5 m
WhistleB (Navex Global)	$10-40K	$50-200K	3-8 m

BLUEPRINT SERIES

SECTION E

Execution Mechanics

Ai6x6™ Transformation Blueprint

Story 1 — The Manhattan Project Coordination Model (1942–1946): The Manhattan Project united scientists, engineers, and military leaders across dispersed sites under extreme secrecy and urgency. Workstreams were interdependent, requiring precise coordination without full visibility. Structured reporting, controlled interfaces, and disciplined compartmentalization enabled progress despite complexity. Human judgment guided trade-offs while machines and labs executed specialized tasks. The project demonstrated how execution succeeds when governance, clarity, and cadence align. It became a template for managing high-stakes, multi-disciplinary programs.

Story 2 — The Berlin Airlift Logistics System (1948–1949): With land routes blocked, Allied forces had to supply an entire city solely by air. Aircraft schedules, loading patterns, and turnaround times were optimized to the minute. Machines provided lift and speed, but humans orchestrated sequencing, prioritization, and contingency planning. Continuous feedback loops refined operations daily. The airlift proved that disciplined execution can overcome structural constraints. It showed how precision, rhythm, and adaptability create operational breakthroughs.

Story 3 — The Interstate Highway Construction Program (1956–1970s): The United States embarked on building a nationwide highway network requiring coordination across states, contractors, and engineering disciplines. Standardized designs ensured consistency while local teams adapted to terrain and conditions. Machines handled heavy construction, but humans managed sequencing, safety, and quality. The program relied on predictable funding, clear specifications, and phased execution. It demonstrated how large-scale delivery succeeds when structure and flexibility coexist. The highways became a symbol of disciplined national execution.

Story 4 — Toyota's Just-in-Time Logistics (1960s–1980s): As Toyota refined its production system, logistics became a critical execution layer. Parts arrived exactly when needed, minimizing inventory and waste. Machines tracked flow, but humans monitored signals, resolved disruptions, and maintained rhythm. Kanban created transparency across the value chain. The system required precision, trust, and continuous calibration. It showed how execution excellence emerges when every component of the workflow reinforces timing, quality, and flow.

Story 5 — FedEx's Overnight Delivery Network (1973–1990s): FedEx built a hub-and-spoke system capable of moving packages across the country overnight. Aircraft, sorting machines, and routing algorithms provided speed, but humans managed exceptions, weather disruptions, and customer commitments. The network operated on tight cycles that demanded flawless coordination. Data informed decisions while people ensured reliability. The system demonstrated how execution mechanics scale when tech and human oversight operate in synchronized cadence.

Story 6 — The Rise of Agile Software Delivery (2001–2015): Software teams struggled with long cycles, rigid plans, and late-stage failures. Agile introduced iterative delivery, cross-functional collaboration, and continuous feedback. Machines automated testing and deployment, while humans prioritized work, clarified requirements, and adapted to change. Execution shifted from linear phases to dynamic cycles. The approach improved predictability and responsiveness. It became a global standard for modern execution, proving that disciplined iteration outperforms rigid planning.

LESSONS FROM A WORKFORCE TRANSFORMATION PERSPECTIVE

Workforce maturity in execution mechanics emerges when people shift from performing isolated tasks to sustaining integrated workflows. Over time, workers learned to operate within structured systems, interpret signals, resolve disruptions, and maintain flow. Execution excellence depends on teams who understand dependencies, uphold standards, and collaborate across boundaries. Machines provide speed and automation, but humans ensure continuity, judgment, and escalation. The workforce becomes a stabilizing force—absorbing variability, reinforcing discipline, and improving processes. Execution maturity is achieved when teams operate with shared rhythm and accountability, enabling the organization to deliver consistently even under stress.

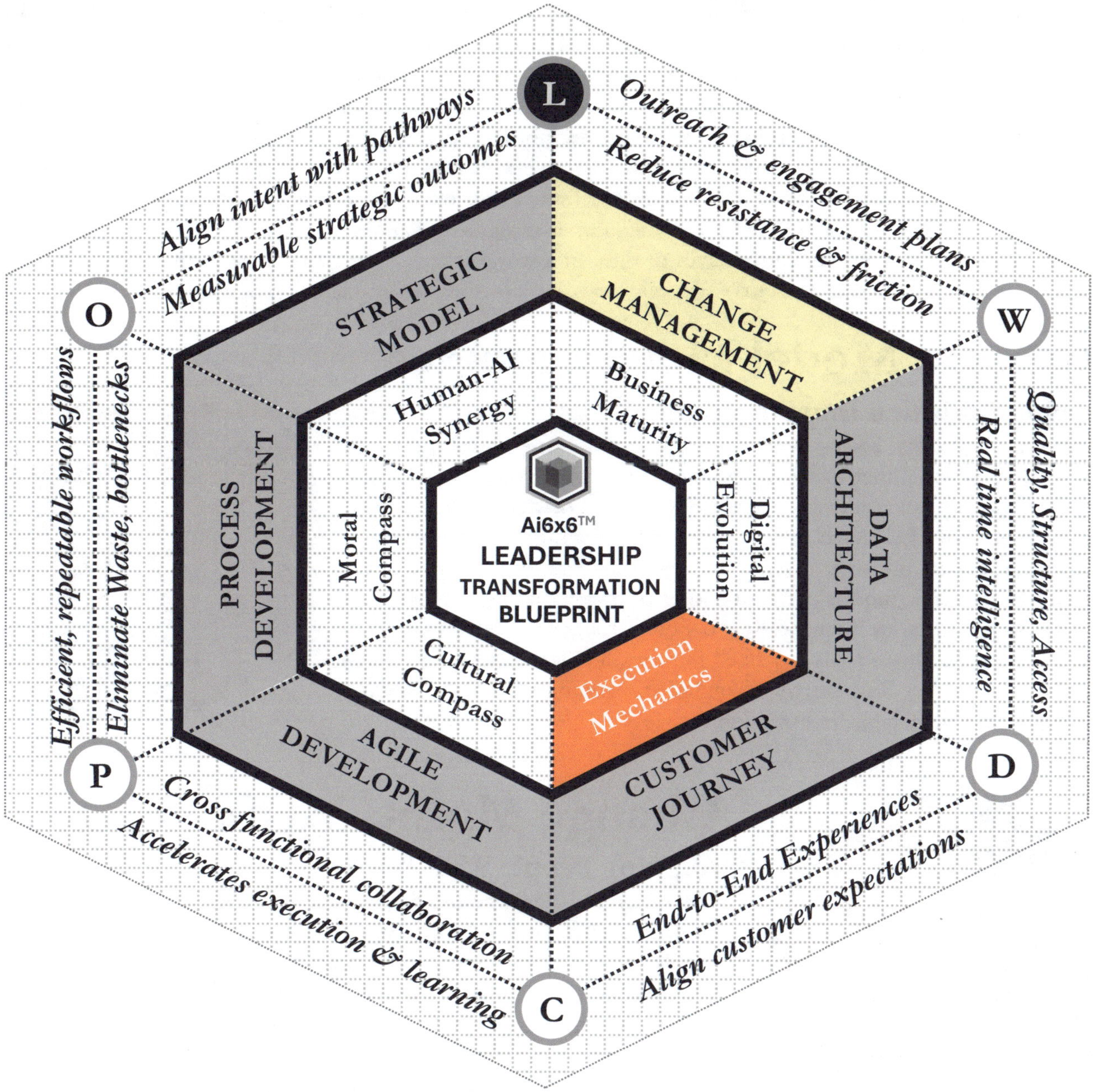

The **Ai6x6 Research Lab** identified this lens as the primary antidote to the "change fatigue" that frequently cripples workforces during radical shifts. We chose it because the human mind requires a disciplined, structured pathway to navigate the emotional stages of transition—from the first shock of readiness to the final state of stabilization. This lens isn't merely about change management tools; it is about the psychological certainty that comes from proven methodologies like ADKAR and Kotter's 8-Step. It is vital because it provides a mental scaffolding, turning the anxiety of the unknown into a series of predictable, manageable transitions.

This lens allows the employee's heart to remain calm while the organizational structures around them move at terminal velocity. By relying on structured application and performance models, the worker replaces resistance with an evidence-based conviction that the change is both necessary and navigable. We believe that change execution is the ultimate form of workforce mindfulness—the practice of honoring the emotional reality of transition while maintaining operational excellence. This lens ensures every shift has a clear roadmap, allowing the workforce to move through Bridges' or Satir's models with dignity. It provides the steady footing necessary for the workforce to leap into new roles with the assurance of a supported professional.

Execution has always been the crucible where strategy proves its worth. Leaders can articulate bold visions, design elegant frameworks, and outline ambitious roadmaps, but none of it matters if the organization cannot execute with precision, coherence, and resilience. In the age of AI and accelerated change, **Execution Mechanics** becomes the lens that determines whether transformation becomes real or remains theoretical. It forces leaders to confront the uncomfortable truth that most failures are not failures of strategy but failures of execution. The gap between intention and impact is rarely philosophical; it is operational. Execution Mechanics exposes that gap and provides the architecture to close it. This lens reframes execution as a system—an interconnected set of methodologies that convert ambition into momentum. It recognizes that execution is not the final step of strategy but the environment in which strategy lives or dies. In a world where AI accelerates decision cycles, compresses timelines, and amplifies both strengths and weaknesses, leaders must design execution as they design vision.

Strategic Model Methodologies

Turning Ambition Into Structure

Execution Mechanics begins with **strategic model methodologies**, the structured approaches that translate ambition into actionable pathways. These methodologies are not academic exercises; they are the scaffolding that ensures strategy is grounded. Leaders who neglect this dimension find themselves improvising under pressure, relying on intuition rather than disciplined frameworks. Strategic models define the logic of execution—how choices are made, how priorities are sequenced, and how resources are allocated. They ensure that every initiative has a defined purpose, a coherent structure, and a measurable outcome. Without this foundation, execution becomes reactive, fragmented, and vulnerable to disruption.

Change Mgmt. Methodologies

Aligning People With Progress

The next dimension, **change management methodologies**, recognizes that execution is ultimately a human endeavor. No transformation succeeds unless people understand, accept, and commit to the change. Leaders who underestimate this reality discover that resistance is not a barrier but a predictable response to uncertainty. Effective change management acknowledges the emotional, cultural, and psychological dimensions of transformation. It ensures that communication is transparent, expectations are clear, and support is available. It turns change from something imposed into something shared. In this dimension, execution becomes relational—rooted in trust, empathy, and the ability to guide people toward possibility.

Data Architecture Methodologies

Building the Backbone of Execution

Data architecture provide the backbone for modern execution. In a world where decisions must be informed by evidence, organizations cannot afford fragmented data, inconsistent definitions, or unreliable pipelines. Data architecture is not a technical concern; it is a strategic one. It determines whether insights are trustworthy, whether systems can scale, and whether AI can operate responsibly. Leaders who invest in strong data architecture create the conditions for intelligence to flow freely across the enterprise. In this dimension, execution becomes intelligent—powered by clarity, consistency, and evidence.

Customer Journey Methodologies

Anchoring Execution in Value

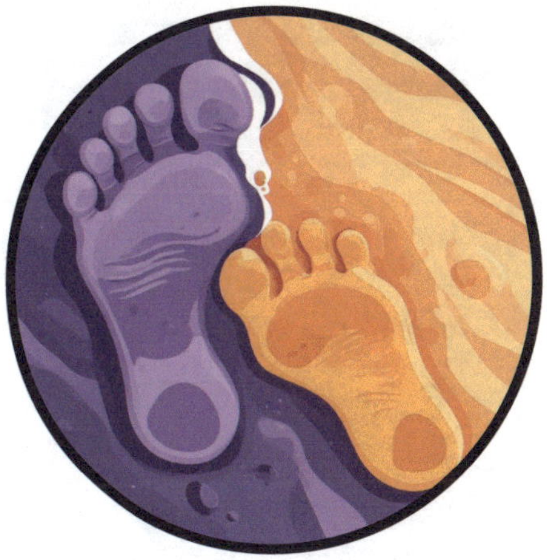

Customer journey ensure that execution remains anchored in value creation. Too many organizations design processes around internal convenience rather than customer experience. Customer journey thinking forces leaders to see the organization from the outside in, identifying friction, gaps, and opportunities that internal perspectives overlook. It ensures that execution is not only efficient but meaningful. It aligns operations with the expectations, needs, and emotions of the people the organization serves. In this dimension, execution becomes empathetic—shaped by the reality of customers rather than the assumptions of leaders.

Agile Development Methodologies

Enabling Adaptation at Speed

Agile development introduce the adaptability required in a world where conditions shift rapidly. Traditional execution models assume stability, predictability, and linear progress. Agile assumes the opposite. It embraces iteration, experimentation, and continuous learning. It allows teams to adjust quickly, respond to feedback, and deliver value incrementally. Leaders who adopt agile mindsets create organizations that move with speed without sacrificing quality. In this dimension, execution becomes dynamic—capable of evolving in real time as new information emerges.

Process Improvement Methodologies

Sustaining Excellence Over Time

The final dimension, **process improvement methodologies**, completes the system by ensuring that execution evolves rather than stagnates. They provide the discipline to identify inefficiencies, eliminate waste, and refine workflows. They ensure that execution is not a one-time effort but a continuous practice. Leaders who embrace process improvement build organizations that get better over time, not simply bigger or busier. In this dimension, execution becomes self-correcting—driven by curiosity, discipline, and a commitment to excellence.

Execution Mechanics is the lens that transforms strategy into reality by ensuring that ambition is supported by structure, that change is supported by people, and that progress is supported by systems. It reveals that execution is not the final step of leadership but the continuous arena in which leadership is tested. In a world where AI accelerates both opportunity and exposure, leaders must design execution as deliberately as they design vision, building the operational muscle that allows the organization to move with coherence under pressure. This lens makes clear that execution is not a matter of effort but of engineering—of constructing the models, architectures, and behaviors that convert intention into repeatable performance. Execution becomes a living capability rather than a periodic push, a discipline that strengthens with use rather than erodes under stress.

EXECUTION MECHANICS ANALOGY – STREAMLINING AIRPORT

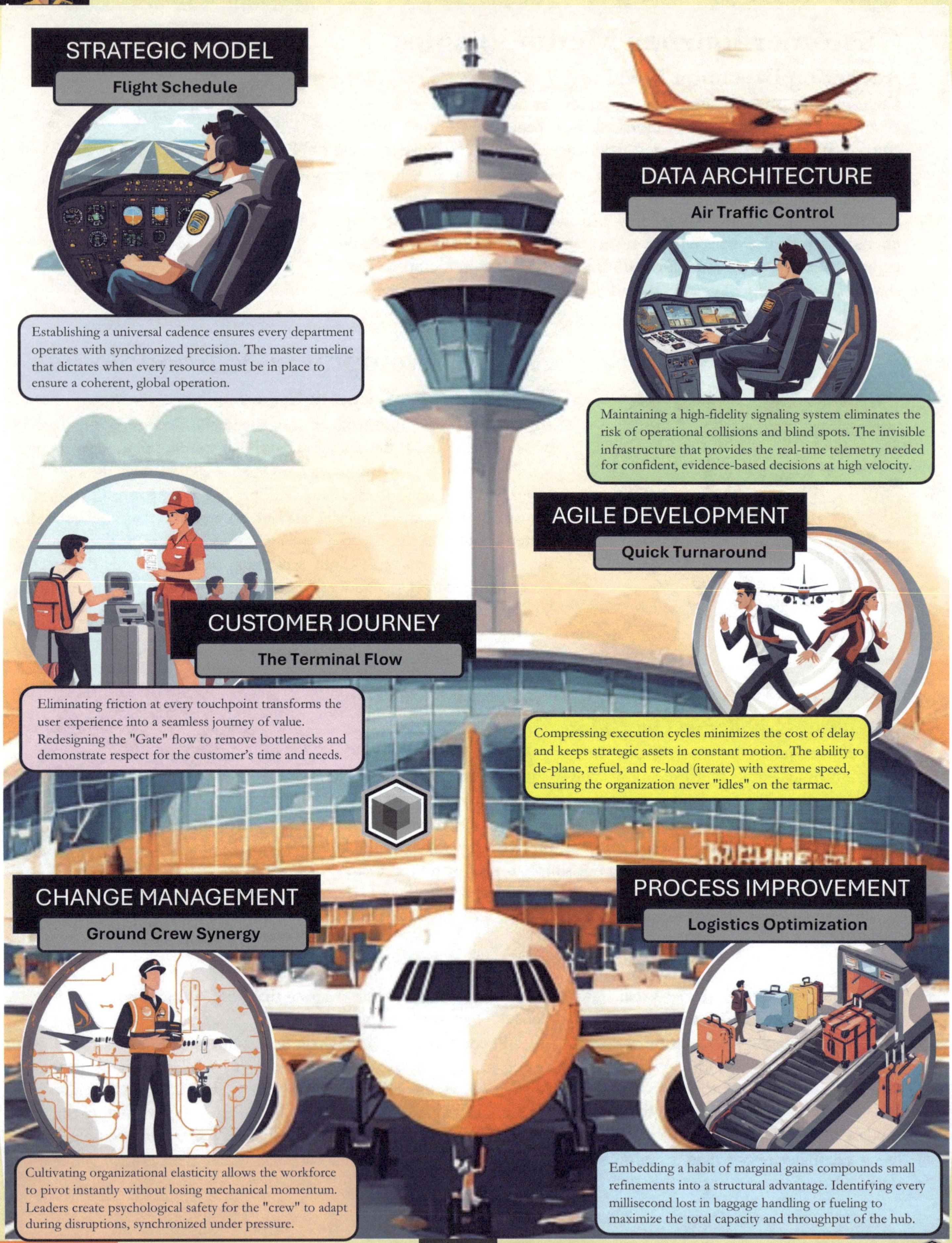

STRATEGIC MODEL
Flight Schedule

Establishing a universal cadence ensures every department operates with synchronized precision. The master timeline that dictates when every resource must be in place to ensure a coherent, global operation.

DATA ARCHITECTURE
Air Traffic Control

Maintaining a high-fidelity signaling system eliminates the risk of operational collisions and blind spots. The invisible infrastructure that provides the real-time telemetry needed for confident, evidence-based decisions at high velocity.

AGILE DEVELOPMENT
Quick Turnaround

Compressing execution cycles minimizes the cost of delay and keeps strategic assets in constant motion. The ability to de-plane, refuel, and re-load (iterate) with extreme speed, ensuring the organization never "idles" on the tarmac.

CUSTOMER JOURNEY
The Terminal Flow

Eliminating friction at every touchpoint transforms the user experience into a seamless journey of value. Redesigning the "Gate" flow to remove bottlenecks and demonstrate respect for the customer's time and needs.

CHANGE MANAGEMENT
Ground Crew Synergy

Cultivating organizational elasticity allows the workforce to pivot instantly without losing mechanical momentum. Leaders create psychological safety for the "crew" to adapt during disruptions, synchronized under pressure.

PROCESS IMPROVEMENT
Logistics Optimization

Embedding a habit of marginal gains compounds small refinements into a structural advantage. Identifying every millisecond lost in baggage handling or fueling to maximize the total capacity and throughput of the hub.

COMMON FRAMEWORK - TOGAF (The Open Group Architecture Framework)

Book 1: Leadership Transformation—Strategic Direction : The frameworks in this pillar ensure that all efforts are strategically aligned and that leaders are focused on the right outcomes for value creation.

1. **McKinsey 7-S Framework:** Aligns the organization's strategy with its structure, systems, staff, and skills to ensure all components support the desired transformation.
2. **Business Model Canvas:** Creates a shared understanding of how value will be created, delivered, and captured, providing a clear map for all teams.
3. **McKinsey Three Horizons of Growth:** Structures innovation efforts across short-term improvements (Horizon 1), scaling existing successes (Horizon 2), and future-proofing the business (Horizon 3), ensuring a sustainable pipeline of value.
4. **Balanced Scorecard:** Translates the strategic vision into measurable performance objectives, connecting operational activities to high-level financial and non-financial goals.
5. **SWOT Analysis:** Provides a structured way to evaluate the strategic context, enabling leaders to leverage strengths, mitigate weaknesses, seize opportunities, and counter threats.
6. **Blue Ocean Strategy:** Drives leaders to look for uncontested market space, shifting the focus from competing for value to creating new value.

Book 2: Workforce Transformation—Adoption and Engagement : This pillar addresses the human element of change, ensuring that employees are prepared, engaged, and able to sustain new ways of working.

1. **Kotter's 8-Step Change Model:** Provides a step-by-step process for leading major change, from creating a sense of urgency to solidifying new approaches in the corporate culture.
2. **ADKAR Model:** Guides change managers in ensuring that individual employees have the Awareness, Desire, Knowledge, Ability, and Reinforcement needed for successful change.
3. **Lewin's Change Management Model:** Conceptualizes change as a three-stage process: "Unfreeze" existing habits, implement the "Change," and "Refreeze" the new processes to make them stick.
4. **Bridges' Transition Model:** Focuses on the psychological journey of individuals, helping them navigate the emotional stages of an ending, a neutral zone, and a new beginning.
5. **Prosci Change Management Methodology:** Provides a structured, research-based methodology for applying change management to any project or initiative.
6. **The Satir Change Model:** Helps leaders understand and anticipate the emotional and performance impacts of change on individuals and teams.

Book 3: Data Transformation—Measurement and Insights - This pillar provides the tools and architectures to manage data as a strategic asset, ensuring that value can be reliably measured and tracked.

1. **Zachman Framework:** Offers a comprehensive, logical structure for classifying and organizing enterprise architecture, ensuring data management is part of the larger business strategy.
2. **DAMA-DMBOK:** Provides a set of best practices for managing data throughout its lifecycle, ensuring the quality and integrity of data used for value measurement.
3. **Data Mesh:** Promotes a decentralized, domain-driven approach to data ownership, empowering teams to create and manage data as a product.
4. **Lambda Architecture (or Kappa Architecture):** Establishes a technical architecture for processing large volumes of data efficiently, supporting real-time value tracking and analytics.
5. **Master Data Management (MDM) Frameworks:** Ensure that an organization's critical business data is consistent, accurate, and provides a "single source of truth" for reliable reporting.
6. **Data Fabric:** Creates a flexible, interconnected data architecture that supports seamless access and integration of data, regardless of its location.

Book 4: Customer Transformation—External Value Delivery: This pillar focuses on defining and delivering value from the customer's perspective, ensuring that transformation efforts translate to measurable customer outcomes.

1. **Customer Journey Mapping:** Visualizes the customer's end-to-end experience, identifying pain points and opportunities for value creation.
2. **Service Blueprinting:** Expands on journey mapping by detailing the front-stage and backstage processes, people, and systems that support a customer experience.
3. **Jobs to be Done (JTBD) Framework:** Helps define "value" from the customer's perspective by focusing on the functional and emotional outcomes they seek.
4. **Design Thinking Methodology:** Provides a human-centered approach to innovation that focuses on empathizing with customers to design meaningful solutions.
5. **CX (Customer Experience) Maturity Models:** Provides a benchmark for tracking and improving customer experience over time, linking strategic initiatives to better customer outcomes.
6. **Persona Development Frameworks:** Creates detailed profiles of target users, ensuring that transformation efforts are grounded in a deep understanding of customer needs.

Book 5: Product Transformation—Internal Value Delivery : This pillar establishes an agile, iterative process for continuously building and improving the products and services that deliver customer value.

1. **Scrum:** Enables cross-functional teams to deliver value incrementally in short sprints, allowing for continuous feedback and adaptation.
2. **Kanban:** Helps manage and visualize workflow, limiting work-in-progress and improving the flow of value to the customer.
3. **Lean Startup:** Applies a "build-measure-learn" feedback loop to product development, ensuring teams are constantly validating their assumptions about value creation.
4. **Scaled Agile Framework (SAFe):** Provides a system for coordinating and aligning multiple agile teams across the enterprise to deliver large-scale solutions.
5. **Feature Flags / Toggles Architecture:** Enables teams to deploy new features and capabilities quickly and safely, allowing for real-time testing of value propositions.
6. **Continuous Integration / Continuous Delivery (CI/CD) Pipeline Architecture:** Automates the software delivery process, allowing teams to deliver new value to customers efficiently and reliably.

Book 6: Operations Transformation—Efficiency and Sustainability
This pillar focuses on optimizing the internal processes and systems that support the entire value chain, ensuring efficiency and sustainability.

1. **Lean Management:** Aims to eliminate waste and optimize the flow of value through the organization's processes, generating cost savings and efficiency gains.
2. **Six Sigma (DMAIC Methodology):** Uses a data-driven approach to improve processes by defining, measuring, analyzing, improving, and controlling them.
3. **BPM (Business Process Management) Lifecycle:** Provides a structured approach to continuously improving business processes, ensuring sustained value delivery over the long term.
4. **SCOR Model (Supply Chain Operations Reference):** Offers a standard framework for managing and improving supply chain efficiency, a key driver of operational value.
5. **Digital Twin Architecture:** Creates virtual models of physical assets or processes, enabling simulation and optimization before real-world changes are made.
6. **Robotic Process Automation (RPA) Frameworks/Methodologies:** Automates repetitive, rule-based tasks, freeing up human resources for higher-value activities.

TOGAF (The Open Group Architecture Framework) is a widely recognized and comprehensive enterprise architecture methodology that provides a structured approach for developing and managing enterprise architecture. It helps organizations design, plan, implement, and govern an enterprise information architecture. At its core is the Architecture Development Method (ADM), a detailed, iterative process for developing various architectural domains: Business, Data, Application, and Technology. For leadership transformation, particularly in achieving digital maturity, TOGAF is indispensable. Digital transformation is not just about implementing new technologies; it requires aligning business strategy with IT capabilities across the entire enterprise. Without a coherent architecture, digital initiatives can become fragmented, leading to integration challenges, redundant systems, security vulnerabilities, and a lack of scalability. TOGAF provides the necessary rigor to ensure that digital investments are strategically aligned, well-integrated, and support the organization's long-term digital vision. It helps leaders understand the current state of their digital landscape, define the desired future state (e.g., cloud-native infrastructure, API-driven integration, AI-enabled applications), and create a roadmap to get there. By fostering a common language and framework for architects and stakeholders, TOGAF enables effective communication and collaboration across complex digital initiatives. It ensures that the digital transformation is not just a series of isolated projects but a cohesive, well-governed evolution of the entire enterprise, a critical component of achieving and sustaining comprehensive digital maturity. It provides the blueprint for how an organization's digital capabilities (applications, data, infrastructure) support its strategic business goals, ensuring that every technological investment contributes to a unified, adaptable, and future-ready digital enterprise. This structured approach helps avoid costly misalignments and ensures that digital initiatives are built on a solid, scalable foundation, allowing the organization to truly harness the power of technology for competitive advantage.

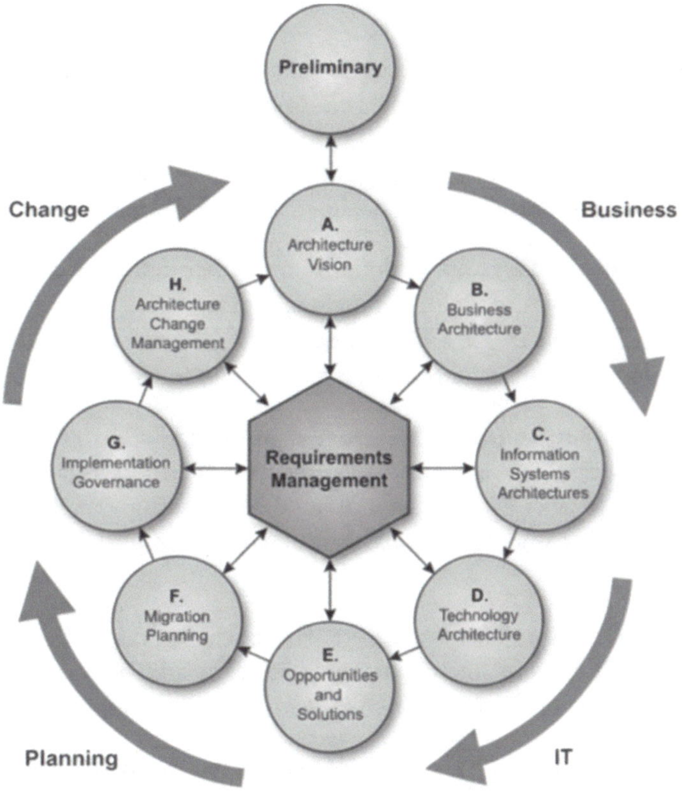

TOGAF Frameworks

TOGAF (THE OPEN GROUP ARCHITECTURE FRAMEWORK) – STEPS

1. **Preliminary Phase: Framework and Principles:** Establish the architecture capability for the enterprise. This involves defining the organizational context for architecture work, identifying key stakeholders, establishing the architecture principles (e.g., "cloud-first," "API-driven," "data-driven decision making"), and defining the scope and governance of the architecture effort.

2. **Phase A: Architecture Vision:** Develop a high-level vision for the target architecture that aligns with business strategy and digital transformation goals. This involves identifying key stakeholders, defining the business value of the architecture work, and securing approval for the architecture development cycle.

3. **Phase B: Business Architecture:** Develop a baseline (current state) and target (future state) Business Architecture that describes the organization's processes, functions, and organizational structure. This ensures that the digital transformation efforts are aligned with and support the core business capabilities and strategic objectives, identifying where digital enablement can optimize business value, streamline workflows, and enhance customer interactions through digital means.

4. **Phase C: Information Systems Architectures (Data & Application):** Develop baseline and target Data Architecture (logical and physical data models, data governance, data flows) and Application Architecture (applications, their interactions, and how they support business functions). This phase is critical for designing integrated digital systems, ensuring data flows seamlessly across digital platforms, and supporting new digital applications that drive business value and customer engagement.

5. **Phase D: Technology Architecture:** Develop baseline and target Technology Architecture that describes the logical software and hardware capabilities required to support the Business, Data, and Application components. This includes defining infrastructure (e.g., cloud platforms, edge computing), platforms (e.g., microservices, AI/ML platforms), and security standards necessary for a robust, scalable, and resilient digital enterprise.

6. **Phase E: Opportunities & Solutions:** Identify and evaluate various implementation options and opportunities to deliver the target architecture. This involves assessing gaps between baseline and target architectures, identifying potential solutions (e.g., Commercial Off-The-Shelf (COTS), custom development, cloud migration, SaaS adoption), and defining high-level implementation projects. For digital maturity, this prioritizes digital solutions that offer the greatest business value and strategic alignment.

7. **Phase F: Migration Planning:** Develop a detailed implementation and migration plan. This involves prioritizing projects, estimating costs, defining a timeline, and outlining the sequence of activities required to transition from the baseline to the target architecture. This ensures a structured and managed approach to digital transformation implementation, minimizing disruption and maximizing efficiency.

8. **Phase G: Implementation Governance:** Oversee the implementation of the architecture. This involves establishing governance processes to ensure that projects adhere to the defined architecture, managing architectural contracts, and providing architectural guidance during development and deployment. This continuous oversight is vital for maintaining architectural integrity and consistency throughout the digital transformation.

9. **Phase H: Architecture Change Management:** Establish processes for managing changes to the architecture. This ensures that the architecture remains relevant and responsive to evolving business needs, technological advancements (e.g., new AI capabilities, quantum computing), and market shifts.

10. **Requirements Management:** This phase runs concurrently throughout all ADM phases. It involves managing the architectural requirements, ensuring they are consistent, complete, and aligned with stakeholder needs. For digital maturity, this ensures that evolving digital requirements (e.g., real-time data access, scalability for new digital services) are captured and integrated into the architectural design and implementation.

Ai6x6™ Transformation Blueprint

Change Management Frameworks

"It is not the strongest of the species that survives, nor the most intelligent that survives. It is the one that is the most adaptable to change."

Charles Darwin. The imperative for the workforce to develop organizational agility and adaptability as the core competitive skill, ensuring survival during disruption.

Change Management Frameworks are the human-centric blueprints essential for ensuring transformation success by systematically addressing the **people side of change**. Unlike architectural models that define the *what* (systems and processes), frameworks like **ADKAR** or the **Kotter 8-Step Process** define the *how*—the phases, activities, and milestones needed to secure employee adoption and usage. This structure is critical because it moves organizational culture and behavioral shifts from abstract concepts to measurable, governed activities. The framework provides tools for assessing organizational readiness, mitigating resistance, ensuring communication is targeted and timely, and defining necessary training. By giving leaders a structured, phased approach to manage transitions, it significantly reduces the human risk associated with digital and AI transformations, ensuring that new processes and systems are not just deployed, but **fully embraced and utilized**, thereby translating technological investment into sustained business value.

The **Change Management Framework** is focused on successfully managing the people side of transformation—ensuring employees are prepared and able to sustain new ways of working—is critical for change longevity. This process is structured by several key frameworks. Kotter's 8-Step Change Model provides a practical, sequential guide, moving from generating urgency to solidifying new behaviors into the corporate culture. At the individual level, the ADKAR Model ensures success by focusing on the necessary Awareness, Desire, Knowledge, Ability, and Reinforcement for each employee. A simpler, conceptual view is offered by Lewin's Change Management Model, which breaks the process into three phases: Unfreeze existing norms, implement the actual Change, and then Refreeze the successful new practices. Recognizing the psychological impact of transition, Bridges' Transition Model addresses the emotional journey of facing an ending, navigating the neutral zone, and embracing the new beginning. Furthermore, the Satir Change Model helps leaders anticipate the emotional and performance curve that teams will experience. Finally, the Prosci Change Management Methodology provides a comprehensive, research-based system for applying these and other tools to drive project success by effectively managing the human element.

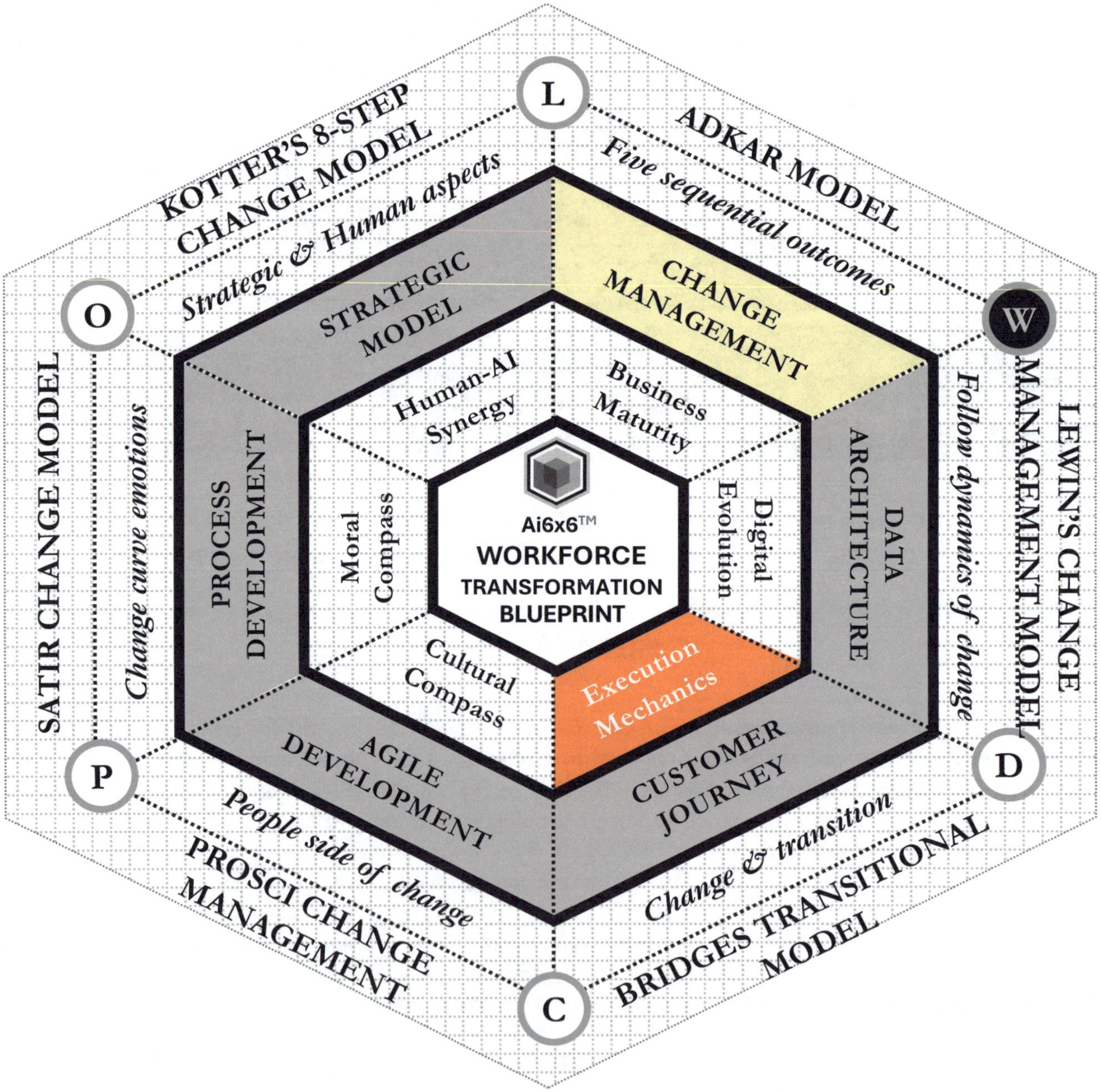

Ai6x6 EXECUTION MECHANICS FRAMEWORK FOR **WORKFORCE LIFECYCLE**

John Kotter's 8-Step Change Model is one of the most widely recognized and applied frameworks for leading successful organizational change. It provides a sequential, step-by-step process that addresses both the strategic and human aspects of transformation. Kotter emphasizes that change often fails not because of a lack of good ideas or technology, but due to a failure in execution and insufficient attention to the human element. For workforce transformation, particularly in the context of digital maturity, this model is highly effective.

Digital transformation often requires significant shifts in employee skills, roles, workflows, and mindsets. Kotter's model guides leaders through the critical stages: from creating a compelling sense of urgency for digital adoption (e.g., highlighting digital competitive threats or opportunities) to building a powerful guiding coalition that champions digital initiatives. It stresses the importance of developing a clear vision for the digitally empowered workforce, communicating this vision broadly and consistently, and empowering employees to act on it by removing barriers like outdated processes or lack of digital tools. Generating short-term wins (e.g., successful pilot of a new digital collaboration tool) helps build momentum and counter skepticism. Finally, consolidating gains and anchoring new digital approaches in the organizational culture ensures that the transformation is sustainable and that digital ways of working become the new norm. By systematically following these steps, leaders can effectively navigate the complexities of digital workforce transformation, minimize resistance, and cultivate a digitally mature workforce that is agile, adaptable, and proficient in leveraging new technologies..

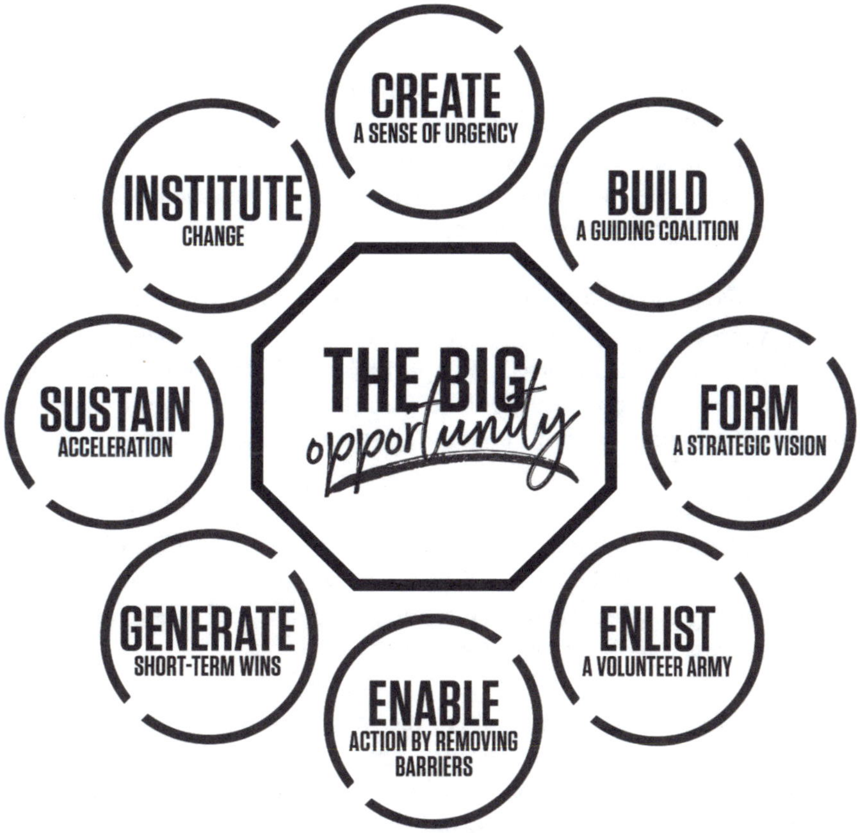

Kotter's 8-Step Change Model

1. **Create Urgency:** Help the organization see the critical need for change and the importance of immediate action. This involves identifying potential crises (e.g., declining market share due to digital disruptors), major opportunities (e.g., new digital market segments), or compelling evidence that the status quo (e.g., manual processes, outdated systems) is unsustainable. Leaders must communicate this urgency widely and genuinely to gain initial buy-in and overcome complacency, especially regarding the imperative for digital transformation and upskilling.

2. **Form Guiding Coalition:** Assemble a powerful, diverse group of leaders and managers who are committed to driving the change. This coalition should have sufficient power, expertise (including digital literacy), credibility, and leadership skills to effectively guide the transformation. They will act as the core team, leading by example and coordinating efforts across the organization, ensuring cross-functional alignment for digital initiatives.

3. Develop Vision & Strategy: Create a clear, compelling, and inspiring future vision for the transformed organization. This vision should articulate the desired future state (e.g., "a digitally empowered workforce delivering seamless customer experiences") and why it's better. Concurrently, develop a strategic plan that outlines the specific steps and initiatives required to achieve that vision, providing a roadmap for the transformation journey, including key digital milestones.

4. **Communicate Vision:** Broadly and frequently share the vision using multiple channels (e.g., town halls, newsletters, internal digital platforms, team meetings, informal conversations) to gain understanding and buy-in from all employees. Communication must be consistent, clear, and address potential concerns about job security or new skill requirements. Leaders must "walk the talk," embodying the vision in their own actions and decisions, particularly by embracing new digital tools themselves.

5. **Empower Action:** Remove obstacles that hinder employees from acting on the vision. This involves identifying and addressing bureaucratic hurdles, outdated processes, or resistant individuals. Leaders must also encourage risk-taking, provide necessary resources (e.g., access to new digital platforms, training budgets), and empower employees at all levels to innovate and implement changes, fostering a sense of ownership and agency in the digital transformation.

6. **Generate Short-Term Wins:** Plan for and celebrate visible, early successes to build momentum and demonstrate tangible progress. These "wins" provide evidence that the change is working, boost morale, and reinforce the efforts of those involved. For workforce digital transformation, this could be the successful rollout of a new collaboration tool, a quick win from an automated process, or a team successfully adopting a new digital workflow, countering skepticism and providing crucial data for learning.

7. **Consolidate Gains & Produce More Change:** Use the credibility gained from early wins to tackle more complex changes and continue the transformation. Avoid declaring victory too early. Leaders must continuously learn from successes and failures, adapt strategies, and ensure that the momentum is sustained by launching new initiatives (e.g., advanced digital training programs, AI adoption) and reinforcing the need for ongoing evolution towards full digital maturity.

8. **Anchor New Approaches in Culture:** Embed the new behaviors, processes, and values into the organization's culture, leadership development programs, and reward systems for lasting impact. This ensures that the transformed way of working – a digital-first mindset, data literacy, continuous learning – becomes the new norm, preventing regression to old habits. Cultural embedding is the final, crucial step to ensure the workforce transformation is sustainable and truly achieves digital maturity.

2. ADKAR MODEL

The ADKAR Model, developed by Prosci, is a goal-oriented change management model that focuses on the individual's journey through change. Unlike models that focus solely on organizational processes, ADKAR identifies five sequential outcomes that an individual must achieve for change to be successful: Awareness, Desire, Knowledge, Ability, and Reinforcement. For workforce transformation, particularly in the context of digital maturity, ADKAR is incredibly powerful because it provides a diagnostic framework.

Digital transformation often requires employees to learn new skills, adopt new tools, and change ingrained habits. If a digital initiative is struggling, ADKAR allows leaders to pinpoint exactly which element is missing for individuals or groups. For example, employees might have the Knowledge of how to use a new digital platform but lack the Desire because they don't understand its personal benefits. Or they might have the Knowledge but not the Ability due to insufficient practice or coaching. By addressing each of these elements systematically, leaders can tailor their change management interventions, such as specific communication campaigns to build Awareness, incentive programs to foster Desire, targeted digital training to build Knowledge, hands-on coaching to develop Ability, and recognition programs to provide Reinforcement. This granular, people-centric approach ensures that the workforce not only understands the digital changes but actively embraces, adopts, and sustains them, accelerating the organization's overall digital maturity by building a digitally proficient and adaptable workforce..

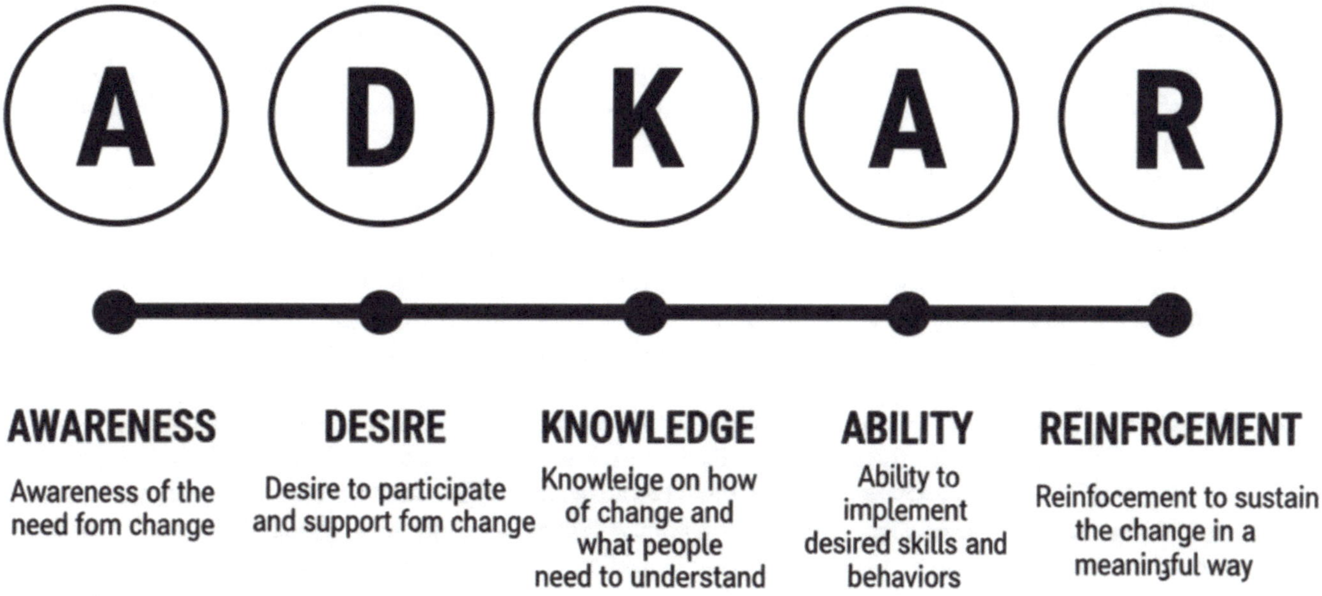

: ADKAR Model | Image source: https://cjni.net

1. **Awareness**: The first step is to ensure employees understand *why* the change is happening. This includes communicating the business reasons for the digital transformation (e.g., market shifts, competitive pressure, new opportunities), the risks of not changing (e.g., becoming obsolete, losing market share), and the benefits of moving to the new digital state (e.g., increased efficiency, new career opportunities, enhanced customer service). Without clear awareness, employees may resist simply because they don't understand the necessity or purpose of adopting new digital tools and processes.

2. **Desire**: Foster a willingness in employees to participate in and support the change. This is the personal choice to engage with the digital transformation. Leaders must address individual motivations, concerns, and potential fears (e.g., fear of job displacement, difficulty learning new tech), showing how the change aligns with personal goals or values (e.g., skill development, career growth, contributing to company success). This step often requires empathetic listening and addressing "what's in it for me?" from the employee's perspective regarding digital adoption.

3. **Knowledge**: Provide employees with the necessary information and training on *how* to change. This includes understanding new digital processes, using new software systems, operating digital tools, and acquiring required digital skills (e.g., data analytics, cloud computing, cybersecurity basics). Knowledge development can take various forms, such as formal digital training programs, workshops, online courses, comprehensive documentation, and access to digital experts, ensuring employees know what they need to do differently in a digitally-enabled environment.

4. **Ability**: Offer coaching, practice, and hands-on experience to build competence and confidence in applying the new digital knowledge and skills. Knowledge alone is often insufficient; employees need opportunities to practice and refine new digital behaviors in a supportive environment. This step bridges the gap between understanding digital concepts and actual performance, ensuring employees can effectively use the new digital tools and processes in their daily work. This practical application is key to digital maturity.

5. **Reinforcement**: Implement mechanisms (e.g., recognition, rewards, feedback, performance management aligned with digital adoption) to ensure the change sticks and becomes the new norm. Reinforcement helps to prevent employees from reverting to old habits and sustains the new digital behaviors. It acknowledges successful adoption of digital tools, celebrates digital champions, and provides ongoing support to embed the digital transformation into daily routines and cultural norms.

6. **Monitor & Diagnose**: Continuously assess individual progress through the ADKAR stages. If an employee is struggling with the digital transformation, leaders can diagnose which ADKAR element is missing (e.g., lack of Awareness about the digital strategy, lack of Desire to embrace new digital tools, insufficient Knowledge of digital systems, or limited Ability to apply new digital skills) and provide targeted, personalized support. This diagnostic capability makes ADKAR a powerful tool for tailoring change management interventions at an individual level, accelerating digital readiness across the workforce.

Kurt Lewin's Change Management Model is a foundational and widely cited framework that simplifies the complex process of organizational change into three distinct phases: Unfreeze, Change, and Refreeze. It provides a simple yet powerful mental model for understanding the dynamics of change. For workforce transformation, particularly in the context of digital maturity, this model is highly relevant.

Digital transformation often requires dismantling old ways of working (manual processes, legacy systems, traditional mindsets) before new digital approaches can be adopted. The "Unfreeze" phase is critical for breaking down resistance to digital change by highlighting the shortcomings of the current state and creating a compelling case for digital adoption. The "Change" phase then involves the actual implementation of new digital tools, workflows, and skills, which can be a period of confusion and reduced productivity as employees adapt. Leaders must provide extensive support and training during this phase. Finally, the "Refreeze" phase is crucial for embedding the new digital behaviors and processes into the organizational culture, ensuring that the investment in digital transformation yields sustainable results and that the workforce becomes truly digitally mature. This involves updating policies, aligning incentives, and consistently reinforcing the new digital norms. By understanding these three phases, leaders can strategically plan their interventions, anticipate challenges, and effectively guide their workforce through the necessary shifts to embrace a digital-first operating model, preventing a relapse into old, less efficient ways of working..

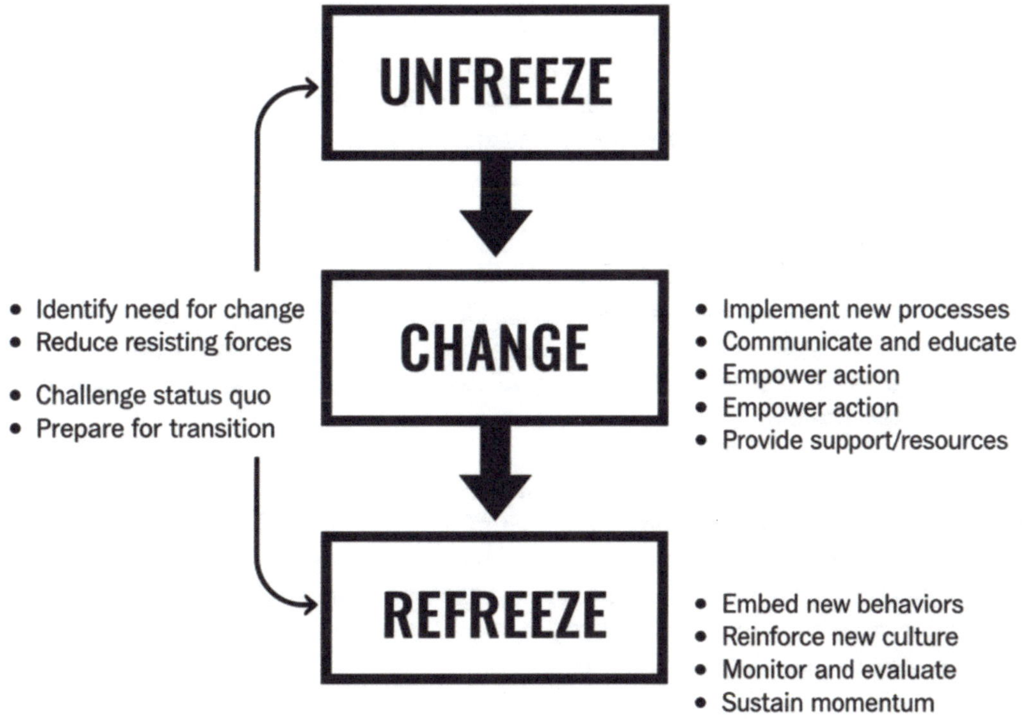

: Lewin's Change Management Model

1. **Unfreeze**: This preparatory phase involves creating a compelling need for change by disrupting the existing equilibrium. Leaders communicate the urgency (e.g., competitive threats from digitally native companies, opportunities from AI), highlight the shortcomings of the current state (e.g., slow manual processes, fragmented data), and challenge existing habits and mindsets (e.g., resistance to new digital tools). The goal is to make people receptive to change by showing why the current way of doing things is no longer sustainable or optimal, reducing resistance by making the imperative for digital transformation undeniable.

2. **Change**: This is the actual implementation phase where the transformation occurs. New behaviors, processes, systems, and organizational structures are introduced. In the context of digital workforce transformation, this means implementing new digital tools (e.g., collaboration platforms, automation software), adopting new digital workflows, and training employees on digital skills. During this period, employees are learning new ways of working, which can lead to confusion, anxiety, and a temporary dip in productivity. Leaders must provide continuous support, clear communication, necessary training, and resources to guide employees through this digital transition.

3. **Refreeze**: In this final phase, the new digital state is solidified and integrated into the organization's culture, policies, and systems to make the changes permanent. The goal is to prevent a return to old habits by reinforcing the new norms through rewards (e.g., recognizing digital champions), updated job descriptions that include digital competencies, and consistent leadership behavior that models digital adoption. This step ensures the digital transformation is sustainable and becomes the new "status quo," fostering a digitally mature workforce that instinctively leverages technology.

4. **Identify Driving Forces**: Before "unfreezing," leaders must identify and understand the forces that push for change. These could include market pressures from digital disruptors, the emergence of new digital technologies (e.g., AI, IoT), competitive threats from digitally agile rivals, new digital opportunities (e.g., e-commerce expansion), or internal inefficiencies caused by outdated manual processes. Recognizing these digital drivers helps build a compelling case for why digital transformation is necessary and urgent.

5. **Identify Restraining Forces**: Simultaneously, leaders need to identify and understand the forces that resist digital change. These often include fear of new technology, inertia towards learning new digital skills, lack of trust in digital solutions, vested interests in old manual processes, or a comfortable status quo that avoids digital disruption. Recognizing these digital restraints allows for proactive strategies to mitigate them (e.g., targeted training, clear communication of benefits) and reduce friction during the digital transformation process.

6. **Force Field Analysis**: Use the identified driving and restraining forces to conduct a force field analysis. This involves visually mapping these forces and their relative strengths. The goal is to plan interventions that either strengthen the digital driving forces (e.g., highlighting success stories of digital adoption) or weaken the digital restraining forces (e.g., providing extensive support for digital skill development) to shift the equilibrium towards the desired digital change. This analysis provides a strategic roadmap for managing resistance and enabling digital movement.

William Bridges' Transition Model distinguishes between "change" (the external event, like a new digital system being implemented) and "transition" (the internal, psychological process that people go through as they come to terms with the new situation). This distinction is critical for workforce transformation, especially in the context of digital maturity.

Digital transformation is not just about installing new software; it fundamentally alters how people work, interact, and perceive their roles. Bridges' model highlights three phases individuals experience: Endings, the Neutral Zone, and New Beginnings. Leaders often focus on the "new beginning" (the desired digital state) but neglect the emotional journey of "endings" (letting go of old, comfortable manual processes or familiar tools) and the challenging "neutral zone" (the period of confusion and uncertainty while adapting to new digital workflows and skills). Failing to manage these psychological transitions effectively can lead to resistance, disengagement, and a slow adoption of new digital capabilities. By understanding and addressing each phase, leaders can provide empathetic support, clear communication about the journey, and the necessary resources (e.g., coaching, training on new digital platforms) to help employees navigate the emotional landscape of digital change. This human-centered approach ensures that the workforce is not just technically capable but also psychologically ready and willing to embrace the new digital reality, accelerating the organization's journey towards true digital maturity..

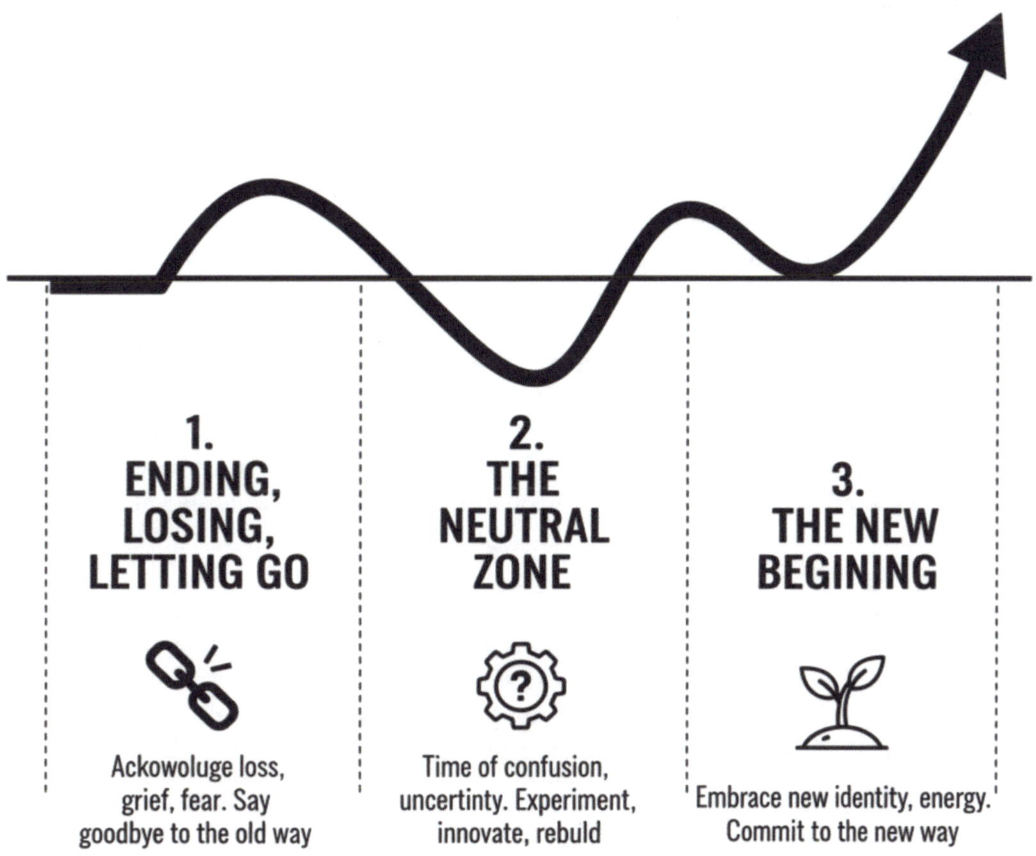

1. **Ending, Losing, Letting Go:** The first stage of transition involves helping individuals acknowledge and process the end of the old way of doing things. This is crucial because people must let go of the past before they can embrace the future. In digital transformation, this means recognizing the "loss" of familiar manual processes, old job roles, or a comfortable technological status quo. Leaders need to identify what is being lost (e.g., a sense of mastery with old tools, established routines) and allow for natural reactions like grief, anger, or resistance. Clear, empathetic communication about what's ending is vital to facilitate this letting go.

2. **The Neutral Zone:** This is often the most challenging stage – a period of uncertainty, confusion, and potential low morale and productivity between the old and the new. Individuals may feel disoriented, anxious, or unmotivated as old habits are gone but new digital ones aren't fully formed. This is where employees are learning new digital systems, adapting to new workflows, and figuring out their new roles in a digitally transformed environment. Leaders must provide clear structure, consistent support, frequent communication, and opportunities for experimentation, celebrating small wins to build momentum through this ambiguous phase.

3. **The New Beginning:** In this final stage, individuals begin to embrace the new way of working and their new identity within the transformed organization. They start to develop new digital skills, adopt new digital behaviors, and see the benefits of the digital change. They gain confidence in using new digital tools and contributing to the digitally-enabled future. Leaders must clearly communicate the new digital vision, reinforce new behaviors through recognition and rewards, and celebrate successes to solidify the new state and ensure the digital transition is complete and sustainable.

4. **Communicate** Continuously: Throughout all stages, provide clear, consistent, and empathetic communication about what is changing, why it's changing (the digital imperative), and how it will impact individuals' roles and daily work. Address concerns openly, listen actively to feedback regarding new digital tools or processes, and be transparent about challenges. Effective communication helps build trust and reduces anxiety during the digital transition, fostering a more informed and engaged workforce.

5. **Provide Support:** Offer comprehensive support to help individuals navigate the digital transition. This includes providing necessary training for new digital skills (e.g., data literacy, cloud proficiency), coaching for new digital behaviors, access to resources (e.g., digital helpdesks, online learning platforms), and creating a supportive environment where employees feel safe to learn and make mistakes with new technologies. Tailored support addresses individual needs and helps build digital capability across the workforce.

6. **Reinforce & Celebrate:** Acknowledge and reward efforts, celebrate milestones, and recognize individuals and teams who successfully adopt new digital behaviors and processes. Reinforcement helps to solidify the digital transition, embed the new ways of working, and prevent regression to old habits. Celebrating successes (e.g., a team's successful adoption of a new digital collaboration tool, a significant efficiency gain from automation) builds positive momentum and reinforces the benefits of the digital transformation, driving towards digital maturity.

The Prosci Change Management Methodology, built around the ADKAR Model, provides a structured and research-based approach to managing the people side of change. It offers a comprehensive set of tools, processes, and best practices for applying change management to projects and initiatives.

For workforce transformation, particularly in the context of digital maturity, Prosci is invaluable because it provides a systematic way to ensure that digital initiatives are not just technically sound but also humanly adopted. Many digital transformations fail not due to technology, but due to insufficient attention to employee readiness and adoption. Prosci's methodology guides leaders through preparing for change (defining strategy, building sponsorship), managing change (executing communication, coaching, training, and resistance management plans), and sustaining outcomes (reinforcing new behaviors and integrating changes into the culture). It emphasizes the critical role of active and visible sponsorship from senior leaders for digital initiatives. By systematically addressing the human elements of digital transformation—from building awareness of the need for new digital skills to reinforcing the use of new digital tools—Prosci helps organizations minimize resistance, accelerate adoption rates, and ensure that the workforce is fully equipped and willing to leverage new digital capabilities. This structured approach to managing the human element of digital change is a key enabler for achieving and sustaining digital maturity across the enterprise..

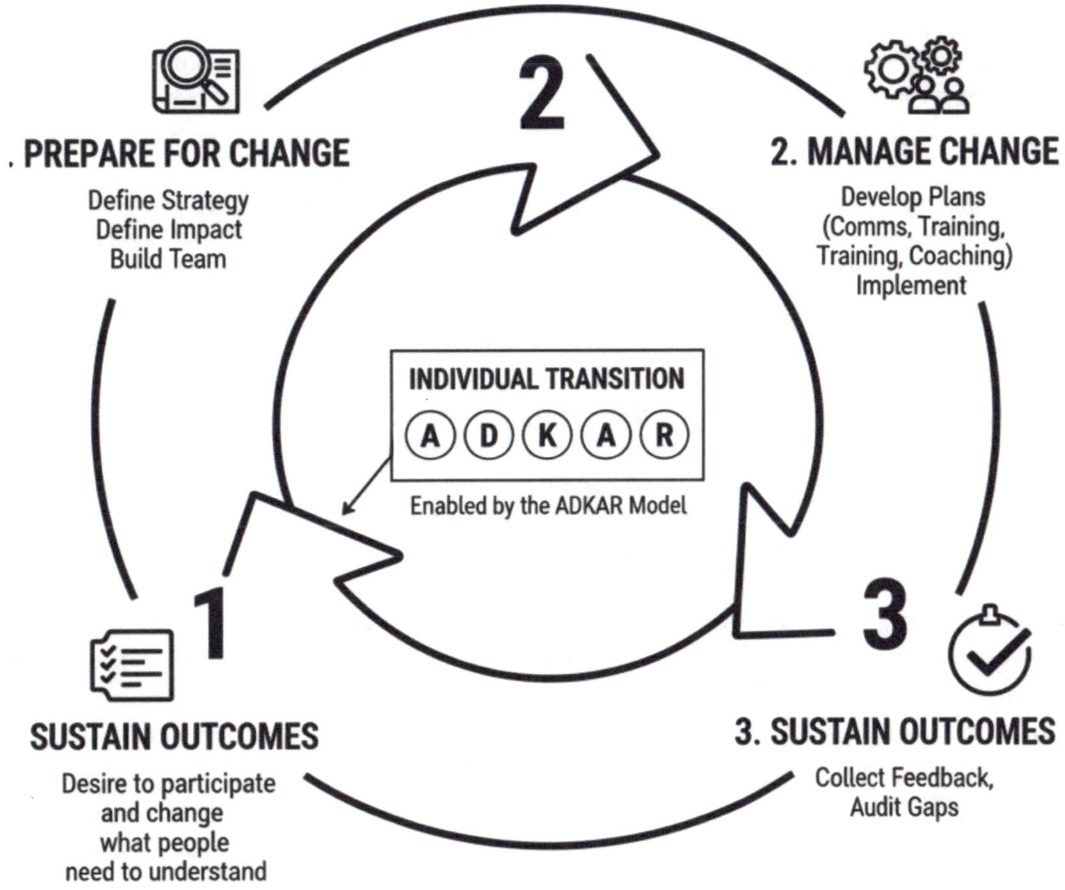

1. **Prepare for Change (Strategy):** This phase focuses on laying the groundwork for successful change. It involves defining the overall change management strategy based on the characteristics of the specific change (e.g., a large-scale digital transformation vs. a small software update), developing a robust sponsorship model to ensure active executive support for digital initiatives, and building a dedicated change management team with the necessary skills and resources to lead the effort. This strategic preparation ensures digital changes are approached systematically.

2. **Manage Change (Execution):** This phase involves the systematic execution of change management plans. It includes developing and delivering comprehensive communication plans to build awareness and desire for digital adoption, creating sponsorship roadmaps to guide leaders in their active and visible support of new technologies, developing coaching plans for managers to support their teams' digital upskilling, delivering targeted training programs for new digital tools, and proactively developing resistance management plans to address anticipated challenges and concerns related to digital transformation.

3. **Sustain Outcomes (Reinforcement):** The final phase focuses on ensuring the change sticks and delivers lasting results. This involves continuously collecting feedback and diagnosing any gaps in digital adoption, implementing corrective actions based on insights (e.g., additional training for a specific digital tool), celebrating successes to reinforce new digital behaviors, and integrating the change into the organization's culture, systems, and processes (e.g., updating performance reviews to include digital competencies). This ensures the workforce's digital maturity is sustained.

4. **Assess Readiness:** Before managing the change, conduct thorough assessments of organizational and individual readiness for the digital transformation. This involves evaluating the organization's capacity for digital change, identifying potential resistance points to new technologies, and understanding the current digital skills and mindsets of the workforce. This data informs the tailoring of change management strategies, ensuring interventions are targeted and effective for digital adoption.

5. **Develop Sponsorship Roadmap:** Create a detailed plan that outlines the specific activities and behaviors expected from key leaders and sponsors throughout the digital transformation. Active and visible sponsorship from the top is consistently identified as the number one success factor in change initiatives. This roadmap ensures sponsors understand and fulfill their critical role in championing digital initiatives, communicating benefits, and removing barriers.

6. **Integrate Change into Culture:** Beyond the project's completion, actively work to embed the new ways of working – particularly digital-first thinking, data literacy, and continuous learning – into the daily operations, systems, and organizational values. This involves updating policies, aligning reward systems, revising job descriptions to reflect digital roles, and ensuring that new behaviors are consistently modeled by leadership, making the transformed digital state the new "business as usual" and fostering deep digital maturity.

6. THE SATIR CHANGE MODEL (VIRGINIA SATIR)

The Satir Change Model, often depicted as a "change curve," provides a powerful framework for understanding the typical emotional and performance stages an organization (and its workforce) goes through during significant change. Developed by family therapist Virginia Satir, it highlights that disruption is a natural part of growth. It describes a progression from the "Late Status Quo" to the introduction of a "Foreign Element," leading to "Chaos," then "Integrating Practice," and finally a "New Status Quo."

For workforce transformation, particularly in the context of digital maturity, this model is invaluable because it helps leaders anticipate and normalize the inevitable dips in productivity and morale that occur when new digital systems or processes are introduced. When a new digital tool is rolled out, it acts as the "foreign element," often pushing the organization into "chaos" as employees struggle with unfamiliar interfaces, new workflows, and skill gaps. Understanding this predictable pattern allows leaders to prepare for the "chaos" phase, provide targeted support (e.g., intensive digital training, empathetic communication), and guide the workforce through the "integrating practice" phase where new digital skills are learned and adopted. By recognizing these stages, leaders can manage expectations, reduce anxiety, and accelerate the transition to a more effective "new status quo" where digital capabilities are fully integrated and optimized. This empathetic and anticipatory approach to managing the human side of digital change is a hallmark of a digitally mature organization..

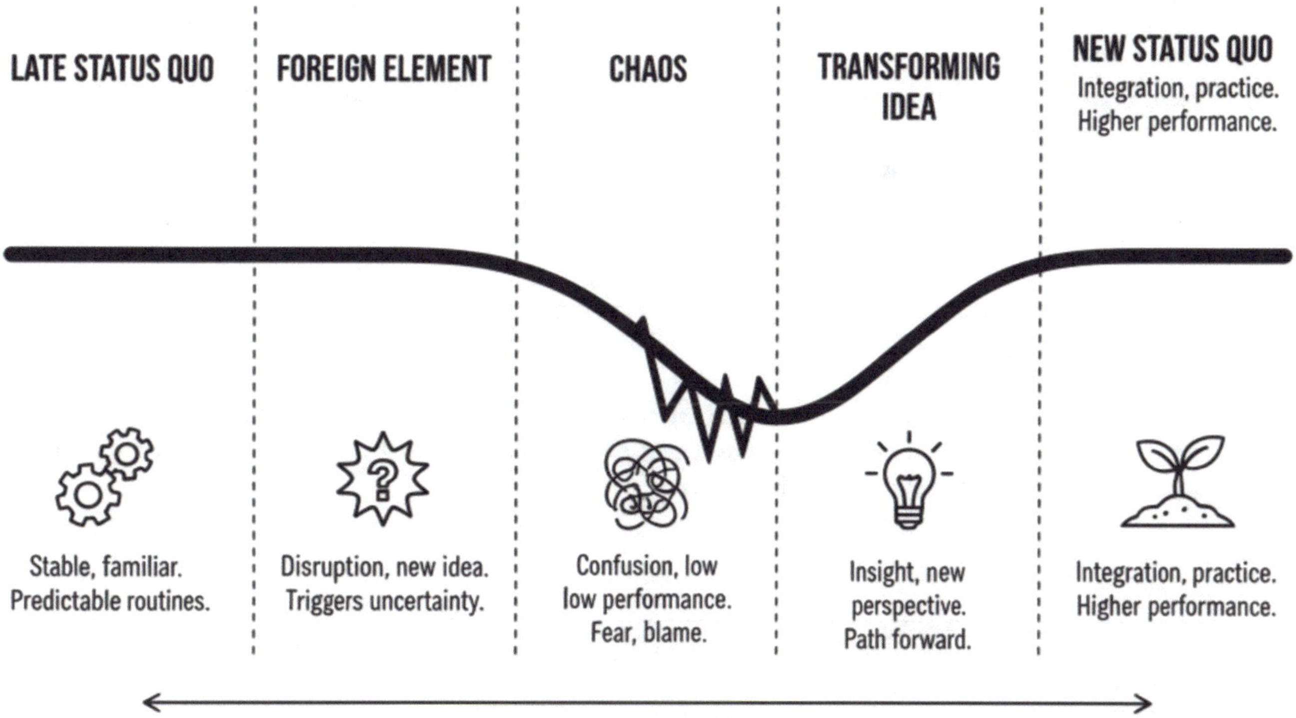

: Satir Change model

1. **Late Status Quo:** This is the period of normal, stable operations before any significant change is introduced. Performance is consistent, and established routines, norms, and expectations are firmly in place. People are comfortable and familiar with their roles and processes, operating within a predictable environment. In a pre-digital transformation context, this might represent reliance on manual processes, legacy systems, and traditional communication methods.

2. **Foreign Element:** A new idea, technology (e.g., a new enterprise software, AI integration), policy, or external event (e.g., a digital disruptor entering the market) acts as the "foreign element" that disrupts the status quo. This is the initial trigger for change, creating awareness that the current way of doing things is no longer sustainable or optimal. It introduces uncertainty and signals that a significant digital shift is imminent, moving the organization out of its comfort zone.

3. **Chaos:** As the foreign element is introduced and the old status quo breaks down, performance dips significantly. Confusion, frustration, anxiety, and resistance are common emotions. This is a critical period where the organization's established systems and processes may no longer work effectively, leading to a temporary state of disarray and reduced productivity. For digital transformation, this often happens when new digital tools are rolled out, and employees struggle with new interfaces, workflows, or skill requirements. Strong, empathetic leadership is vital here to navigate the digital disruption.

4. **Integrating Practice:** The organization begins to experiment with new ways of working, new behaviors, and new systems in response to the chaos. Through trial and error, learning, and adaptation, individuals and teams start to find solutions and integrate new digital practices. Performance slowly begins to improve as understanding grows and new routines are formed. This phase involves extensive training, coaching, and support to help the workforce master new digital tools and processes.

5. **New Status Quo:** The new ways of working become the norm. Performance stabilizes at a new, often higher, level, and the organization regains a sense of stability and predictability. The new behaviors, processes, and systems (now digitally enabled) are embedded into the daily operations, becoming the accepted and expected way of functioning. This represents a significant step towards digital maturity, where the new digital capabilities are fully integrated.

6. **Monitor & Reinforce:** Continuously track performance in the new status quo and provide reinforcement to ensure the changes are sustained and further optimized. This involves celebrating successes (e.g., achieving new digital efficiency targets), providing ongoing support for digital tools, and making minor adjustments as needed to ensure the new norms are firmly established and the organization remains adaptable for future digital changes. This continuous feedback loop is essential for maintaining and advancing digital maturity.

"The future belongs to those who learn more skills and combine them in creative ways."
— Michael S. Dell, Founder, Dell Technologies

Ai6x6™ Transformation Blueprint

The Unlearning Rate (UR)

"The illiterate of the 21st century will not be those who cannot read and write, but those who cannot learn, unlearn, and relearn." --- **Alvin Toffler,** the futurist who predicted the "Information Overload" of the modern era defined the ability to discard legacy mental models as the core skill of the post-industrial age. For the **Workforce**, the Unlearning Rate is the metric of the "Momentum Builder." As AI renders traditional skills obsolete, Toffler's insight frames "Unlearning" as a competitive advantage, forcing teams to shed high-carbon habits to make room for regenerative innovation.

In the traditional industrial model, a "skilled" worker was a repository of static knowledge. We hired for "experience," which was essentially a measure of how many times a human had repeated a specific task. However, in the AI-augmented era, the value of static knowledge is depreciating at an exponential rate. The machine now holds the "functional" truth of how to code, how to calculate, and how to analyze. The new bottleneck isn't a lack of skills; it is the presence of old ones. This is the Identity Chasm of the workforce: the psychological friction of letting go of a "legacy self" to make room for an "augmented self." To survive this transition, we must shift the culture from rewarding "Output Volume" to rewarding "Cognitive Velocity." The primary metric for this evolution is the Unlearning Rate (UR).

In the workforce, the chasm exists between **Task-Based Identity** and **Strategic Agency**. Employees often wear their manual skills like armor, believing that their value lies in the "doing." When AI automates the "doing," the worker feels existentially exposed. Identifying this chasm is essential to move teams from "Skill-Armor" to "Cognitive Velocity." It shifts the worker's purpose from being a repository of static knowledge to becoming a pilot of infinite intelligence, fulfilling their raison d'être to strive.

The greatest barrier to workforce transformation is Skill-Armor. Employees often tie their self-worth and job security to the "manual touch"—the specific, repetitive tasks they have mastered over decades. When AI automates those tasks, the employee doesn't just see a more efficient process; they see an existential threat. They cling to "Ghost Processes"—manual workarounds that serve no purpose other than to make the human feel "necessary."

This creates a state of "Functional Stagnation." If the machine can handle the execution, but the human refuses to let go of the steering wheel, the organization pays for the machine's speed but continues to move at a human crawl. To bridge this chasm, the workforce must undergo a psychological shift: realizing that their value is not in the task, but in their Moral Authority over the machine's output. This is the move from being a "Process Follower" to being a "System Architect."

THE RAISON D'ÊTRE:

"TO STRIVE"

By mastering the UR, the workforce fulfills its true purpose: To Strive. When humans are freed from the "mechanical" aspects of their jobs, they are pushed into the frontier of mastery. Unlearning is not about loss; it is about Liberation. It allows the human to move from the "how" (function) to the "why" (purpose). In the post-AI era, the most valuable employees aren't the ones who know the most; they are the ones who can "empty their cup" the fastest. We are not replacing the worker; we are replacing the drudgery that prevents the worker from reaching their full potential.

DEFINING THE UR: THE LOGIC OF SKILL-SHEDDING

The Unlearning Rate is a measurement of Role Migration. It asks one simple question: How much of yesterday's identity have you successfully retired to claim today's opportunity?

In a stagnant organization, the UR is near zero. People are doing the same things they did five years ago, just with an AI tool sitting unused on their desktop. In a high-velocity culture, the UR is high. Roles are constantly being "refactored." A junior analyst might unlearn "data cleaning" (now handled by AI) to master "scenario synthesis" (the human agency). The goal is to maximize the speed at which "Legacy Tasks" are decommissioned so that "Human Agency" can be redeployed toward higher-value creation.

THE OBJECTIVE: TOWARD COGNITIVE AGENCY

The goal of the UR intervention is to reach a state of Equilibrium we call Cognitive Agency. In this state, the workforce operates as a fleet of "Strategic Pilots." They understand that their primary job is to direct the intelligence of the machine toward the most valuable problems.

Cognitive Agency allows the workforce to scale its impact without scaling its effort. By increasing the UR, you aren't just "upskilling"; you are "right-skilling" for a world of infinite intelligence. You are allowing the "Logic" of the machine to handle the "Work," so the "Spirit" of the worker can handle the "Creation." This is the only way to maintain human relevance in a world where tasks are becoming a commodity.

SAMPLE CASE STUDIES ON MASTERY CHASM

CASE STUDY	THE CHASM REALITY	ACTION & RESULT
1: A GLOBAL CPG COMPANY spent millions on localized marketing copy. The team spent 70% of their time on basic translation and formatting, leaving no room for creative strategy or cultural nuance.	The team was trapped in a **Task-Based Identity**, fearing that machine-generated drafts made their language skills redundant. They resisted AI to protect their perceived value.	Leadership launched a **"Skill-Shedding Salon,"** incentivizing staff to automate 100% of first drafts. This pivoted the team to "Cultural Orchestrators," increasing campaign performance by 22%.
2: A TECH FIRM was drowning in technical debt. Senior engineers spent most of their week fixing minor bugs in legacy code, viewing their knowledge of the old system's "quirks" as their primary job security.	This was an **Expert Wall** scenario. Engineers refused to feed their "secret sauce" into AI pair-programming tools, viewing the machine as a threat to their status rather than a tool for liberation.	The firm enacted **Collaborative Prompting**, rewarding engineers who trained AI to handle 80% of routine fixes. Veterans became "System Architects," accelerating the release cycle by 3x.
3: A NATIONAL INSURANCE PROVIDER struggled with slow turnaround for property claims. Field adjusters spent hours manually measuring damage and typing reports, a process they felt defined their professional rigor.	Adjusters wore **Skill-Armor**, "ghosting" new AI vision tools because they believed only "looking with their own eyes" ensured accuracy and maintained their professional worth.	Management implemented **Legacy Decommissioning** by removing manual forms. Adjusters retrained as "Empathy Specialists" for complex cases. Processing time dropped from 12 days to 36 hours.
4: A PUBLIC UNIVERSITY was paralyzed by manual enrollment paperwork. Admissions staff spent months cross-referencing transcripts, viewing this "paper-pushing" as essential to protecting academic standards.	Staff suffered from **Task-Based Identity**. They feared an algorithm would make their roles vanish, leading to "Shadow Paperwork" where they privately double-checked the machine's work.	The university held a **"Retirement Ritual"** for manual filing. Staff were pivoted to "Student Success Coaches," focusing on retention. This increased student satisfaction scores by 40%.
5: A GLOBAL FASHION RETAILER noticed store managers spent 15 hours weekly on inventory spreadsheets. Managers took pride in this "manual control" even as stockouts and overages increased.	Managers used the **Busywork Shield**. By staying buried in spreadsheets, they avoided high-stakes human tasks like coaching staff or engaging customers—tasks the AI could not perform.	Leadership mandated **Autonomous Inventory Management**. Managers were forced back to the floor as "Brand Ambassadors," leading to a 15% sales increase and a 30% reduction in staff turnover.
6: A TOP-TIER LAW FIRM struggled with the high cost of document discovery. Paralegals spent thousands of billable hours per case searching for keywords—a grueling task viewed as a "rite of passage."	The firm faced **Seniority Stagnation**. Partners felt "the grind" built character, while paralegals feared AI would make junior roles obsolete, leading to "Feature Ghosting" of discovery software.	The firm shifted to **Agency-First Training**, showing paralegals how to orchestrate AI for legal strategy. Paralegals became "Strategy Analysts," winning cases with 40% less overhead.

ASSIGNMENT : THE WORKFORCE CULTURE DIAGNOSTIC AUDIT

The transition from a "Process Follower" to a "Strategic Agent" requires a ruthless inventory of hidden resistance. We often mistake "hard work" for "valuable work," but in the AI era, holding onto manual habits is a form of organizational debt. To bridge the Identity Chasm, you must be able to recognize the specific symptoms of Skill-Armor and apply the precise architectural intervention to liberate the human potential within your teams. This audit is designed to move your workforce from legacy stagnation to cognitive velocity. On the left are ten common Symptoms found in organizations with a low Unlearning Rate. On the right are ten Strategic Interventions from the Ai6x6 playbook.

The Assignment: Match the symptom observed in your current workforce culture to the intervention required to reach Equilibrium.

THE SYMPTOMS (THE DIAGNOSTIC)	MATCH	THE INTERVENTIONS (THE REMEDY)
1. Shadow Paperwork: Staff keeping manual logs or offline spreadsheets "just in case" the AI is wrong.		**A. Capability Mapping:** Redefining job descriptions based on "Human-Only" outcomes like Empathy and Ethics.
2. Feature Ghosting: High-cost AI tools are deployed, but staff only use them for basic "Google-style" searches.		**B. The Retirement Ritual:** Publicly "retiring" legacy processes to signal that it is safe (and expected) to let go.
3. Seniority Stagnation: The most experienced staff are the loudest critics of new technology, fearing loss of status.		**C. Skill-Shedding Incentives:** Rewarding employees for automating their own tasks and moving to higher-value work.
4. The Busywork Shield: Employees filling their day with low-value manual tasks to avoid the "threat" of free time.		**D. Safe-to-Fail Sandboxes:** Creating low-stakes environments where staff can "play" with AI without performance pressure.
5. Task-Based Identity: Employees describe their job by *what they do* (the function) rather than *the value they create.*		**E. "Unlearning" KPIs:** Measuring managers on how quickly their teams decommission legacy workflows.
6. The Expert Wall: Specialists refusing to share data or prompts because they fear "training their replacement."		**F. Agency-First Training:** Shifting from "How to use the tool" to "How to direct the system as a strategist."
7. Automation Resentment: A palpable "Us vs. Them" mentality regarding the machine's role in the daily workflow.		**G. The Transparency Compact:** A guarantee that "Unlearning" leads to role evolution, not just headcount reduction.
8. Cognitive Overload: Staff trying to do their "old job" AND their "new AI job" simultaneously without dropping tasks.		**H. Legacy Decommissioning:** Forcefully removing old software and paper processes to prevent "Ghosting."
9. The Feedback Void: Workers see AI errors but don't report them, secretly hoping the system fails to prove their worth.		**I. Collaborative Prompting:** Making "Model Optimization" a core part of every employee's performance review.
10. Incrementalism Bias: Staff using AI only to do the same old things 5% faster, rather than exploring new capabilities.		**J. Radical Role Refactoring:** Dissolving old job titles and creating new ones that reflect human-machine partnership.

ANSWER MATCH: 1-H, 2-F, 3-B, 4-C, 5-A, 6-I, 7-G, 8-J, 9-D, 10-E

SECTION F

BLUEPRINT SERIES

Cultural Compass

Ai6x6™ Transformation Blueprint

Story 1 — The Post-War Japanese Industrial Ethos (1945–1960): Japan emerged from the war with limited resources but a collective commitment to national rebuilding. Companies embraced a culture of discipline, apprenticeship, and shared responsibility. Workers identified deeply with their organizations, reinforcing loyalty and craftsmanship. Leaders cultivated humility and long-term thinking. This cultural foundation enabled rapid industrial recovery. It demonstrated how shared values can accelerate capability even before systems mature.

Story 2 — NASA's Mission Culture During Apollo (1961–1972): NASA built a culture defined by precision, accountability, and collective purpose. Engineers and astronauts operated with a shared belief that failure was unacceptable, yet learning was essential. Communication was open, direct, and cross-disciplinary. Rituals, reviews, and norms reinforced psychological safety and technical rigor. This culture enabled teams to navigate uncertainty and complexity. It became a global example of mission-driven cohesion.

Story 3 — The Civil Rights Movement's Organizing Ethos (1954–1968): The movement relied on a culture of discipline, nonviolence, and moral clarity. Communities trained in de-escalation, collective action, and strategic communication. Leaders modeled courage and restraint, shaping norms that guided behavior under extreme pressure. Shared values created unity across diverse groups. This cultural coherence amplified impact far beyond organizational structure. It showed how culture can mobilize change at societal scale.

Story 4 — The Nordic Social Trust Model (1970s–1990s): Nordic countries cultivated cultures rooted in transparency, equality, and civic trust. Institutions operated with high accountability and low hierarchy. Citizens expected fairness and contributed to collective welfare. This cultural fabric enabled strong public services, stable governance, and economic resilience. Trust reduced friction in decision-making and collaboration. It demonstrated how cultural norms can become strategic national assets.

Story 5 — The Rise of Silicon Valley's Innovation Culture (1980s–2000s): Silicon Valley developed a culture that celebrated experimentation, risk-taking, and rapid iteration. Informal networks, open information flow, and meritocratic ideals shaped behavior. Failure was reframed as learning, attracting talent willing to push boundaries. Venture capital reinforced speed and ambition. This culture accelerated technological breakthroughs and entrepreneurial growth. It showed how cultural norms can amplify innovation beyond structural advantages.

Story 6 — The Globalization of Corporate Diversity & Inclusion (2000s–2020s): Organizations worldwide recognized that diverse teams outperform homogeneous ones when supported by inclusive cultures. Companies invested in psychological safety, equitable processes, and representation. Employees gained platforms to voice perspectives and challenge norms. Inclusion became a driver of creativity, retention, and ethical integrity. This cultural shift reshaped leadership expectations and organizational identity. It demonstrated how culture evolves to meet societal and competitive demands.

LESSONS FROM A WORKFORCE TRANSFORMATION PERSPECTIVE

Workforce maturity in culture emerges when employees internalize shared values that guide behavior across situations. Over time, organizations learned that culture enables coordination, reduces friction, and strengthens collective identity. Mature workforces operate with norms that support collaboration, learning, and ethical conduct. They uphold standards not because they are mandated, but because they are believed. Culture empowers workers to act autonomously while remaining aligned with organizational purpose. As institutions evolve, the workforce becomes a carrier of cultural continuity—transmitting practices, reinforcing expectations, and sustaining cohesion. Maturity is reached when culture becomes a lived experience, not a documented aspiration.

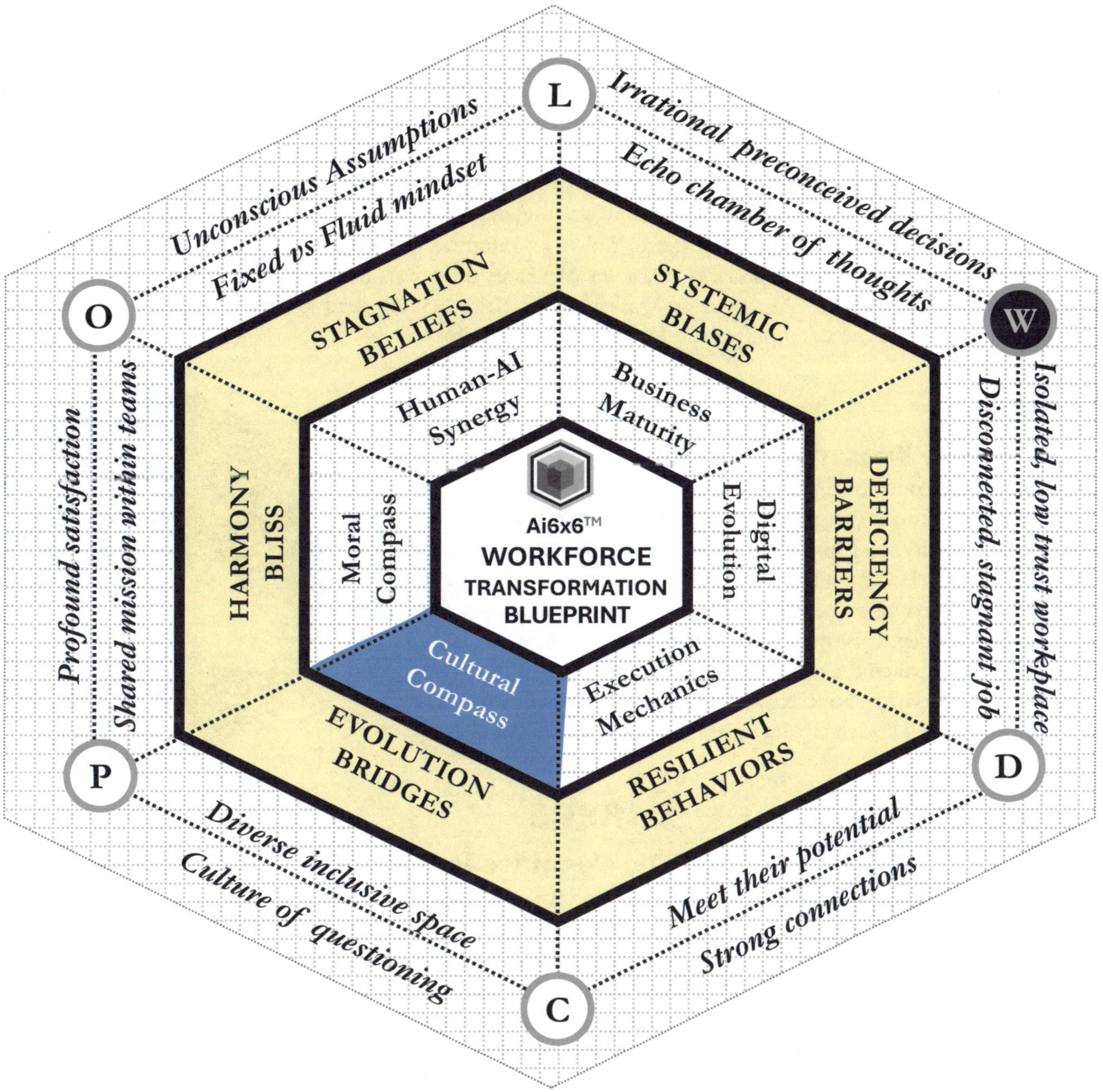

The **Ai6x6 Research Lab** chose this lens because the ultimate vitality of a workforce is found in its collective resonance and psychological safety. We reframed this dimension as the stewardship of "Workforce Harmony," recognizing that culture is the invisible force that either fuels or fractures the transformation engine. It is vital because it addresses the invisible shadows—the deep-seated stagnation, systemic biases, and deficiency mindsets—that live in the minds of the people. This lens is the bridge between the organization's high-tech goals and the workforce's human reality, moving the needle from fear to resilient behaviors.

We selected it to ensure leaders build an evolving and harmonious community rather than just a productive headcount. By dismantling systemic barriers and building "Evolution Bridges," the leader creates a "greenhouse" for human potential where every worker feels equipped to grow alongside the machine. The Lab believes that in an age of automation, human resonance is the only self-sustaining source of organizational energy. This lens forces a look at how beliefs about stagnation impact the collective output. When culture is treated as a discipline, the workforce achieves a state of "bliss"—a high-fidelity hum where every individual feels empowered, valued, and fundamentally connected to the shared mission.

Culture has always been the invisible architecture of an organization, shaping how people think, behave, and make decisions long before any formal strategy takes hold. In the age of AI, this architecture becomes even more consequential. Technology may accelerate capability, but culture determines whether that capability becomes meaningful progress or wasted potential. The **Cultural Compass** lens forces leaders to confront the truth that transformation is not sustained by systems alone; it is sustained by the beliefs people hold, the biases they carry, the barriers they navigate, the behaviors they exhibit, the bridges they build, and the moments of profound satisfaction that anchor their commitment. Without cultural alignment, even the most sophisticated AI strategy collapses under the weight of human resistance. This lens reframes culture as a dynamic system rather than a static identity. It recognizes that culture is not a set of values on a wall but a living pattern of interpretation, interaction, and intention. As AI reshapes workflows, decision cycles, and power dynamics, leaders must understand how cultural forces amplify or undermine transformation.

Beliefs

The Narratives Beneath the Surface

Beliefs and assumptions form the first dimension of the Cultural Compass. They are the quiet forces that shape how people interpret change, how they respond to uncertainty, and how they judge the intentions behind leadership decisions. When beliefs are outdated or misaligned, they distort perception and create friction. Leaders who ignore this dimension discover that people do not resist technology—they resist the meaning they attach to it. Addressing beliefs requires leaders to surface the narratives that operate beneath the surface and reshape them with clarity, honesty, and purpose. In this dimension, cultural truth-telling becomes a leadership act rather than a communication tactic.

Biases

The Cognitive Patterns That Shape Judgment

Biases and mental shortcuts represent the second dimension. These cognitive patterns influence how people evaluate information, assign trust, and make decisions. In an AI-enabled environment, biases can distort both human judgment and the interpretation of machine-generated insights. Addressing them requires intentional design, transparent dialogue, and environments where people feel safe challenging their own assumptions. Leaders must create conditions where cognitive unlearning becomes possible. In this dimension, cultural evolution is driven by awareness, humility, and the willingness to question long-held mental models.

Barriers

The Structural Forces That Restrict Change

Barriers and systemic impediments form the third dimension. These are the structural forces—policies, hierarchies, incentives, and legacy processes—that constrain behavior even when people are willing to change. Leaders often misinterpret resistance as emotional when it is structural. Removing barriers is not about demanding more effort; it is about redesigning the environment so that the desired behaviors become the natural ones. In this dimension, organizational de-friction becomes a strategic priority rather than an operational afterthought, enabling people to move with clarity rather than struggle against constraints.

Behaviors

The Visible Expression of Culture

Behaviors and observable actions represent the fourth dimension. Culture is not what people say; it is what they do consistently. In the AI era, behaviors must evolve to reflect curiosity, adaptability, collaboration, and responsible experimentation. Leaders must model these behaviors visibly and repeatedly. They must demonstrate that learning is valued, that questioning is encouraged, and that accountability is shared. In this dimension, behavioral signaling becomes a powerful tool for shaping norms, reinforcing the idea that culture is enacted through daily choices rather than declared through slogans.

Bridges

The Pathways That Connect People and Purpose

Bridges and cultural pathways form the fifth dimension. These are the connections that allow ideas, trust, and collaboration to flow across the organization. Bridges are built through shared language, shared rituals, and shared purpose. They transform isolated teams into aligned communities. Leaders who invest in cultural pathways create environments where people feel connected to something larger than their role. In this dimension, collective identity-building becomes a catalyst for cohesion, enabling the organization to move as a unified whole rather than a collection of disconnected parts.

Bliss

The Emotional Anchor of Commitment

Bliss and profound satisfaction complete the compass. Bliss is not about comfort; it is about meaning. It is the sense of fulfillment that emerges when people feel their work matters, their contributions are recognized, and their growth is supported. Bliss is the emotional anchor that sustains commitment through uncertainty. Leaders who cultivate it create cultures where people do not merely comply—they care. In this dimension, purposeful fulfillment becomes a strategic asset, strengthening resilience and deepening engagement.

The Cultural Compass lens reminds leaders that transformation is not sustained by technology or process alone. It is sustained by the human experience of change. Culture determines whether people interpret transformation as threat or opportunity, whether they respond with resistance or curiosity, and whether they move in alignment or fragmentation. In a world where AI accelerates both progress and pressure, leaders must treat culture as the primary operating system of the enterprise. When leaders align beliefs, challenge biases, remove barriers, shape behaviors, build bridges, and cultivate meaning, they create cultures capable of evolving continuously. They build environments where people feel psychologically safe, intellectually stimulated, and emotionally invested. Culture becomes the force that amplifies strategy rather than undermining it. It becomes the stabilizing anchor in moments of disruption and the accelerating engine in moments of opportunity.

CULTURAL COMPASS ANALOGY – CULTIVATING JAPANESE GARDEN

The False Climate — BELIEFS

Dismantling "Weather Myths" ensures that unrealistic, universal expectations don't stifle individual growth. Exposing the false belief that all "plants" require the same sun, water, or timing—allowing for specialized care.

The Invasive Weeds — BIASES

Identifying and removing hidden mental shortcuts prevents old habits from choking out innovation. Surgically exposing the leadership judgment and systemic biases that create "shadow" cultures and block progress.

The Hard Clay — BARRIERS

Breaking through structural resistance allows the organizational roots to expand and stabilize. Removing the deeply embedded obstacles like data, automation and AI anxiety that act as physical walls against digital growth.

Calibrated Care — BEHAVIORS

Executing specialized nurturing routines ensures each distinct "species" receives what it needs to thrive. Moving away from "one-size-fits-all" mode to model specific watering, feeding, and light-balancing behaviors required for a diverse ecosystem.

The Trellis System — BRIDGES

Constructing pathways for upward growth provides the support needed for teams to reach their full potential. Building the specific leadership and development paths that allow the workforce to climb through the maturity stages.

The Full Harvest — BLISS

Securing a state of permanent organizational harmony results in a self-sustaining ecosystem of excellence. Reaching the ultimate destination where the "Cultural Soul" operates in a high-fidelity hum, delivering constant value.

Ai6x6™ Transformation Blueprint

Workforce Stagnation Beliefs

"If you hire people just because they can do a job, they'll work for your money. But if you hire people who believe what you believe, they'll work for you with blood, sweat, and tears." - Simon Sinek

Simon Sinek (b. 1973) is a British-American author and motivational speaker. He is best known for his work on the concept of "Start With Why," which he introduced in his famous TED Talk on leadership.

Workforce stagnation beliefs are negative, ingrained assumptions about an organization's employees that prevent a culture of growth and innovation. These beliefs can be held by leaders, managers, or employees. A classic example is Aptitude Beliefs, which suggest an employee's skills and potential are fixed. A manager with this belief will never invest in employee development, creating a self-fulfilling prophecy of stagnation. Similarly, outdated Work Ethic Beliefs or a pervasive skepticism about training can create a rigid, one-size-fits-all environment. The journey to a transformed workforce begins with challenging these limiting beliefs and cultivating a new mindset that views every employee as a valuable source of potential, capable of continuous learning and growth.

This Chapter establishes the Relational Core by focusing on Workforce Stagnation Beliefs, the collective mindset that determines change Adoption and builds the essential Cultural Compass. It examines how shared group convictions—the unwritten rules of the organization—either accelerate or sabotage large-scale transformation efforts. The core is built on Collaborative Beliefs, demonstrating that shared success, cross-functional teamwork, and silo demolition are non-negotiable. This is supported by Innovation Beliefs, which foster a psychologically safe environment where new ideas are welcomed, tested, and risk is managed, not feared. The book then tackles performance and trust through Accountability Beliefs and Transparent Beliefs, ensuring clear expectations, honest feedback, and open communication are institutionalized during times of uncertainty and rapid change. Furthermore, it promotes Continuous Learning Beliefs, positioning skill development as a lifelong organizational imperative, and culminates with Ownership Beliefs, cultivating a stakeholder mentality where every employee acts as a strategic owner, thereby securing lasting cultural change and sustainable results.

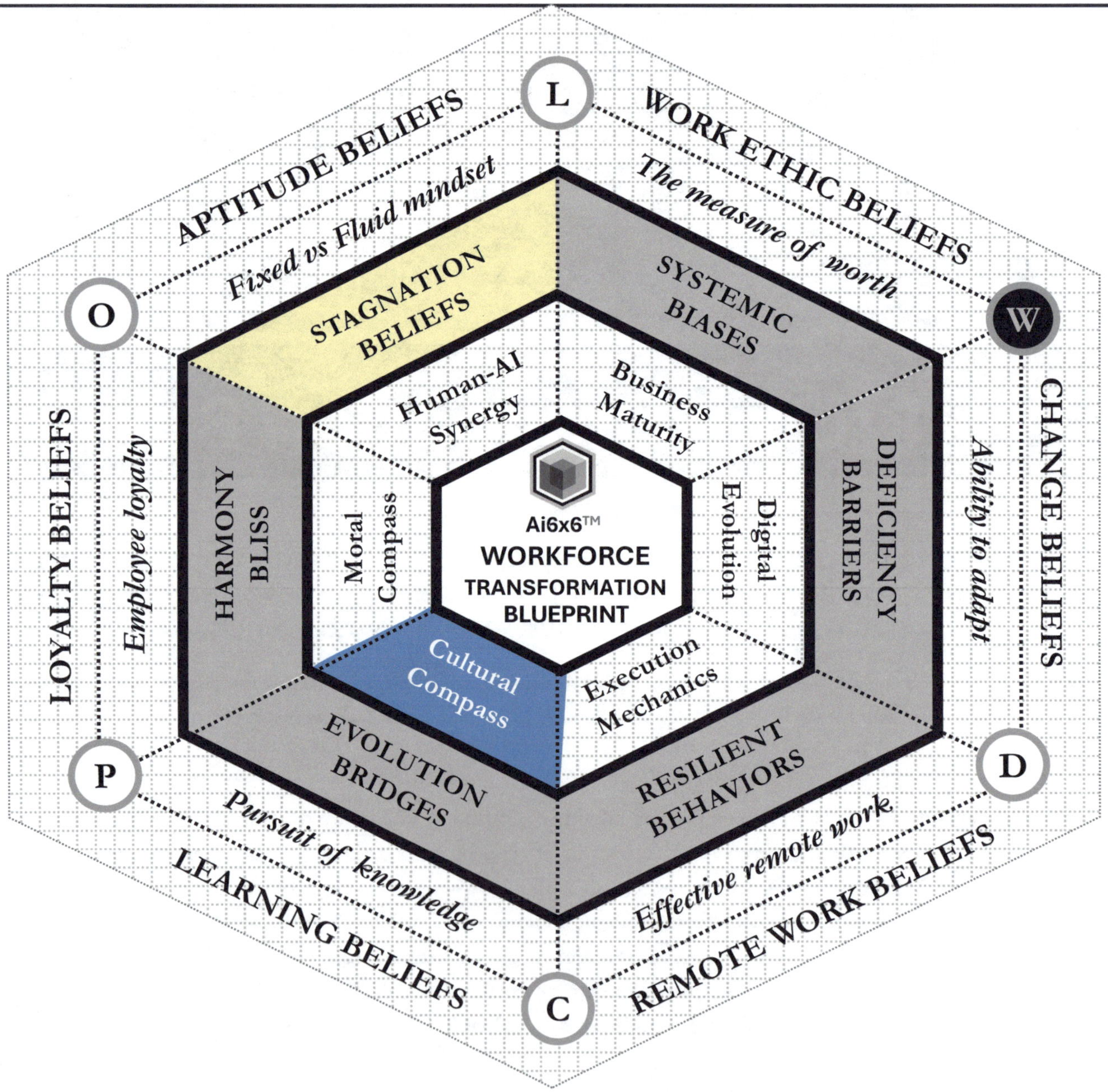

Ai6x6 CULTURAL COMPASS FOR BELIEFS | WORKFORCE TRANSFORMATION

1. APTITUDE BELIEFS | The Fixed vs. Fluid Mindset

If you believe talent is "born, not made," you treat your workforce like a fixed asset rather than a growing one. In the context of transformation, this belief kills your ROI on upskilling initiatives. Executives who view aptitude as static will constantly look outside for new talent instead of unlocking the latent potential within their own ranks. This creates a stagnant culture where employees stop trying to evolve because they feel their ceiling is already pre-determined by your rigid assessment. Real transformation requires believing that with the right environment and incentives, your current workforce can acquire the digital and strategic fluencies required for the future. Without this belief, you are essentially managing a depreciating asset, whereas a "growth mindset" belief turns your workforce into a compounding investment.

DO'S FOR THE LEADER	DON'TS FOR THE LEADER
Believe in a growth mindset.	Believe skills are fixed and can't change.
Invest in employee training.	Assume people can't learn new skills.
Provide feedback for development.	Don't give feedback on growth.
Encourage everyone to experiment.	Only promote "naturally talented" people.
Acknowledge that learning takes time.	Expect instant expertise from people.
Highlight the importance of effort.	Focus only on innate abilities.

SITUATION	SUGGESTED ACTION
Leaders assume some employees “aren't wired” for AI work and avoid offering development opportunities	Design structured learning paths, emphasize growth potential, and showcase internal upskilling examples to shift beliefs from fixed aptitude toward expandable capability through practice
Employees believe technical skills are innate, avoiding engagement with new AI tools and platforms	Create low-pressure learning labs, normalize early mistakes, and highlight visible progress to reframe technical competence as learnable rather than predetermined
Managers label individuals as “non-technical,” excluding them from strategic digital or AI initiatives	Rotate employees through guided AI-adjacent tasks, pair them with mentors, and track improvement to demonstrate capability growth over time
Teams assume only specialists can meaningfully work with AI systems or automation workflows	Offer cross-functional AI literacy sessions, demystify core concepts, and reinforce shared responsibility for digital fluency across roles and levels
Employees fear failure when learning AI tools, reinforcing beliefs that improvement is unlikely or embarrassing	Celebrate incremental progress publicly, spotlight learning journeys, and frame early missteps as essential steps in building durable capability
Leaders rely heavily on past performance to predict future digital capability and advancement potential	Incorporate learning agility, curiosity, and experimentation into talent assessments, rewarding demonstrated growth rather than static historical skill profiles

Viewing "work ethic" strictly through the lens of hours logged or physical presence is a legacy belief that sabotages modern productivity. In a transformed workforce, value is found in outcomes and cognitive intensity, not duration. If your leadership team still prizes "first in, last out," you are likely rewarding performative busyness rather than actual impact. This belief alienates your most efficient high-performers and forces a culture of presenteeism. To transform, you must shift your belief to value the quality of contribution and the ability to solve complex problems, regardless of the traditional "9-to-5" markers of effort. By redefining a "good" work ethic as the ability to drive results and innovate, you free your best talent to work in ways that maximize their actual output.

DO'S FOR THE LEADER	DON'TS FOR THE LEADER
Define a good work ethic clearly.	Make assumptions about work ethic.
Judge people by output, not hours.	Value "face time" over results.
Trust your team to manage their time.	Micromanage people's daily activities.
Provide autonomy over how work is done.	Dictate every step of the work process.
Focus on achieving shared goals.	Prioritize individual tasks and effort.
Reward results and collaboration.	Only reward long hours and sacrifice.

SITUATION	SUGGESTED ACTION
Leaders equate long hours with commitment, undervaluing efficiency, automation, and outcome quality	Redefine performance expectations around impact, intelligent tool use, and sustainable delivery rather than visible time spent or constant presence
Employees believe constant availability signals dedication, leading to burnout and declining performance	Establish clear boundaries, model healthy disconnection, and reinforce that recovery, focus, and thoughtful pacing strengthen long-term contribution and reliability
Managers assume quiet or introverted employees lack drive or strong work ethic	Evaluate work ethic using reliability, ownership, and outcomes, decoupling perceived energy levels from actual contribution and professional commitment
Teams view asking for help as weakness, equating independence with superior work ethic	Normalize collaborative problem-solving, reward knowledge-sharing, and position effective escalation as a mature, high-performance behavior rather than deficiency
Leaders reward visible busyness instead of strategic prioritization and automation adoption	Recognize those who streamline workflows, leverage AI effectively, and focus on high-value activities that materially advance organizational goals
Employees believe multitasking demonstrates strong work ethic, despite reduced focus and quality	Promote single-tasking, structured work blocks, and AI-assisted task management to improve accuracy, reduce errors, and elevate overall performance standards

The assumption that "people naturally resist change" becomes a self-fulfilling prophecy. If you believe your employees are inherently rigid, you will approach transformation with a defensive, top-down mandate that triggers the very resistance you fear. In reality, people don't resist change; they resist being changed without agency. When you hold this negative belief, you fail to design inclusive transition paths, leading to a workforce that feels "done to" rather than "part of." Transforming this belief involves recognizing that your team is capable of incredible adaptation if they understand the 'why' and feel supported through the 'how.' When you start believing in your team's inherent adaptability, you stop being a "enforcer" of change and become an "architect" of a naturally evolving organization.

DO'S FOR THE LEADER	DON'TS FOR THE LEADER
Communicate the reasons for change.	Spring major changes on employees.
Get employee buy-in before changes.	Force change without explanation.
Provide training for new methods.	Assume employees will adapt instantly.
Listen to concerns about change.	Dismiss resistance as negativity.
Focus on the benefits of new ways.	Only talk about problems of the old.
Celebrate adapting to new systems.	Punish people for not changing fast.

SITUATION	SUGGESTED ACTION
Leaders assume certain employees "can't handle change," excluding them from AI transformation initiatives	Provide structured change support, coaching, and phased exposure, then highlight their progress to challenge limiting assumptions about adaptability
Teams believe new AI-enabled processes will disrupt routines and reduce efficiency	Run small pilots demonstrating time savings and quality improvements, then scale gradually while sharing concrete before-and-after comparisons
Employees fear AI tools will replace their roles, resisting adoption and experimentation	Communicate role evolution clearly, emphasize augmentation, and co-design new responsibilities that elevate human judgment and relationship-driven work
Managers assume older employees struggle more with new systems and digital workflows	Offer inclusive, paced training with peer support, and track adoption data to demonstrate adaptability across age groups objectively
Teams believe change is chaotic, unpredictable, and primarily driven by external pressure	Establish predictable change cadences, transparent roadmaps, and feedback loops that frame change as structured, intentional, and navigable
Employees assume new methods will increase workload rather than streamline tasks	Show how automation removes repetitive work, measure reclaimed time, and reassign capacity toward higher-value, more engaging responsibilities

Believing that physical proximity is the only way to ensure accountability or culture-building is a major barrier to attracting global talent. If you equate remote work with "slacking," you create a surveillance-heavy environment that destroys psychological safety. For an executive, this belief limits your scale. A transformed workforce leverages asynchronous communication and digital trust to operate across borders. If you cling to the belief that eyes-on-glass is the only way to lead, you will lose your best people to competitors who offer the flexibility and trust that modern professionals now consider a non-negotiable standard for engagement. Remote work isn't just a perk; it's a strategic capability that allows you to access the best minds in the world, unconstrained by geography.

DO'S FOR THE LEADER	DON'TS FOR THE LEADER
Trust employees to be productive.	Assume remote work is less productive.
Provide tools for remote collaboration.	Fail to invest in remote work tech.
Set clear goals and performance metrics.	Judge people by their office presence.
Foster an inclusive remote culture.	Create a two-tier system for work.
Hold regular, focused virtual meetings.	Let meetings become unfocused and long.
Prioritize work-life balance for all.	Expect constant availability from staff.

SITUATION	SUGGESTED ACTION
Leaders assume remote employees are less productive because work is less physically visible	Implement outcome-based metrics, review deliverables regularly, and decouple perceived effort from physical presence or online availability indicators
Teams believe collaboration suffers without frequent in-person interaction and hallway conversations	Establish structured rituals—check-ins, design sessions, retrospectives—and use digital whiteboards to maintain creative, synchronous, and asynchronous collaboration
Managers assume remote work reduces accountability and ownership of commitments	Define clear deliverables, deadlines, and decision rights, supported by transparent task boards and shared progress visibility across stakeholders
Employees fear remote work limits career advancement and leadership visibility	Create explicit sponsorship mechanisms, virtual showcase forums, and promotion criteria that recognize impact regardless of physical location
Leaders believe innovation only happens when people share the same physical space	Facilitate global virtual ideation sprints, leverage diverse time zones, and capture ideas asynchronously to broaden participation and creative range
Teams assume remote work inevitably weakens culture and shared identity	Design intentional digital culture practices—rituals, recognition, storytelling—that reinforce belonging, values, and collective purpose across distributed environments

5. LEARNING BELIEFS | The Pursuit of Knowledge

Assuming that learning only happens in formal, "classroom" settings leads to a workforce that is perpetually behind the market curve. If you believe employees learn best through infrequent, mandatory training sessions, you are missing the 90% of development that happens through peer-to-peer sharing and on-the-job experimentation. Transformation requires a belief in "continuous flow" learning. Executives who hold onto old pedagogical beliefs will see their teams' skills decay in real-time. You must believe that learning is a social, daily habit, incentivizing your workforce to treat every project as a lab for new skill acquisition. When learning is woven into the daily flow of work, the organization becomes an "intelligence engine" that stays ahead of the competition by constantly integrating new information and modern methodologies.

DO'S FOR THE LEADER	DON'TS FOR THE LEADER
Encourage a culture of lifelong learning.	Believe learning stops after training.
Offer diverse learning opportunities.	Only use one training method.
Create a safe place for new ideas.	Punish people who make learning mistakes.
Reward those who share knowledge.	Create competition around who knows more.
Invest in ongoing education.	Cut training budgets to save money.
Mentor new employees.	Assume new hires can fend for themselves.

SITUATION	SUGGESTED ACTION
Leaders assume employees prefer traditional classroom training, limiting experimentation with modern learning formats	Offer blended learning options—micro-learning, simulations, peer labs—and track engagement to refine formats based on demonstrated effectiveness
Teams believe learning must be formal and scheduled to be valuable or legitimate	Highlight informal learning moments, embed short learning nudges into workflows, and recognize self-directed skill development publicly
Employees assume they must master AI tools before applying them in real work	Encourage learning-by-doing with supervised practice, emphasizing iterative improvement and safe experimentation within defined guardrails
Managers believe only top performers deserve advanced AI or strategic training opportunities	Provide broad access to foundational programs, then layer advanced tracks based on interest, progress, and demonstrated learning commitment
Teams assume learning is a one-time event rather than an ongoing capability-building process	Establish recurring learning cycles, reflection checkpoints, and skill refreshers that normalize continuous development as part of work
Employees believe meaningful learning requires long, uninterrupted time blocks they rarely have	Promote short, focused learning bursts integrated into daily routines, supported by AI-curated content and just-in-time resources

6. LOYALTY BELIEFS | The Bond of Belonging

The belief that loyalty is driven by tenure or "gratitude" for a paycheck is dangerously outdated. In a transformed workforce, loyalty is a byproduct of growth and alignment. If you assume employees owe you their career just because you hired them, you will be blindsided by the "Great Resignation" cycles. Modern loyalty is a two-way street; it's about the "career equity" you help an employee build. When you shift this belief, you stop trying to "trap" talent with golden handcuffs and start "attracting" it through a compelling mission and continuous development. Loyalty today is found in the shared journey toward a meaningful goal. By believing that loyalty is earned through the value you provide to your employees, you build a much more resilient and dedicated core.

DO'S FOR THE LEADER	DON'TS FOR THE LEADER
Earn loyalty through trust and respect.	Expect blind loyalty without earning it.
Invest in employee well-being.	Value the company more than people.
Reward loyalty with recognition.	Let hard work go unnoticed.
Build a positive company culture.	Tolerate a toxic or unhealthy environment.
Be transparent and honest with staff.	Keep important info from employees.
Provide opportunities for advancement.	Expect loyalty from a dead-end job.

SITUATION	SUGGESTED ACTION
Leaders assume loyalty is primarily driven by tenure, stability, and job security	Reinforce loyalty through meaningful work, transparent communication, and visible investment in employee growth and future relevance
Teams believe loyalty requires sacrificing personal boundaries and always saying yes	Model healthy boundaries, respect capacity limits, and show that sustainable performance strengthens long-term commitment and trust
Managers assume compensation alone determines whether employees stay or leave	Elevate purpose, recognition, development, and fair treatment as core retention levers alongside competitive, transparent compensation structures
Employees believe loyalty means staying silent about concerns or misalignment	Create safe feedback channels, act visibly on input, and reinforce that constructive honesty deepens trust and mutual commitment
Leaders assume loyalty, once damaged, cannot be rebuilt meaningfully	Rebuild trust through consistent behavior, clear commitments, and follow-through that demonstrates reliability over time, not slogans
Teams believe loyalty means resisting change to protect legacy ways of working	Reframe loyalty as supporting evolution that keeps the organization relevant, resilient, and capable of serving stakeholders better

CHAPTER F2

Ai6x6™ Transformation Blueprint

Workforce Systemic Biases

"Diversity is being invited to the party; inclusion is being asked to dance." - Verna Myers

Verna Myers is an American lawyer, diversity consultant, and author. She is a renowned speaker on implicit bias and is a thought leader on fostering inclusive workplaces. She currently serves as Netflix's VP of Inclusion Strategy.

Workforce systemic biases are unconscious patterns of thought that silently erode fairness, collaboration, and diversity. These biases are often invisible, yet they profoundly impact everything from hiring to promotions. A prime example is In-Group Bias, where managers unconsciously favor people with similar backgrounds or work styles. This creates a sense of exclusion and prevents the organization from benefiting from diverse perspectives. Groupthink Bias pressures employees to conform to a dominant opinion, stifling creativity and discouraging unconventional ideas. Addressing these biases is not about eliminating natural human preference but about creating formal, objective processes that ensure everyone is evaluated fairly. As Verna Myers says, "Diversity is being invited to the party; inclusion is being asked to dance." Transformation is about building an environment where every employee feels included, valued, and empowered to contribute their unique perspective.

This chapter establishes the Relational domain by exploring Workforce Systemic Biases, the collective group dynamics and interpersonal perceptions that undermine change Adoption and distort the Cultural Compass. It examines how unconscious prejudices and group affiliations sabotage cross-functional collaboration and trust, the twin engines of transformation. The chapter details the resistance created by In-Group Bias, which favors one's own team and resists collaboration with other departments, and the dangerous intellectual stagnation caused by Conformity/Groupthink Bias, where the suppression of dissent leads to poor decisions and failed execution. It addresses the unfairness in talent management caused by the Halo/Horns Effect Bias (a single trait influencing overall perception) and the hierarchy-preserving effect of Authority Bias (undue deference to leadership, stifling bottom-up innovation). Finally, it details the pervasive impact of Affinity Bias (favoring similar people in hiring and promotion) and Stereotyping Bias (generalizing group characteristics), providing tools to dismantle these barriers and create a truly inclusive, adaptive culture.

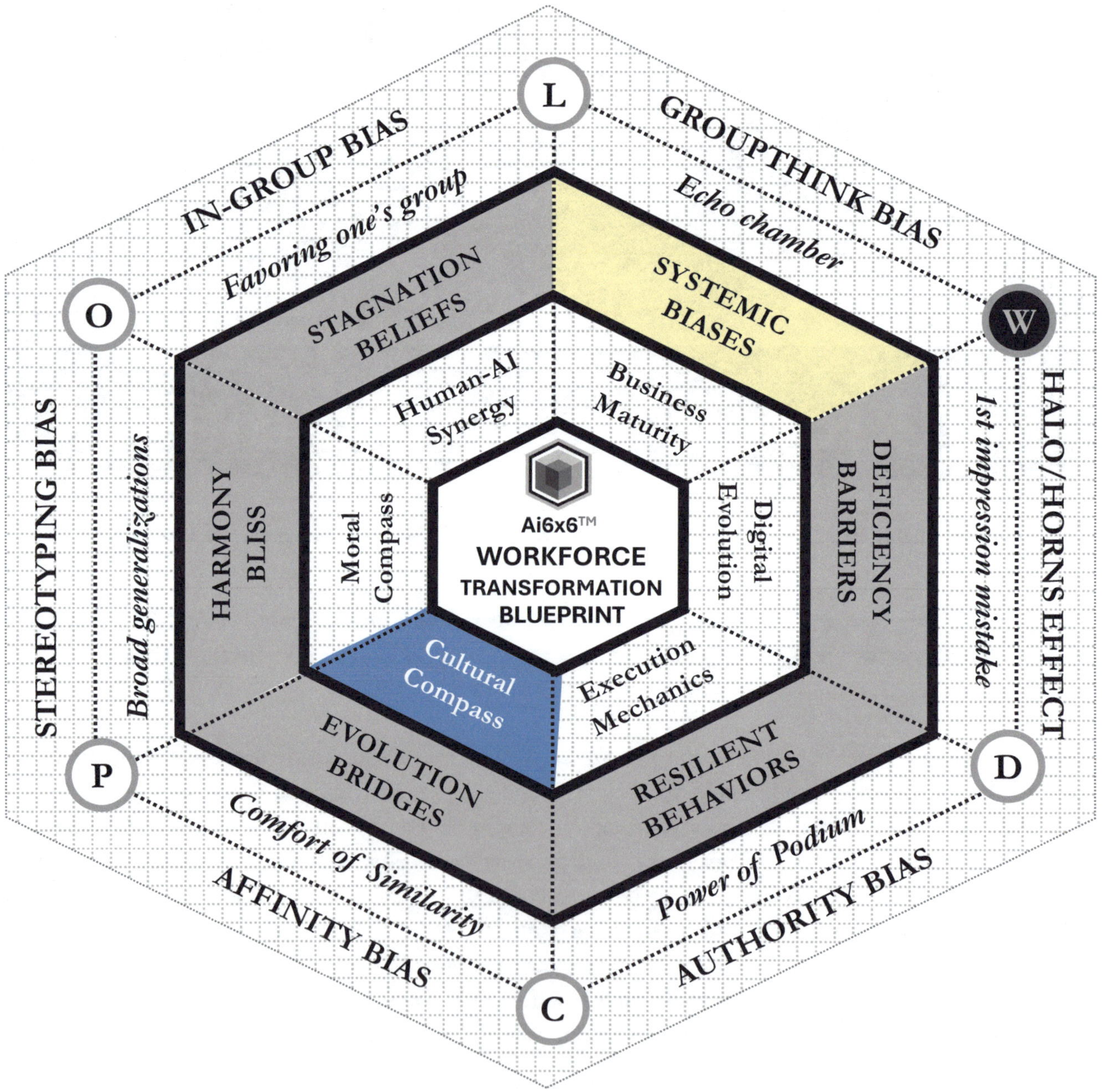

CULTURAL COMPASS FOR BIASES | WORKFORCE TRANSFORMATION

1. IN-GROUP BIAS | The Inner Circle

This bias creates invisible walls between departments, where "HQ" is favored over regional offices, or "Product" is favored over "Support." In a transformation effort, this leads to a fragmented strategy where different teams prioritize their own success over the company's evolution. As an executive, you might find yourself unconsciously approving budgets or resources for the groups you "know" best, further starving the departments that might actually be the engines of your future growth. Breaking this bias is essential to creating a unified workforce where everyone feels they are playing for the same team, regardless of their specific badge or location. When you dismantle in-group bias, you unlock a cross-functional synergy that allows the organization to move as a single, powerful unit against external market forces.

DO'S FOR THE LEADER	DON'TS FOR THE LEADER
Value diverse backgrounds and views.	Favor people in your own group.
Rotate people on different teams.	Create tight knit, exclusive groups.
Celebrate cross-functional success.	Only celebrate your own team's wins.
Promote based on merit, not connections.	Promote people based on friendship.
Create common goals for all to work toward.	Let teams compete against one another.
Listen to everyone's ideas.	Ignore ideas from outside your group.

SITUATION	SUGGESTED ACTION
A leader repeatedly assigns high-visibility projects to long-time insiders, overlooking newer contributors	Publish clear selection criteria, rotate ownership intentionally, and review opportunity distribution regularly to ensure visibility and advancement are not confined to established inner circles
A core project team always includes the same familiar members, excluding capable colleagues from other functions	Require cross-functional representation on strategic initiatives and track participation patterns to prevent closed groups from monopolizing influence, information, and development opportunities
Informal decisions are made in private chats among friends before formal meetings even occur	Move key decisions into transparent forums, document rationales, and ensure impacted stakeholders participate directly rather than relying on pre-aligned in-group agreements
Performance feedback is richer and more developmental for people socially close to the manager	Standardize feedback cadence and structure, using shared templates and calibration sessions to equalize developmental input across all employees, not just trusted insiders
Social events and networking opportunities are organized around interests shared mainly by one dominant subgroup	Design inclusive networking formats, rotate themes intentionally, and invite employee input so relationship-building is not limited to one cultural or social reference point
Critical information about upcoming changes circulates first within a favored inner circle, reaching others much later	Establish consistent communication channels and timing, ensuring all affected employees receive the same information simultaneously, regardless of proximity to informal power networks

2. GROUPTHINK BIAS | The Echo Chamber

When the desire for harmony outweighs the need for critical evaluation, transformation stalls. Groupthink leads to "vanity projects" and the avoidance of hard truths. In your daily executive meetings, if no one is disagreeing with the transformation roadmap, you are likely in a groupthink trap. This bias suppresses the "early warning signals" that employees on the front lines are trying to send. To transform successfully, you must actively dismantle the pressure to conform, encouraging a "duty to dissent" that ensures your strategic decisions are stress-tested by the diverse perspectives of your entire workforce. True innovation requires the clashing of different ideas; if everyone is nodding in agreement, you aren't leading an evolution—you are presiding over a decline. Encouraging healthy debate is your primary defense against strategic stagnation.

DO'S FOR THE LEADER	DON'TS FOR THE LEADER
Encourage respectful disagreement.	Pressure people to conform to opinions.
Appoint a "devil's advocate."	Silence dissenting voices to keep the peace.
Use anonymous feedback channels.	Punish people who speak their mind.
Allow time for independent thought.	Force people to make quick decisions.
Seek input from outside experts.	Stay within the group's point of view.
Reward creative, new ideas.	Only reward conformity and consensus.

SITUATION	SUGGESTED ACTION
A project team quickly endorses the first proposal, avoiding deeper analysis to maintain harmony	Assign a rotating "challenge role," require at least two alternative options, and document trade-offs before any decision is finalized, normalizing constructive dissent as standard practice
Junior employees remain silent during strategy reviews because senior colleagues appear fully aligned	Use anonymous pre-reads, collect written perspectives beforehand, and explicitly invite junior viewpoints during meetings, signaling that disagreement is valued rather than career-limiting
A leadership group repeatedly reuses last year's plan, assuming consensus equals correctness	Schedule structured assumption-testing sessions, bring in external perspectives, and require evidence for continuing legacy approaches instead of defaulting to familiar group comfort
Teams rush to meet deadlines by skipping risk discussions that might slow agreement	Build mandatory risk checkpoints into project timelines and measure leaders on quality of challenge, not just speed of consensus and delivery
Employees privately question a major AI rollout but publicly echo enthusiastic messaging	Create confidential feedback channels, share aggregated concerns transparently, and adjust plans visibly so employees see that voicing doubts influences real decisions
A steering committee rarely records dissenting opinions, presenting decisions as unanimously supported	Capture minority views in decision logs, review them during retrospectives, and highlight instances where early dissent later proved directionally correct

3. HALO/HORNS EFFECT BIAS | The First Impression Fallacy

Judging an entire employee's potential based on a single success (Halo) or one mistake (Horns) creates a wildly inaccurate map of your talent. In a transformation, you might over-rely on a "Halo" employee who isn't actually suited for the new digital landscape, or overlook a "Horns" employee who has exactly the skills you need but once missed a deadline. This bias leads to poor succession planning and misaligned roles. As a leader, you must look for holistic data. Otherwise, you risk building your future around individuals rather than capabilities. By neutralizing this bias, you ensure that you are making talent decisions based on a comprehensive view of performance and potential, which is the only way to build a robust, future-proof leadership pipeline that isn't vulnerable to individual failures.

DO'S FOR THE LEADER	DON'TS FOR THE LEADER
Evaluate people based on all traits.	Judge people based on a single trait.
Use clear, objective performance metrics.	Rely on vague or subjective feedback.
Focus on specific behaviors.	Make broad, general judgments.
Provide balanced feedback.	Give overly positive or negative reviews.
Base promotions on a full history.	Promote someone based on a single success.
Give everyone a fresh chance.	Let a bad first impression stick.

SITUATION	SUGGESTED ACTION
A charismatic presenter is assumed highly competent across domains, receiving disproportionate responsibilities	Separate presentation skills from domain expertise in evaluations, require evidence of performance in each area, and distribute responsibilities based on demonstrated capability rather than charisma
An employee who once missed a critical deadline is continually viewed as unreliable	Reassess performance using recent data, discuss context openly, and update expectations based on current behavior rather than a single historical failure
A technically brilliant specialist is excused from collaboration expectations because of perceived exceptional value	Define role expectations holistically, including teamwork and communication, and incorporate peer feedback into performance reviews to balance technical excellence with collaborative impact
An early positive impression during onboarding shapes a manager's view despite later mixed results	Use periodic, structured check-ins with objective metrics and peer input to recalibrate perceptions, ensuring early impressions do not permanently anchor evaluations
A single awkward interaction leads colleagues to avoid involving someone in cross-functional initiatives	Encourage direct clarification conversations, provide coaching if needed, and evaluate suitability for projects using skills and outcomes rather than isolated interpersonal moments
A leader's favorite employee receives consistently generous ratings despite average measurable performance	Introduce calibration sessions with comparative data, require justification for ratings, and flag discrepancies between narrative praise and objective results for review

4. AUTHORITY BIAS | The Power of the Podium

When employees overvalue your opinion simply because you are the boss, they stop thinking for themselves. This bias is the enemy of an agile, transformed workforce. If your team waits for your "OK" for every minor adjustment, you have become the bottleneck for transformation. Authority bias prevents the "bottom-up" innovation that is crucial for staying ahead of market shifts. Your daily challenge is to intentionally downplay your own authority to create space for others to lead. If you don't, you will be surrounded by followers who execute your mistakes with perfect precision, rather than partners who prevent them. To transform, you must cultivate a culture where the "best idea wins," regardless of who said it, ensuring that your organization is guided by collective wisdom rather than hierarchy.

DO'S FOR THE LEADER	DON'TS FOR THE LEADER
Encourage critical inquiry.	Overvalue opinions of authority figures.
Empower teams to question leaders.	Expect blind obedience to authority.
Value evidence over position.	Let a leader's opinion override facts.
Create a flat organizational structure.	Create a rigid hierarchy.
Give everyone a voice in decisions.	Let one person make all the decisions.
Reward independent thinking.	Punish people who think for themselves.

SITUATION	SUGGESTED ACTION
A senior executive suggests a solution and the team immediately stops exploring alternatives	Require option comparison documents, including pros, cons, and data, before endorsement; treat executive ideas as inputs, not conclusions, within the decision process
Analysts soften data that contradicts leadership's preferred narrative to avoid perceived confrontation	Establish protections for evidence-based challenge, review unfiltered data in dedicated sessions, and recognize individuals who surface uncomfortable truths constructively
Project teams delay decisions, waiting for explicit executive approval on issues within their remit	Clarify decision rights, publish delegation boundaries, and hold teams accountable for timely choices rather than escalation by default to senior leaders
A manager's opinion dominates cross-functional meetings, with others deferring even when expertise differs	Use neutral facilitation, time-boxed speaking turns, and explicit expert-voice prioritization when topics fall within specific technical or operational domains
Employees prioritize tasks requested by senior leaders over agreed strategic priorities	Implement portfolio-level prioritization, require leaders to route requests through existing governance, and align personal asks with documented organizational priorities
A respected founder's historical preferences continue guiding decisions despite changed market conditions	Conduct periodic strategy resets, explicitly question legacy assumptions, and use current data to validate whether traditional directives still create advantage

5. AFFINITY BIAS | The Comfort of Similarity

Unconsciously favoring people who share your background, education, or hobbies leads to a "mirrored" workforce that lacks cognitive diversity. In transformation, this bias is lethal; you end up solving new-world problems with old-world thinking. If your "top talent" list looks remarkably like you, you have an affinity bias problem. This limits the range of solutions your team can generate and alienates the very diverse talent you need to understand a global, fragmented customer base. Overcoming this requires building systemic checks into hiring and promotion that prioritize "culture add" over "culture fit," ensuring your workforce evolves beyond your own limitations. By intentionally seeking out people who think differently, you expand the organization's "intellectual footprint" and its ability to innovate in markets you personally don't fully understand yet.

DO'S FOR THE LEADER	DON'TS FOR THE LEADER
Use blind resume reviews.	Unconsciously favor people like you.
Create diverse interview panels.	Have only one person interview candidates.
Focus on skills and qualifications.	Hire for "culture fit" over talent.
Challenge your own assumptions.	Only hire people from your network.
Acknowledge your own biases.	Assume you are unbiased.
Provide training on unconscious bias.	Ignore bias in your hiring process.

SITUATION	SUGGESTED ACTION
A manager frequently socializes with employees who share similar backgrounds, offering them informal opportunities	Track informal and formal opportunity distribution, schedule inclusive touchpoints intentionally, and hold leaders accountable for balanced relationship-building across demographic and stylistic differences
Performance reviews are more generous for employees who mirror the manager's communication style	Use standardized criteria, include multi-rater feedback, and audit rating patterns for correlation with style similarity rather than objective contribution or outcomes
Interview panels consistently favor candidates who attended similar schools or worked at familiar companies	Blind certain résumé fields where possible, diversify panels, and emphasize structured scoring rubrics over subjective "fit" impressions tied to shared history
High-potential nominations disproportionately feature employees with similar career paths to current leaders	Require data-backed nominations, review talent slates for diversity, and challenge leaders to sponsor individuals whose profiles differ meaningfully from their own
Informal brainstorming sessions occur with a small circle of like-minded colleagues before broader engagement	Move ideation into open forums, invite contrasting perspectives deliberately, and rotate participants to avoid reinforcing narrow affinity-based echo chambers
Coaching conversations are deeper and more frequent with employees who share personal interests with their manager	Standardize coaching frequency, use shared agendas, and monitor participation to ensure developmental depth is not reserved for affinity-aligned relationships

6. STEREOTYPING BIAS | The Mental Shortcut

Generalizing about "Millennials," "Gen Z," or "Legacy Employees" creates resentment and limits individual potential. If you believe "older workers can't learn tech" or "younger workers lack grit," you are mismanaging your human capital based on myths. This bias prevents you from seeing the unique value each person brings to the transformation journey. In your daily leadership, stereotyping leads to misaligned training and poor communication. A transformed workforce is one where individuals are seen as unique contributors. By discarding these labels, you unlock the ability to pair the wisdom of experience with the energy of new perspectives, creating a powerful, intergenerational engine. True transformation recognizes that talent is individual, not generational, and that your greatest advantage lies in harnessing the unique strengths of every person on your team.

DO'S FOR THE LEADER	DON'TS FOR THE LEADER
Treat all people as unique individuals.	Make generalizations based on identity.
Get to know people personally.	Judge people based on a group they belong to.
Focus on individual skills and talents.	Assign roles based on stereotypes.
Promote an inclusive workplace.	Allow stereotypes to guide your decisions.
Challenge stereotypes when you hear them.	Tolerate stereotypes in the workplace.
Give everyone a fair and equal chance.	Assume people lack certain skills.

SITUATION	SUGGESTED ACTION
A leader assumes certain demographic groups are less suited for client-facing roles	Use objective performance data, client feedback, and interest surveys to assign roles, explicitly rejecting assumptions tied to identity rather than capability
New hires from specific regions are presumed weaker in communication before any interaction occurs	Provide equal onboarding, observe actual behavior, and address biased comments directly, reinforcing that evaluation must follow evidence, not cultural shortcuts
Technical aptitude is assumed higher for some groups, leading to skewed training investments	Offer uniform access to advanced programs, monitor participation by segment, and adjust outreach to correct disparities driven by stereotype-based expectations
Employees from underrepresented groups are repeatedly asked to speak for their entire community	Set guidelines for inclusive dialogue, discourage tokenization, and encourage individuals to share personal perspectives rather than represent broad categories
Certain age groups are stereotyped as resistant to AI, limiting their involvement in pilots	Invite mixed-age cohorts into experiments, track adoption objectively, and share success stories that contradict age-based assumptions about adaptability
Accent or language style leads colleagues to underestimate someone's strategic thinking	Focus evaluations on written analyses, decisions, and outcomes; provide communication support where needed without equating fluency style with intellectual capability

BLUEPRINT SERIES

Ai6x6™ Transformation Blueprint

Workforce Deficiency Barriers

"Obstacles don't have to stop you. If you run into a wall, don't turn around and give up. Figure out how to climb it, go through it, or work around it." - Michael Jordan

Michael Jordan (b. 1963) is a former professional basketball player. He is widely regarded as the greatest player in the history of the sport, known for his relentless work ethic, competitive drive, and his success in the business world.

Workforce deficiency barriers are structural or cultural impediments that prevent employees from feeling engaged, supported, and empowered. These barriers are the root cause of low morale and a lack of innovation. A Lack of Trust Barrier creates an environment where micromanagement is the norm, slowing down work and making employees feel disrespected. A Lack of Growth Barrier, where career opportunities are scarce, leads to stagnation and high turnover. A Lack of Voice Barrier makes employees feel powerless, preventing valuable feedback and innovative ideas from reaching leadership. Overcoming these barriers requires a focused effort to rebuild a culture of trust, create clear pathways for advancement, and empower every employee to contribute their ideas without fear. As Michael Jordan suggests, obstacles don't have to stop you.

This component establishes the Relational domain by exploring Workforce Deficiency Barriers, the collective group dynamics and organizational friction that undermine change Adoption and distort the Cultural Compass. A primary obstacle is Communication Silos, where departments or teams are unwilling to share information, duplicating effort and preventing cross-functional efficiency. This section details the burnout caused by Change Fatigue, where employees become emotionally and professionally exhausted by continuous, poorly managed initiatives, leading to apathy and resistance. A critical structural issue is Skill Gaps, the lack of necessary technical, analytical, or behavioral abilities required to operate the transformed enterprise, turning new systems into expensive ornaments. Cultural Resistance is explored as a deep attachment to old ways of working, often rooted in tribal memory and fear of the unknown. The component stresses the importance of fostering Psychological Safety, the absence of which means employees fear speaking up, admitting mistakes, or offering contrary opinions, thereby suppressing crucial feedback. Finally, it addresses the fundamental Lack of Buy-In, where employees do not see or accept the value proposition of the transformation for themselves or the company.

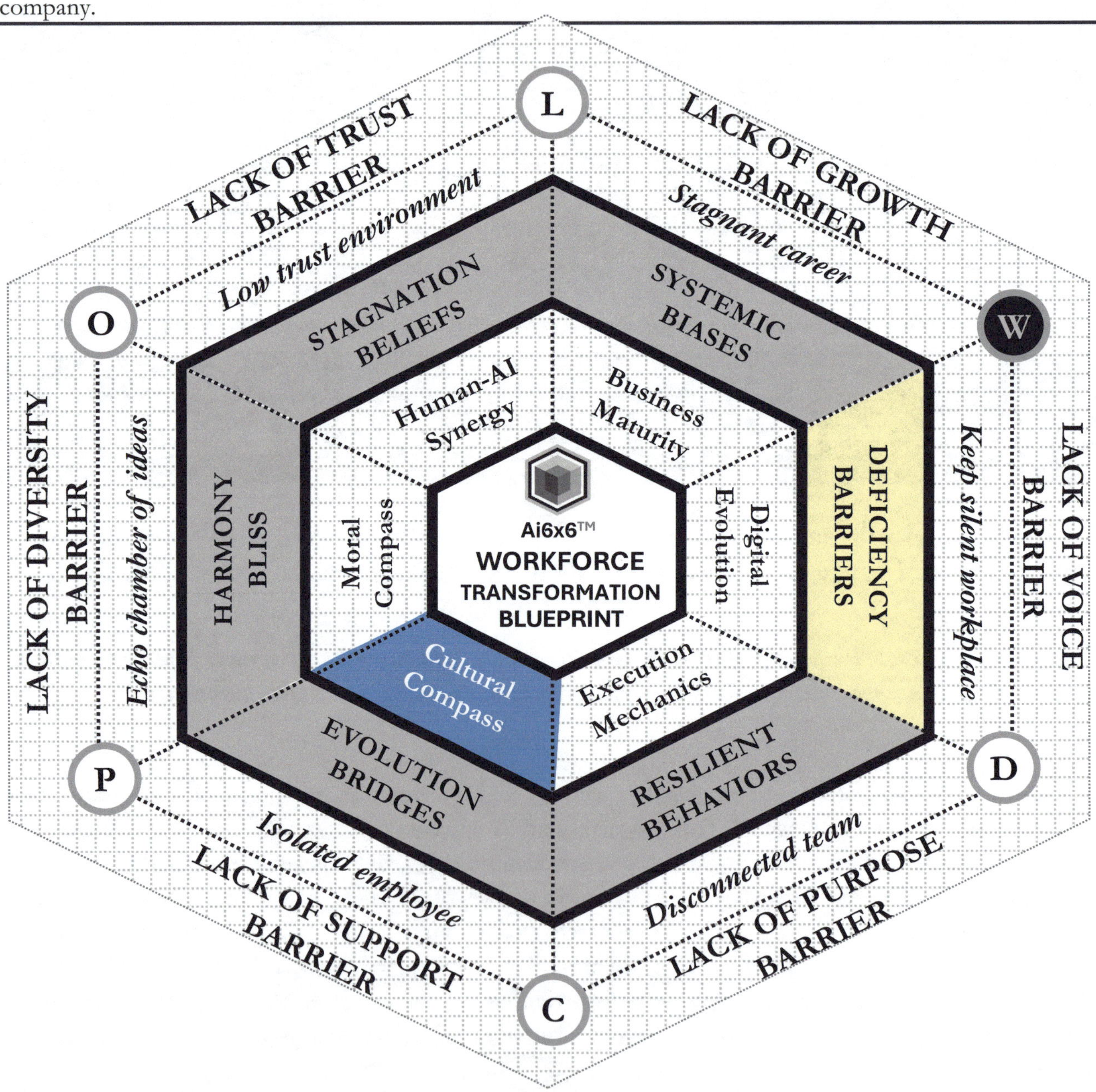

Ai6x6 CULTURAL COMPASS FOR BARRIERS | WORKFORCE TRANSFORMATION

1. LACK OF TRUST BARRIER | The Missing Foundation

Without trust, the "transaction costs" of doing business skyrocket. In a low-trust workforce, every new transformation initiative is viewed with suspicion—as a potential "downsizing" or "monitoring" tactic. Employees won't share their best ideas or admit to mistakes, which are the two things you need most during a shift. This barrier turns your workplace into a political minefield where people spend more energy protecting themselves than advancing the mission. Transforming this requires you to be the first to show vulnerability, proving that trust is not a luxury, but the foundational infrastructure upon which all high-stakes organizational change is built. When trust is established, the speed of execution increases dramatically because the need for constant validation and suspicion-driven "fact-checking" simply disappears from your daily operations.

DO'S FOR THE LEADER	DON'TS FOR THE LEADER
Foster an environment of mutual trust.	Let a low-trust environment fester.
Be transparent with all information.	Hide information from your team.
Delegate decisions and projects.	Micromanage and take back control.
Empower people to do their jobs.	Check on employees constantly.
Acknowledge mistakes and failures.	Blame others and never admit you're wrong.
Be consistent and reliable.	Be unpredictable and unreliable.

SITUATION	SUGGESTED ACTION
Teams withhold information because they assume colleagues may misuse it or undermine them	Establish transparent communication norms, model vulnerability, and reinforce shared accountability to rebuild trust and strengthen collaborative behavior across the organization
Employees avoid raising concerns due to fear their input will be judged harshly or dismissed	Create psychologically safe forums, acknowledge contributions publicly, and demonstrate consistent follow-through to show that honest input is valued and protected
Managers double-check every task, signaling doubt in employee capability and reliability	Shift to outcome-based expectations, provide autonomy within clear boundaries, and recognize dependable execution to reinforce mutual trust and confidence
Cross-functional groups hesitate to share resources because they expect unequal reciprocation	Implement reciprocal agreements, track commitments, and celebrate balanced collaboration to normalize fair exchange and reduce territorial behavior
Employees assume leadership decisions are made secretly without considering broader impact	Communicate decision rationales clearly, invite clarifying questions, and share context proactively to strengthen transparency and organizational trust
Team members interpret silence as disapproval, creating unnecessary tension and misalignment	Encourage explicit acknowledgment, clarify intent regularly, and train teams to replace assumptions with direct communication to reduce mistrust

If employees don't see a future for themselves within your transformation, they will find one elsewhere. A lack of clear growth pathways leads to "quiet quitting" and a brain drain of your most ambitious talent. You cannot transform a workforce that feels it is in a dead-end. This barrier manifests as a lack of internal mobility and a rigid "ladder" that no longer fits the modern career "lattice." To overcome this, you must transform your HR systems to support continuous skill-building and project-based roles, ensuring that as the company evolves, every employee feels their personal market value is increasing. By providing clear avenues for advancement, you turn the workplace into a "career incubator," where the best talent is incentivized to stay because their professional growth is inextricably linked to the company's success.

DO'S FOR THE LEADER	DON'TS FOR THE LEADER
Create opportunities for advancement.	Provide few chances for career growth.
Invest in training and new skills.	Cut training budgets to save money.
Offer a clear career path.	Have no path for employee development.
Promote from within when possible.	Always hire externally for new roles.
Have regular growth conversations.	Only talk about performance reviews.
Reward learning and new skills.	Only reward day-to-day tasks.

SITUATION	SUGGESTED ACTION
Employees feel stuck because advancement paths are unclear and rarely communicated effectively	Publish transparent career frameworks, outline required competencies, and provide regular development conversations to clarify growth possibilities and expectations
High-potential employees leave after realizing skill development opportunities are inconsistent or inaccessible	Build equitable learning pathways, track participation, and ensure managers actively support development planning for all employees, not just select individuals
Teams repeat the same tasks without exposure to new responsibilities or stretch assignments	Rotate responsibilities intentionally, introduce project-based learning, and assign growth-oriented challenges aligned with individual aspirations and organizational needs
Managers prioritize immediate delivery over long-term employee development, limiting capability expansion	Integrate development goals into performance plans, measure leaders on talent growth, and allocate time for structured learning activities
Employees assume promotions depend on tenure rather than demonstrated capability or readiness	Use competency-based evaluations, communicate criteria clearly, and highlight examples of advancement driven by performance and learning agility
Career conversations occur only during annual reviews, leaving employees uncertain about future direction	Establish quarterly development check-ins, provide actionable guidance, and track progress to maintain momentum and clarity throughout the year

3. LACK OF VOICE BARRIER | The Silent Majority

When employees feel they have no say in how work is done, they disengage. In many transformation efforts, the "front line" knows exactly where the friction points are, but they are never asked. This barrier creates a disconnect between executive strategy and operational reality. If your feedback loops are one-way (top-down), you are flying blind. To break this barrier, you must create "radical transparency" channels where the lowest-level employee can flag a systemic issue without fear. Empowering the workforce voice ensures that your transformation is grounded and fueled by the collective intelligence of those doing the work. When everyone has a voice, the organization benefits from thousands of eyes and ears on the ground, making your strategic adjustments far more accurate and timelier.

DO'S FOR THE LEADER	DON'TS FOR THE LEADER
Empower employees to share their ideas.	Make people afraid to speak up.
Listen to all employee feedback.	Dismiss ideas from junior staff.
Create channels for anonymous feedback.	Punish people who criticize the company.
Welcome dissent and disagreement.	Create a culture of forced consensus.
Involve employees in decision-making.	Make decisions in a vacuum.
Act on employee suggestions.	Let feedback go into a black hole.

SITUATION	SUGGESTED ACTION
Employees stay silent during meetings because past suggestions were ignored or dismissed quickly	Implement structured idea-capture processes, acknowledge contributions publicly, and demonstrate visible action on employee input to rebuild confidence in speaking up
Junior employees hesitate to challenge assumptions due to fear of negative career consequences	Create protected challenge mechanisms, encourage respectful dissent, and recognize individuals who raise thoughtful questions that strengthen decision quality
Teams rely on a few dominant voices, limiting participation from quieter contributors	Use facilitated turn-taking, anonymous input tools, and rotating discussion leads to ensure balanced participation across all team members
Employees avoid reporting issues because previous concerns resulted in no meaningful change	Close the loop on all feedback, communicate actions taken, and set expectations for response timelines to reinforce credibility
Managers unintentionally shut down ideas by reacting too quickly with criticism or skepticism	Train leaders in appreciative inquiry, require exploratory questioning before evaluation, and model curiosity to encourage open contribution
Employees feel their lived experiences are undervalued, reducing willingness to share insights	Integrate diverse perspectives into decision processes, highlight contributions from varied backgrounds, and reinforce that all viewpoints matter

A workforce that doesn't understand the "why" behind the transformation will quickly lose motivation. If people feel like they are just "cogs in a machine" working to increase shareholder value, they will give you the bare minimum. This barrier is especially high for younger generations who demand alignment with their values. Transforming this means moving beyond a mission statement on the wall to a "shared purpose" that is felt in daily tasks. As an executive, your primary job is to connect the dots between the company's digital evolution and a meaningful impact on the world, giving your team a reason to lean in when things get difficult. When work has purpose, it stops being a "job" and becomes a "mission," which is the highest level of engagement any workforce can achieve.

DO'S FOR THE LEADER	DON'TS FOR THE LEADER
Connect work to the company's mission.	Let employees feel disconnected.
Communicate the bigger picture.	Focus on only the small details.
Show how each role contributes.	Make work feel meaningless or unimportant.
Celebrate how work impacts customers.	Ignore the ultimate impact of your work.
Ask employees what gives them purpose.	Assume everyone is motivated by money.
Foster a sense of shared purpose.	Let people feel isolated in their role.

SITUATION	SUGGESTED ACTION
Employees complete tasks mechanically without understanding how their work contributes to broader goals	Connect daily responsibilities to strategic outcomes, share impact stories, and reinforce how individual contributions advance organizational purpose
Teams feel disconnected from leadership's vision because messaging is abstract and infrequent	Communicate purpose consistently, translate vision into practical examples, and involve teams in shaping how mission shows up operationally
Employees struggle to stay motivated because priorities shift without clear explanation	Provide context for changes, outline decision drivers, and reinforce continuity between evolving priorities and long-term mission
Managers focus solely on metrics, neglecting meaning and human connection in their communication	Integrate purpose-driven narratives into performance discussions, highlighting how work supports customers, communities, and organizational aspirations
Employees perceive their roles as transactional rather than meaningful or impactful	Redesign roles to include purpose-aligned responsibilities and recognize contributions that advance mission-critical outcomes
Teams feel excluded from strategic conversations, weakening their sense of belonging and purpose	Invite employees into planning sessions, gather input early, and show how their insights shape strategic direction

Transformation is exhausting. If employees feel isolated—tasked with "doing more with less" without new tools or emotional support—they will burn out. This barrier is often hidden in the "middle management" layer, where leaders are squeezed between executive demands and team needs.5 A lack of support manifests as high stress, poor mental health, and declining quality. Overcoming this requires investing in the "infrastructure of care": better tools, wellness resources, and a culture that recognizes the human cost of change. When employees feel backed by the organization, they are far more likely to take the risks necessary to drive the transformation forward. Providing genuine support isn't a "soft" benefit; it is the "life support system" that allows your workforce to sustain high performance through intense periods of change.

DO'S FOR THE LEADER	DON'TS FOR THE LEADER
Create a strong support system.	Let employees feel isolated and alone.
Encourage peer-to-peer support.	Create competition between employees.
Train leaders to be empathetic.	Train leaders to be distant or robotic.
Provide mental health resources.	Ignore signs of stress and burnout.
Be flexible with personal issues.	Be rigid about personal needs.
Check in on your team often.	Assume your team is always fine.

SITUATION	SUGGESTED ACTION
Employees hesitate to ask for help because they fear appearing incompetent or unprepared	Normalize help-seeking, celebrate collaborative problem-solving, and train leaders to respond supportively when employees request assistance
New hires struggle because onboarding lacks structure and consistent guidance from managers	Implement standardized onboarding plans, assign mentors, and schedule regular check-ins to ensure new employees feel supported and connected
Teams feel overwhelmed because workload distribution is uneven and rarely reviewed	Conduct workload audits, rebalance responsibilities transparently, and provide resources to prevent burnout and strengthen team cohesion
Employees feel isolated in hybrid environments due to inconsistent communication rhythms	Establish predictable touchpoints, use intentional connection rituals, and ensure managers maintain regular one-to-one engagement
Managers overlook emotional strain during high-pressure periods, focusing solely on output	Train leaders to recognize stress signals, encourage open dialogue, and provide access to supportive resources during demanding cycles
Employees lack clarity on where to seek guidance for complex or unfamiliar tasks	Create centralized support channels, document expert resources, and ensure employees know exactly where to find help when needed

A homogeneous workforce is a fragile workforce. If everyone thinks the same way, you have a collective blind spot. A lack of diversity is a structural barrier to innovation; you will keep producing the same safe, incremental ideas. In a global economy, a lack of varied perspectives means you will fail to see the opportunities or threats emerging from different markets or demographics. Transformation requires "cognitive friction"—the clashing of different viewpoints to create a better spark. To break this barrier, you must move beyond diversity as a metric and embrace it as a strategic necessity. A diverse workforce is more resilient, more creative, and more capable of understanding the complex, multifaceted customer base that defines the modern economy, giving you a decisive competitive edge.

DO'S FOR THE LEADER	DON'TS FOR THE LEADER
Actively recruit from diverse sources.	Only hire from a narrow talent pool.
Create an inclusive culture.	Foster a homogeneous culture.
Value different backgrounds and ideas.	Ignore the benefits of varied perspectives.
Promote based on merit, not sameness.	Promote people based on familiarity.
Challenge your own biases.	Assume a diverse team is not needed.
Create a fair and equal workplace.	Let some people have an easier time.

SITUATION	SUGGESTED ACTION
Hiring pipelines consistently produce candidates from similar backgrounds, limiting cognitive and cultural diversity	Expand sourcing channels, partner with diverse institutions, and use structured evaluation criteria to broaden representation meaningfully
Teams rely on familiar perspectives, reducing creativity and innovation in problem-solving	Introduce mixed-perspective working groups, rotate roles, and encourage contrasting viewpoints to strengthen solution quality
Leadership roles are dominated by one demographic group, limiting aspirational visibility for others	Build diverse succession slates, provide equitable sponsorship, and track advancement patterns to correct systemic imbalances
Employees hesitate to express unique viewpoints because dominant norms shape acceptable behavior	Reinforce inclusive communication norms, celebrate divergent thinking, and ensure leaders model openness to varied perspectives
Project decisions reflect narrow experiences, overlooking customer segments with different needs	Integrate diverse voices into design processes, conduct inclusive research, and validate assumptions across multiple demographic groups
Cultural celebrations and recognition programs reflect only a narrow set of identities	Co-create inclusive cultural calendars, rotate representation, and ensure recognition practices honor the full spectrum of employee backgrounds

Ai6x6™ Transformation Blueprint

Workforce Resilient Behaviors

"Excellence then, is not an act, but a habit." - Aristotle

Aristotle (384–322 BCE) was an ancient Greek philosopher and polymath. He is one of the most influential figures in Western philosophy and is known for his work in logic, science, and ethics, particularly his ideas on virtue and character.

Workforce resilient behaviors are positive, proactive actions that directly counter negative beliefs, biases, and barriers. These are the practices that build a dynamic, adaptable, and high-performing culture. Growth Behavior is the active pursuit of professional development by employees and the organization, a direct antidote to stagnation. Shared Voice Behavior is a culture where every employee feels heard and valued, a direct counter to the lack of a voice barrier. Diverse Team Behavior is the practice of actively building inclusive teams where varied perspectives are valued, which effectively counters in-group bias. As Aristotle reminds us, "Excellence then, is not an act, but a habit." These behaviors, when adopted, transform the organization into a cohesive, collaborative, and high-performing community.

This framework establishes the Relational domain by exploring Workforce Resilient Behaviors, the interpersonal actions and group dynamics that successfully drive change Adoption and refine the Cultural Compass. A primary required behavior is Collaborative Behavior, ensuring teams work toward shared, enterprise-wide goals and actively support colleagues across functional and departmental lines. The organization must foster Open-Minded Behavior, where individuals are receptive to new ideas and challenging perspectives, moving beyond the complacency of internal tradition. Employees must exhibit Feedback-Seeking Behavior, proactively soliciting and acting on input regarding their performance and the transformation's progress from peers and managers. This is supported by Supportive Behavior, where colleagues actively help and encourage each other through difficult transitions and skill development. The framework encourages Proactive Behavior, taking initiative to solve latent problems and improve processes before issues escalate or are formally mandated. Finally, success in change adoption requires Empathetic Behavior, understanding and respecting others' feelings, particularly during periods of organizational disruption and uncertainty.

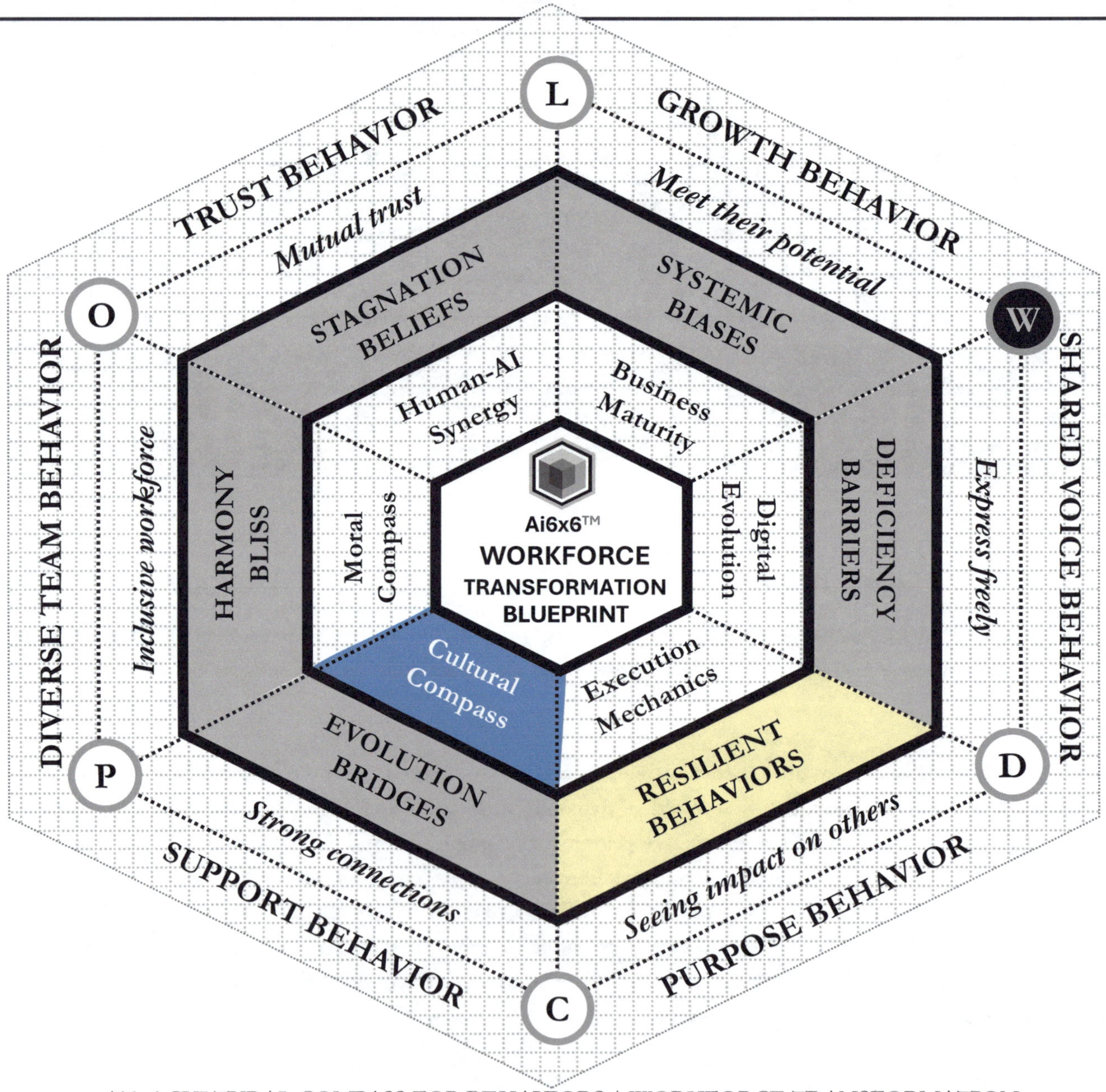

Ai6x6 CULTURAL COMPASS FOR BEHAVIORS | WORKFORCE TRANSFORMATION

Mutual trust is the "high-speed rail" of workforce transformation. When employees trust each other and their leaders, they collaborate without hesitation and share information freely. This behavior eliminates the need for excessive "check-ins" and bureaucratic oversight, allowing for a much more agile response to market shifts. In your daily life as an executive, you'll notice that trusted teams resolve their own conflicts and take more initiative. By modeling trust—starting with assuming positive intent—you create a "safety net" that encourages your workforce to reach for higher goals and experiment with more bold, transformative ideas without the paralyzing fear of being blamed for mistakes. Trust transforms the workplace from a series of contracts into a powerful, living ecosystem of shared responsibility and high-velocity execution.

DO'S FOR THE LEADER	DON'TS FOR THE LEADER
Create a culture of mutual trust.	Allow a low-trust environment.
Be open and transparent with staff.	Withhold information from your team.
Give employees autonomy.	Micromanage everything they do.
Be consistent and reliable.	Be unpredictable and unreliable.
Acknowledge mistakes and fix them.	Blame others and never admit you're wrong.
Empower teams to make decisions.	Take control back from the team.

SITUATION	SUGGESTED ACTION
Employees openly share information because they trust colleagues to use it responsibly and collaboratively	Reinforce transparent communication norms, celebrate responsible knowledge-sharing, and highlight examples where trust accelerated execution and strengthened team cohesion across functions
Teams confidently delegate work knowing others will deliver reliably without excessive oversight or micromanagement	Recognize dependable follow-through, model autonomy-supportive leadership, and maintain clear expectations that reinforce mutual confidence and shared accountability
Employees raise concerns early because they trust leaders to respond constructively and without retaliation	Provide timely acknowledgment, act visibly on input, and reinforce psychological safety as a core expectation for all leaders and teams
Cross-functional groups collaborate smoothly because they trust each other's expertise and intentions	Establish shared goals, clarify interdependencies, and spotlight successful partnerships that demonstrate the value of trusting cross-team relationships
Managers assume positive intent when misunderstandings occur, reducing unnecessary conflict and defensiveness	Encourage clarifying dialogue, train teams to check assumptions, and reinforce norms that prioritize curiosity over blame
Employees feel comfortable admitting mistakes because they trust the organization to treat errors as learning opportunities	Normalize reflective debriefs, highlight lessons learned, and reward transparency to strengthen a culture of continuous improvement

In a transformed workforce, learning is not an event; it's a constant behavior. Employees who exhibit "growth behavior" are proactively looking for ways to improve their skills and the company's processes. This creates a "pull" rather than a "push" for development. As an executive, you support this by providing the time, resources, and recognition for continuous upskilling. When growth becomes a cultural habit, your workforce becomes self-correcting and future-proof. You no longer have to worry about your team's skills becoming obsolete because they are already moving toward the next horizon, ensuring the organization's capability always stays one step ahead. This behavior turns the organization into a "learning machine" that treats every new market challenge as an opportunity to gain a new, actionable capability rather than a threat.

DO'S FOR THE LEADER	DON'TS FOR THE LEADER
Offer professional development.	Cut training budgets and opportunities.
Encourage learning and new skills.	Punish people for making mistakes.
Provide clear career paths.	Keep career paths vague and unclear.
Promote from within the company.	Always hire new people from outside.
Celebrate learning and new skills.	Only reward people for old tasks.
Give regular feedback for growth.	Never give feedback on development.

SITUATION	SUGGESTED ACTION
Employees proactively request stretch assignments to expand their skills and accelerate development	Provide structured opportunities, pair them with experienced mentors, and track progress to reinforce growth as a shared responsibility
Teams regularly participate in learning programs because development is viewed as essential, not optional	Integrate learning into workflows, celebrate participation, and align development activities with strategic capability needs
Employees openly discuss career aspirations with managers, seeking guidance on advancement pathways	Train leaders to hold meaningful development conversations and provide actionable steps that support long-term growth
Managers encourage employees to explore new skills beyond their current roles or functions	Offer cross-functional rotations, skill-building projects, and curated learning resources that broaden capability and perspective
Employees share newly acquired knowledge with peers, strengthening collective capability across the organization	Create knowledge-sharing rituals, recognize contributors, and embed learning transfer into team routines
Teams experiment with new tools and methods to continuously improve performance and adaptability	Support safe experimentation, provide resources, and highlight successful innovations that emerged from proactive learning

3. SHARED VOICE BEHAVIOR | The Power of Every Voice

A culture of "shared voice" ensures that the best ideas rise to the top, regardless of where they come from. When every employee feels valued and heard, you unlock a massive reservoir of creative energy. Daily, this looks like meetings where the most junior person speaks first, and leaders listen more than they talk. This behavior democratizes innovation and ensures that transformation is a collective effort rather than a top-down mandate. It builds a deep sense of "psychological ownership" across the workforce, where people aren't just following a plan—they are building their own future. When everyone shares their voice, you gain a more complete picture of the market and your own operations, leading to more robust decisions and a much higher level of commitment to execution.

DO'S FOR THE LEADER	DON'TS FOR THE LEADER
Ensure every employee feels heard.	Dismiss feedback from junior staff.
Create a safe place to share ideas.	Punish people who speak their mind.
Involve all staff in decisions.	Make decisions in secret meetings.
Listen to all feedback and act on it.	Let feedback go into a black hole.
Encourage respectful disagreement.	Promote a culture of forced agreement.
Use anonymous suggestion boxes.	Only listen to a few trusted people.

SITUATION	SUGGESTED ACTION
Employees contribute ideas confidently because they know their perspectives will be considered seriously	Acknowledge contributions publicly, integrate ideas into decisions, and demonstrate visible follow-through to reinforce credibility
Teams use structured discussions ensuring quieter voices participate equally alongside more vocal colleagues	Facilitate balanced turn-taking, use anonymous input tools, and rotate discussion leads to broaden participation
Employees feel comfortable challenging assumptions respectfully, strengthening decision quality and innovation	Train teams in constructive dissent, reward thoughtful challenge, and model openness to alternative viewpoints
Managers regularly solicit feedback before finalizing decisions, ensuring employees feel included and respected	Use pre-decision surveys, share themes transparently, and show how input influenced outcomes
Employees raise concerns early because they trust leaders to respond thoughtfully and without defensiveness	Provide timely acknowledgment, clarify next steps, and reinforce that honest feedback strengthens organizational performance
Teams celebrate contributions from all members, reinforcing shared ownership and collective success	Highlight diverse inputs during reviews, recognize collaborative wins, and reinforce inclusive contribution norms

When employees have a deep understanding of their work's meaning, they exhibit "purpose behavior"—going above and beyond because they believe in the goal. This is the ultimate "force multiplier" for transformation. Purpose-driven employees are more resilient to stress and more committed to long-term objectives. For you, this means your leadership task shifts from "management" to "inspiration." You don't have to monitor their output because their internal motivation is already aligned with the organization's mission. By consistently connecting daily tasks to the bigger picture, you create a workforce that is fueled by a sense of contribution. This behavior ensures that the organization’s "energy level" remains high, even during the most difficult phases of change, as everyone is anchored by a shared, inspiring "why."

DO’S FOR THE LEADER	DON’TS FOR THE LEADER
Help staff understand their work's meaning.	Let people feel disconnected.
Communicate the company's mission.	Keep the mission vague or unclear.
Show how each role contributes.	Make work feel pointless or boring.
Share customer success stories.	Never connect work to customer joy.
Foster a sense of shared purpose.	Let people feel isolated in their job.
Celebrate achieving meaningful goals.	Only celebrate achieving metrics.

SITUATION	SUGGESTED ACTION
Employees connect daily tasks to broader organizational goals, strengthening motivation and commitment	Communicate strategic context regularly, share impact stories, and reinforce how individual work advances mission-critical outcomes
Teams prioritize work aligned with purpose rather than reacting to shifting demands without clarity	Clarify strategic priorities, explain rationale, and ensure teams understand how choices support long-term direction
Employees feel energized because they understand how their contributions benefit customers and communities	Share customer insights, highlight real-world impact, and integrate purpose narratives into team rituals
Managers discuss meaning and impact during performance conversations, not just metrics and deliverables	Incorporate purpose-aligned goals, explore personal motivations, and connect achievements to organizational aspirations
Employees advocate for improvements that strengthen mission alignment and long-term value	Encourage purpose-driven suggestions, evaluate proposals transparently, and recognize contributions that enhance mission clarity
Teams maintain resilience during challenges because they remain anchored in shared purpose and direction	Reinforce mission during difficult periods, celebrate progress, and remind teams of the broader significance of their work

5. SUPPORT BEHAVIOR | The Safety Net

Support behavior creates a "we've got your back" culture that is essential for navigating the turbulence of transformation. This looks like peers helping each other through a learning curve and leaders prioritizing their team's well-being over a single deadline. When support is a standard behavior, the organization's "stress tolerance" increases significantly. You'll see less burnout and a higher quality of collaborative work. As an executive, when you foster this environment, you build a "high-care, high-performance" workforce that can sustain its energy over the long haul. This collective support system acts as a buffer against the inevitable setbacks of change, ensuring the transformation stays on track even when the pressure is at its peak. Support behavior is the "glue" that keeps the team together when things get messy.

DO'S FOR THE LEADER	DON'TS FOR THE LEADER
Create strong peer support systems.	Create a cutthroat work environment.
Train leaders to be great mentors.	Have leaders who are distant or cold.
Provide mental health resources.	Ignore signs of stress or burnout.
Be flexible with personal needs.	Be rigid and uncaring about people.
Check in on employee well-being.	Assume your team is always okay.
Encourage teamwork and collaboration.	Foster a solo-worker mentality.

SITUATION	SUGGESTED ACTION
Employees readily assist colleagues facing heavy workloads or unfamiliar tasks	Encourage reciprocal support, recognize collaborative behavior, and ensure leaders model shared responsibility
Teams check in regularly on each other's well-being, strengthening connection and trust	Establish consistent touchpoints, normalize emotional awareness, and integrate well-being discussions into team routines
Managers provide guidance proactively rather than waiting for employees to struggle silently	Train leaders to anticipate needs, offer timely coaching, and maintain open communication channels
Employees feel comfortable asking for help without fear of judgment or negative perception	Reinforce help-seeking as a strength, celebrate collaborative problem-solving, and remove stigma around requesting support
Teams share resources and expertise freely, accelerating collective performance and capability	Create shared knowledge hubs, encourage peer mentoring, and highlight examples of effective resource-sharing
Leaders advocate for employees during cross-functional discussions, ensuring fair consideration and support	Promote sponsorship behaviors, track advocacy patterns, and reinforce leader accountability for equitable representation

6. DIVERSE TEAM BEHAVIOR | The Mosaic of Minds

Actively valuing varied backgrounds and perspectives is a behavior that leads to "cognitive diversity in action." In a transformed workforce, diverse teams are the primary units of problem-solving. They are better at spotting risks and identifying new opportunities that a more uniform team would miss. Daily, this means you are intentional about who is in the room and whose voice is being amplified. This behavior goes beyond "inclusion" to "integration," where the power of difference is harnessed to drive better outcomes. By leading this way, you ensure that your workforce is as multifaceted as the world it serves, making your organization more innovative, empathetic, and successful. Diversity isn't just about representation; it's about a daily behavior of seeking out and integrating different points of view to create a superior strategy.

DO'S FOR THE LEADER	DON'TS FOR THE LEADER
Create an inclusive workplace.	Foster a homogeneous work culture.
Value varied backgrounds and views.	Dismiss ideas that are different.
Promote based on merit, not familiarity.	Promote people who are just like you.
Actively recruit diverse talent.	Rely on referrals from a narrow circle.
Listen to all different perspectives.	Let only a few people's voices be heard.
Be a fair and equal opportunity employer.	Show favoritism to certain people.

SITUATION	SUGGESTED ACTION
Teams actively seek diverse viewpoints to strengthen problem-solving and innovation	Encourage mixed-perspective groups, highlight benefits of diversity, and integrate inclusive practices into decision processes
Employees feel comfortable expressing cultural perspectives without fear of misunderstanding or dismissal	Provide inclusive communication training, model curiosity, and reinforce respect for varied experiences
Managers intentionally build teams with diverse backgrounds, skills, and thinking styles	Use structured hiring criteria, broaden sourcing channels, and track representation across roles
Teams challenge assumptions that limit inclusion, ensuring all voices influence outcomes meaningfully	Facilitate bias-aware discussions, encourage reflective questioning, and reinforce inclusive decision-making norms
Employees celebrate cultural differences, strengthening belonging and organizational identity	Create inclusive rituals, rotate cultural spotlights, and ensure recognition practices reflect diverse contributions
Leaders evaluate ideas based on merit rather than similarity to their own experiences	Use structured evaluation frameworks, diversify review panels, and monitor decision patterns for fairness

Ai6x6™ Transformation Blueprint

Workforce Evolution Bridges

"Our very survival depends on our ability to stay awake, to adjust to new ideas, to remain vigilant and to face the challenge of change." - Martin Luther King Jr.

Martin Luther King Jr. (1929–1968) was an American Baptist minister and activist. He was the most visible leader of the Civil Rights Movement and is famous for his use of nonviolent civil disobedience to achieve equality.

Workforce evolution bridges are the practical mechanisms that connect a stagnant present to a resilient and thriving future. They are strategic tools that allow an organization to move beyond negative patterns and create a new reality where fairness, trust, and growth are the norm. The Psychological Safety Bridge is a formal framework for creating a safe environment where open communication and respectful disagreement are encouraged, which is a direct response to the lack of voice and groupthink bias. The Holistic Evaluation Bridge is a system for assessing an employee's performance comprehensively, looking beyond a single trait to provide a fair and balanced view, thereby mitigating the halo/horns effect. As Martin Luther King Jr.'s quote suggests, our survival depends on our ability to "stay awake, to adjust to new ideas... and to face the challenge of change." These bridges are the actionable tools that provide a clear road map for navigating the difficult journey of transformation.

This framework establishes the interpersonal and structural connections that successfully drive change Adoption and refine the Cultural Compass. Cross-Functional Bridges are essential, enabling collaboration between traditionally siloed departments through shared goals and mutual accountability, accelerating end-to-end value delivery. Knowledge-Sharing Bridges mandate the creation of formal systems and cultural practices (e.g., internal wikis, brown bag sessions) for reliably transferring crucial information and specialized skills across the organization. Psychological Safety Bridges establish a culture where people feel safe to speak up, admit mistakes, and take calculated risks without fear of punishment, which is foundational to innovation. The framework highlights Team-Building Bridges, intentional activities and structures designed to foster camaraderie, mutual respect, and social cohesion among dispersed or new teams. Empathy Bridges focus on training and processes that help employees understand and respect others' perspectives, leading to smoother inter-team relations and reduced friction during stressful transitions. Finally, Conflict-Resolution Bridges provide structured, neutral processes for promptly and effectively resolving interpersonal and inter-departmental disputes, preventing localized conflicts within initiatives.

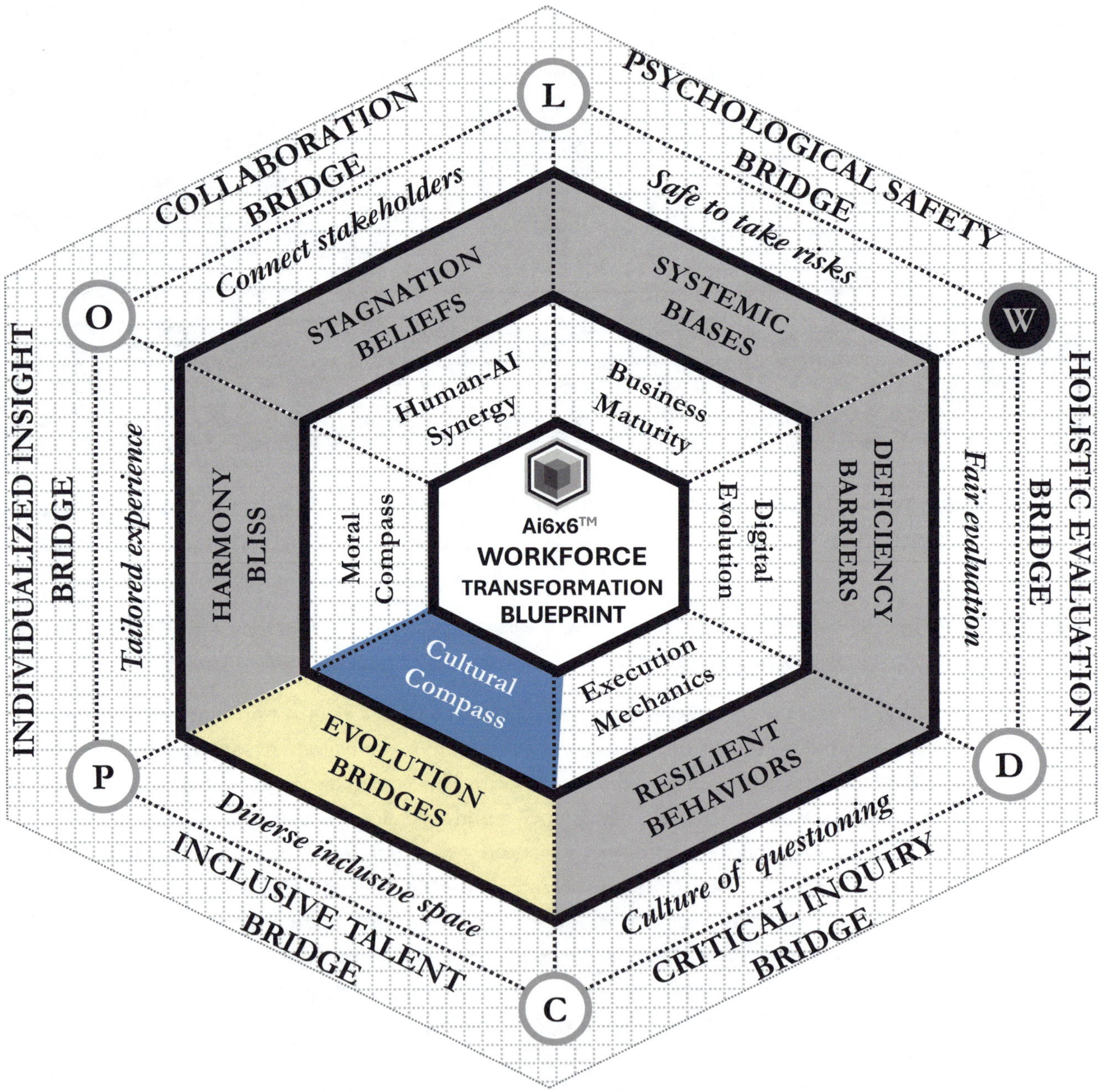

CULTURAL COMPASS FOR BRIDGES | WORKFORCE TRANSFORMATION

1. COLLABORATION BRIDGE | The Weaving of Connections

This bridge connects the "silos" that naturally form in large organizations, helping to overcome in-group bias. It provides the tools and the incentives for cross-functional teams to work together on shared goals. In your daily operations, this bridge manifests as "unified project platforms" and "inter-departmental rotation programs." It allows for the free flow of ideas and talent across the whole workforce. By building this bridge, you turn a collection of isolated departments into a single, integrated network. This connectivity is essential for transformation, as it ensures that the "right hand" always knows what the "left hand" is doing, leading to much faster and more coherent execution of your strategy. Collaboration becomes a seamless, natural part of the workday rather than a difficult, forced effort across departmental lines.

DO'S FOR THE LEADER	DON'TS FOR THE LEADER
Connect teams to overcome silos.	Let teams operate in isolation.
Create cross-functional projects.	Keep people stuck in their department.
Foster an environment of teamwork.	Create a competitive environment.
Reward group success, not just personal.	Only reward individual achievements.
Use tools that connect people.	Have separate systems for each team.
Encourage sharing knowledge openly.	Hoard knowledge in one department.

SITUATION	SUGGESTED ACTION
Teams work in isolated clusters, limiting shared understanding and reinforcing internal silos	Establish cross-team working groups, rotate ownership of initiatives, and create shared goals that require collaboration across functional and relational boundaries
Employees rely heavily on familiar colleagues, avoiding broader organizational partnerships	Introduce structured collaboration rituals, pair diverse contributors intentionally, and highlight successful cross-team outcomes to normalize broader cooperation
Critical information circulates only within small groups, slowing alignment and decision-making	Implement transparent communication channels, require cross-functional updates, and reinforce expectations for open information flow across all teams
Project teams repeatedly select the same members, excluding capable contributors from other areas	Use inclusive staffing processes, track participation patterns, and ensure diverse representation on high-impact initiatives
Teams hesitate to share resources because they expect unequal reciprocation from other groups	Create reciprocal agreements, clarify shared responsibilities, and celebrate balanced collaboration that benefits multiple teams
Employees struggle to understand how their work connects to other functions' priorities	Map interdependencies clearly, host cross-team briefings, and reinforce shared outcomes that require coordinated execution

Creating a safe environment for open communication is the bridge that allows your workforce to cross from "fear" to "innovation."7 When people feel safe to speak up without fear of punishment, the quality of your information increases instantly. You'll hear about the "hidden risks" before they become crises. This bridge is built through your own vulnerability and the way you handle mistakes. In a transformed workforce, psychological safety is the foundation of "fail fast, learn faster". It allows your team to navigate the uncertainty of change with a sense of security, ensuring that they remain focused on the solution rather than their own self-preservation. This bridge is what turns a group of anxious individuals into a confident, high-performing team that isn't afraid to take the bold risks required for true evolution.

DO'S FOR THE LEADER	DON'TS FOR THE LEADER
Create a safe place for open talk.	Punish people for making mistakes.
Encourage respectful disagreement.	Silence dissent and criticism.
Reward honesty and transparency.	Punish people who are truthful.
Let people take risks without fear.	Create a fear of failure.
Admit your own mistakes as a leader.	Always pretend you're right.
Listen to all concerns with empathy.	Dismiss people's feelings and fears.

SITUATION	SUGGESTED ACTION
Employees hesitate to speak openly because past feedback was dismissed or criticized harshly	Train leaders in supportive responses, acknowledge contributions publicly, and demonstrate consistent follow-through to rebuild confidence in open communication
Teams avoid raising risks early due to fear of negative consequences or blame	Establish blameless review practices, encourage early escalation, and reinforce that transparency strengthens outcomes rather than exposing weakness
Employees remain silent during discussions because dominant voices overshadow quieter contributors	Facilitate balanced participation, use structured turn-taking, and create anonymous input channels to ensure all perspectives are heard
Managers unintentionally shut down ideas by reacting too quickly with skepticism	Require exploratory questioning before evaluation, model curiosity, and reinforce norms that value thoughtful consideration of all contributions
Employees fear admitting mistakes, worrying it will damage credibility or advancement	Normalize reflective debriefs, highlight learning value, and reward transparency that strengthens team capability
Teams avoid difficult conversations, allowing issues to grow silently and reduce trust	Introduce guided dialogue frameworks, train facilitators, and reinforce that respectful honesty is essential for healthy team dynamics

Moving beyond "annual reviews" to a comprehensive system for assessing performance is a critical bridge for workforce transformation. This bridge allows you to see the "whole person"—their potential, their soft skills, and their contribution to the culture—not just their latest KPIs. It provides the data you need for more accurate talent placement and growth planning. In your daily life, this bridge means more meaningful "check-ins" and real-time feedback. By evaluating employees holistically, you show that you value their total contribution, which increases engagement and ensures that you are developing the diverse range of talents the organization will need to thrive. This bridge ensures that talent is recognized and nurtured wherever it exists in the organization, preventing valuable skills from going unnoticed or underutilized.

DO'S FOR THE LEADER	DON'TS FOR THE LEADER
Assess performance comprehensively.	Judge people on one single metric.
Use clear, objective performance metrics.	Use vague, subjective evaluations.
Gather feedback from multiple sources.	Rely on only one person's feedback.
Focus on growth, not just performance.	Only focus on hitting a number.
Tie rewards to holistic performance.	Reward a single, narrow achievement.
Review goals and performance often.	Have only one review per year.

SITUATION	SUGGESTED ACTION
Performance reviews focus narrowly on output metrics, overlooking collaboration and long-term growth	Expand evaluation criteria, incorporate peer input, and balance results with behaviors that strengthen organizational capability
Managers rely heavily on recent events, ignoring broader performance patterns across the year	Use structured evidence logs, review longitudinal data, and calibrate assessments to reduce recency bias
Employees feel evaluations depend more on personal impressions than objective contributions	Implement transparent scoring rubrics, share expectations early, and provide examples of performance at each level
Teams receive inconsistent feedback because managers apply different standards across similar roles	Standardize evaluation frameworks, conduct calibration sessions, and ensure leaders apply criteria consistently
Employees struggle to understand how development areas connect to future opportunities	Link feedback to career pathways, outline specific growth actions, and track progress over time
Managers overlook contextual factors that influence performance, such as workload or resource constraints	Require contextual analysis, review environmental factors, and adjust expectations to ensure fairness and accuracy

4. CRITICAL INQUIRY BRIDGE | The Culture of Questioning

This bridge encourages questioning and independent thinking, helping to dismantle authority bias and groupthink. It's a process that invites employees to challenge "the way we've always done it." Daily, this looks like "red-teaming" your own strategies or holding "question-only" brainstorming sessions. By building this bridge, you create a workforce of "thinkers" rather than "doers." This is essential for transformation because it ensures that your evolution is constantly being refined by the smartest people in the organization. It builds a culture of intellectual rigor and continuous improvement, where every process is subject to a healthy level of scrutiny. This bridge ensures that the organization remains self-correcting, preventing you from going too far down a wrong path because no one felt empowered to ask a difficult question.

DO'S FOR THE LEADER	DON'TS FOR THE LEADER
Encourage questioning and thinking.	Punish people who challenge the status quo.
Ask "why" questions constantly.	Assume the current way is the best way.
Promote a culture of curiosity.	Make people feel stupid for asking.
Use data to challenge assumptions.	Rely on gut feelings instead of facts.
Create a safe place for new ideas.	Reject new ideas without thought.
Train people to solve problems.	Give people problems with only one answer.

SITUATION	SUGGESTED ACTION
Teams accept initial assumptions without examining alternatives or challenging underlying logic	Introduce structured questioning frameworks, assign challenge roles, and require evidence reviews before decisions
Employees hesitate to question established processes, assuming legacy methods are always correct	Encourage process audits, reward constructive challenge, and highlight improvements driven by fresh perspectives
Managers discourage debate by prioritizing speed over thoughtful evaluation	Build time for inquiry into workflows, reinforce quality over haste, and model openness to deeper analysis
Teams rely heavily on expert opinions without validating data or exploring independent viewpoints	Require data triangulation, diversify input sources, and encourage independent assessment before alignment
Employees avoid raising concerns about potential risks because they fear appearing negative	Normalize risk identification, integrate it into planning, and recognize individuals who surface issues early
Teams struggle to innovate because they rarely explore unconventional ideas or alternative approaches	Facilitate ideation sessions, use divergent thinking techniques, and celebrate creative contributions that challenge norms

A system that ensures fair opportunities for everyone is the bridge to a truly diverse and equitable workforce. This bridge involves removing the systemic biases from your hiring, promotion, and development processes. It ensures that "merit" is defined broadly and fairly. In your daily role, this bridge means you have a more robust and talented leadership pipeline. By ensuring that everyone has an equal shot at success, you attract and retain the best talent from every background. This bridge doesn't just "feel good"—it's a strategic imperative that ensures your organization has access to 100% of the available talent pool. It ensures that your transformation is powered by the best possible minds, providing a significant competitive advantage in an increasingly global and highly competitive labor market.

DO'S FOR THE LEADER	DON'TS FOR THE LEADER
Ensure fair opportunities for all.	Give some people better opportunities.
Use unbiased hiring and promotion.	Rely on gut feelings in hiring.
Create objective criteria for success.	Use subjective, vague criteria.
Recruit from a wide range of schools.	Only hire from a small number of places.
Provide a mentor for everyone.	Only mentor a select few.
Support employee resource groups.	Don't support groups for minorities.

SITUATION	SUGGESTED ACTION
High-visibility assignments consistently go to the same small group of employees	Track opportunity distribution, rotate ownership intentionally, and ensure equitable access to career-advancing work
Promotion decisions rely heavily on informal impressions rather than transparent criteria	Publish advancement requirements, use structured evaluations, and calibrate decisions across leaders to ensure fairness
Employees from underrepresented groups receive less sponsorship and advocacy from senior leaders	Establish sponsorship programs, monitor participation, and hold leaders accountable for equitable advocacy
Talent reviews overlook individuals who contribute strongly but lack visibility or self-promotion	Use multi-rater input, review objective outcomes, and ensure recognition is not tied solely to visibility
Development resources are allocated unevenly, favoring employees with existing relationships	Standardize access to learning programs, track utilization, and adjust outreach to correct disparities
Hiring panels favor candidates who resemble current team members, limiting diversity	Diversify interview panels, use structured scoring, and broaden sourcing channels to expand representation

Understanding each employee as a unique individual is the bridge to personalized management and higher engagement. This bridge uses data and empathy to tailor growth paths, communication styles, and work arrangements to the specific needs of each person. Daily, this allows you to lead with more precision and impact. It recognizes that "one-size-fits-all" leadership is dead. By building this bridge, you create a workforce where people feel "seen" and "valued" for who they are, not just what they do. This deep level of connection leads to much higher loyalty and performance, as employees are far more likely to give their best to an organization that truly understands and supports their individual journey. It transforms the relationship between the worker and the firm into a personalized, mutually beneficial partnership.

DO'S FOR THE LEADER	DON'TS FOR THE LEADER
Understand each employee as unique.	Treat all employees exactly the same.
Get to know people personally.	Be a distant, impersonal leader.
Provide personalized development plans.	Use a generic development plan for all.
Tailor feedback to each person's needs.	Give everyone the same feedback.
Ask employees about their strengths.	Assume everyone has the same strengths.
Empower people to follow their passion.	Force people to work on things they hate.

SITUATION	SUGGESTED ACTION
Managers assume all employees are motivated by similar rewards or career paths	Conduct individualized motivation assessments, explore personal goals, and tailor development plans accordingly
Employees feel misunderstood because leaders rely on generalized assumptions rather than personal context	Hold regular one-to-one conversations, document preferences, and adjust leadership approaches to reflect individual needs
Teams apply uniform work styles without considering differences in strengths or cognitive preferences	Use strengths assessments, adapt workflows, and assign tasks aligned with individual capabilities
Managers overlook unique constraints affecting employee performance, such as workload or personal circumstances	Review context regularly, adjust expectations, and provide targeted support that reflects individual realities
Employees feel disconnected because their personal aspirations are not integrated into team planning	Incorporate personal goals into project assignments, track alignment, and revisit aspirations during development discussions
Leaders struggle to understand how different employees learn best, limiting development effectiveness	Offer varied learning modalities, gather feedback, and tailor learning plans to individual preferences and growth patterns

CHAPTER
F6

BLUEPRINT SERIES

Ai6x6™ Transformation Blueprint

Workforce Harmony Bliss

"Teamwork is the ability to work together toward a common vision. The ability to direct individual accomplishments toward organizational objectives. It is the fuel that allows common people to attain uncommon results." - Andrew Carnegie

Andrew Carnegie (1835–1919) was a Scottish-American industrialist and philanthropist. He led the expansion of the American steel industry and is known for his vast philanthropic efforts and his book The Gospel of Wealth.

Workforce harmony bliss is the ultimate state where the collective feels a profound sense of purpose, achievement, and belonging. It is the pinnacle of a successful workforce transformation, where employees are not just productive but deeply engaged. This state is characterized by several key elements. Blissful Talent Cultivation is the deep satisfaction of building a culture that develops and nurtures its people. Blissful Dedication is the joy of having employees who are committed to their work because they understand its meaning. Blissful Innovation is the pleasure of a workforce that embraces change and constantly generates new ideas, driven by a culture of psychological safety and shared voice. As Andrew Carnegie suggests, "Teamwork is the fuel that allows common people to attain uncommon results." The harmony of a transformed workforce is a state of collective brilliance where the organization operates as a cohesive, dynamic, and effective whole.

This framework establishes the Relational domain by exploring Workforce Harmony Bliss, the collective outcomes that validate successful change Adoption and refine the Cultural Compass. The primary result is High Engagement, characterized by a highly motivated, committed, and enthusiastic workforce that actively contributes to organizational goals and discretionary effort. This is built upon Psychological Safety, a culture of mutual trust and respect where individuals feel secure enough to speak up, admit mistakes, and take calculated risks, which is foundational for rapid learning and innovation. Harmony is secured by Low Turnover, indicating that employees are satisfied, feel valued, and choose to remain with the company long-term, dramatically reducing recruitment and training costs. The framework targets Collaborative Excellence, where teams work together seamlessly across functions, achieving superior results through shared effort, breaking down historical siloes. Success is seen in Empowered Teams, which possess the autonomy, resources, and authority to make critical decisions at the point of action, accelerating decision velocity and speed-to-market. Finally, the economic result is High Productivity, a workforce that consistently delivers high-quality work efficiently and effectively, maximizing the human capital return on investment (ROI).

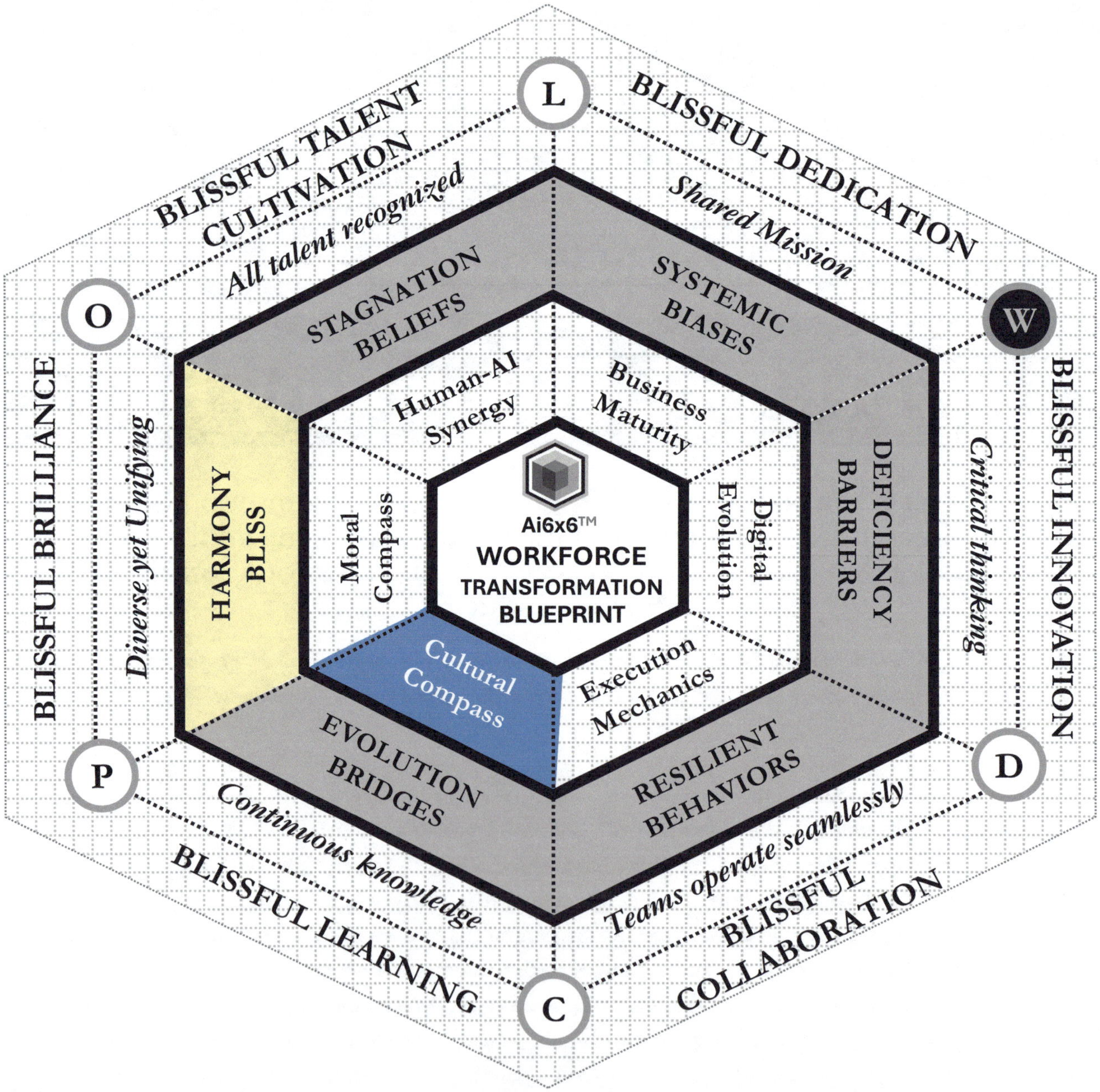

CULTURAL COMPASS FOR BLISS | WORKFORCE TRANSFORMATION

1. BLISSFUL TALENT CULTIVATION | The Garden of Talent:

There is a profound joy in building a culture that successfully develops and nurtures talent. This bliss comes from seeing your "upskilling" programs turn into real career breakthroughs for your employees. It transforms your daily work into a legacy of growth. In a transformed workforce, you are no longer just managing output; you are gardening potential. This state of harmony is reached when the organization's growth and the employee's growth are perfectly aligned. It creates a vibrant environment where people are excited to come to work because they know they are becoming "better versions of themselves" every day. This bliss is the feeling of leading an organization that doesn't just produce profit, but produces more capable, successful human beings, creating a powerful sense of collective achievement.

DO'S FOR THE LEADER	DON'TS FOR THE LEADER
Build a culture that nurtures talent.	Let talent stagnate and get bored.
Invest in people's development.	Ignore your employees' potential.
Create an environment of growth.	Have a static, unchanging workplace.
Give people challenges to grow.	Assign only simple, repetitive tasks.
Celebrate learning and new skills.	Only celebrate achieving deadlines.
Mentor new employees.	Assume new people will figure it out.

SITUATION	SUGGESTED ACTION
Employees feel motivated because leaders consistently invest in their long-term growth and development	Provide structured learning pathways, celebrate progress milestones, and reinforce that cultivating talent is a shared organizational responsibility supported by leadership
Teams actively mentor one another, strengthening capability and accelerating collective advancement across functions	Formalize peer-mentoring programs, recognize effective mentors, and integrate knowledge-sharing rituals into team routines to reinforce continuous capability building across groups
Managers identify emerging strengths early and create opportunities for employees to apply them meaningfully	Use strengths assessments, assign stretch responsibilities, and track progress consistently to reinforce confidence, capability expansion, and long-term development momentum
Employees volunteer for skill-building projects because they trust the organization will support their growth	Provide coaching, allocate resources, and highlight successful development journeys to reinforce confidence in long-term capability expansion and meaningful professional advancement
Teams celebrate learning achievements, reinforcing a culture where growth is valued and consistently visible	Create recognition rituals, share development stories, and integrate learning wins into team communications to strengthen motivation and reinforce continuous improvement behaviors
Leaders prioritize talent cultivation as a strategic imperative rather than an optional organizational activity	Embed development expectations into leadership goals, measure progress consistently, and reward leaders who grow talent effectively across diverse teams and evolving business needs

2. BLISSFUL DEDICATION | The Shared Mission:

The satisfaction of leading a workforce that is genuinely dedicated to meaningful work is the ultimate executive reward. This bliss comes from the knowledge that your team isn't just working for a paycheck; they are working for a purpose they believe in. It creates a "flow state" at the organizational level. Daily, this means you lead with inspiration rather than pressure. You can feel the collective "pull" toward the goal. This bliss provides a deep sense of calm and confidence, as you know that even in the face of immense challenge, your team's shared commitment will carry you through. It turns the hard work of transformation into a fulfilling, collective mission, leaving you with a workforce that is intrinsically motivated to do their best work every single day.

DO'S FOR THE LEADER	DON'TS FOR THE LEADER
Connect work to a meaningful purpose.	Let work feel pointless and boring.
Communicate the mission and vision.	Keep your goals a secret.
Show how work impacts customers.	Never share customer success stories.
Foster a sense of shared dedication.	Let people feel isolated in their role.
Be transparent about company results.	Keep people in the dark about success.
Thank employees for their hard work.	Let hard work go unnoticed.

SITUATION	SUGGESTED ACTION
Employees consistently exceed expectations because they feel deeply connected to their work's purpose and impact	Reinforce mission alignment, share meaningful impact stories, and recognize contributions that advance long-term organizational goals and strengthen employee commitment
Teams maintain high standards during challenging periods because they believe their work genuinely matters	Provide context for shifting priorities, celebrate resilience, and ensure leaders acknowledge sustained dedication throughout demanding cycles
Employees take ownership of outcomes, demonstrating pride in delivering consistently high-quality results	Highlight strong ownership examples, provide autonomy, and reinforce accountability as a positive expression of meaningful commitment
Managers inspire dedication by connecting daily tasks to broader organizational goals and long-term direction	Communicate strategic relevance clearly, explain decision drivers, and show how work contributes to sustained organizational success
Employees proactively solve problems because they care about delivering value to customers and colleagues	Recognize proactive behavior, share customer insights, and reinforce the importance of initiative and thoughtful action across teams
Teams support each other during demanding cycles, driven by shared commitment to meaningful work	Encourage collaborative problem-solving, acknowledge collective effort, and maintain transparent communication during high-pressure periods

3. BLISSFUL INNOVATION | The Creative Engine

Watching your workforce embrace change and constantly innovate is a state of creative bliss. It's the feeling of leading a team that is "ahead of the curve." This bliss comes from the realization that you have successfully built a "resilient engine" of evolution. Daily, this looks like a constant stream of new ideas and improvements bubbling up from every level. It reduces the stress of competition because you know your organization is the one setting the pace. This harmony between strategy and execution allows you to lead with a sense of wonder and excitement, as you witness the collective brilliance of your team solving the future's problems today. This bliss is the realization that your organization has moved from reacting to the future to creating it.

DO'S FOR THE LEADER	DON'TS FOR THE LEADER
Create a workforce that innovates.	Punish people who try new things.
Encourage new ideas and change.	Stick to the way things have always been.
Provide resources for new projects.	Keep resources locked up and controlled.
Foster a safe space for mistakes.	Create a fear of failure.
Challenge old processes and ideas.	Assume the status quo is the best.
Reward creativity and new thinking.	Only reward conformity.

SITUATION	SUGGESTED ACTION
Employees experiment confidently because innovation is celebrated rather than judged prematurely or harshly by colleagues	Provide safe experimentation spaces, highlight learning value, and reward thoughtful risk-taking that strengthens long-term capability and organizational adaptability
Teams continuously refine processes, seeking better ways to deliver value and operational efficiency	Encourage iterative improvement, share successful innovations, and integrate experimentation into regular workflows to reinforce progress and collective learning
Employees propose new ideas frequently because they trust leaders will consider them seriously and fairly	Establish idea-capture systems, evaluate proposals transparently, and communicate outcomes clearly to maintain credibility and encourage continued contribution
Managers champion innovation by modeling curiosity and openness to unconventional approaches and emerging possibilities	Recognize leaders who support experimentation, provide resources, and reinforce innovation as a core leadership expectation across the organization
Teams collaborate across functions to generate creative solutions for complex and evolving challenges	Facilitate cross-functional ideation sessions, rotate participants, and celebrate diverse contributions that strengthen innovation and problem-solving capability
Employees stay ahead of industry trends by exploring emerging tools, methods, and technologies regularly	Provide access to innovation resources, encourage exploration, and integrate trend insights into planning and decision-making processes

The satisfaction of working seamlessly with diverse colleagues across silos and borders is the "symphony" of modern leadership. This bliss comes from the friction-less flow of ideas and effort. It's the joy of seeing "1 + 1 = 3." When your workforce is truly collaborative, the complexity of transformation becomes manageable. Daily, this bliss manifests as efficient, high-energy interactions where everyone is focused on the outcome rather than their ego. It provides a deep sense of connection and belonging, reminding you that while the challenges are great, you are part of a powerful, unified team that can overcome anything through the strength of its collective intelligence. This bliss is the feeling of an organization where the whole is truly greater than the sum of its diverse and talented parts.

DO'S FOR THE LEADER	DON'TS FOR THE LEADER
Let teams work seamlessly together.	Create silos between departments.
Promote an environment of teamwork.	Create a competitive workplace.
Reward groups for their success.	Only reward individuals for their work.
Use tools to connect people.	Have separate systems for each team.
Encourage open communication.	Discourage communication between teams.
Celebrate diversity and new ideas.	Only value ideas from a single source.

SITUATION	SUGGESTED ACTION
Teams coordinate effortlessly because they understand each other's strengths and preferred working styles clearly	Use strengths assessments, align responsibilities intentionally, and reinforce shared ownership of outcomes across diverse teams
Employees build strong relationships across functions, enabling smooth execution and shared organizational success	Facilitate cross-team interactions, celebrate collaborative wins, and maintain transparent communication to strengthen cohesion and trust
Managers encourage inclusive participation, ensuring all voices contribute meaningfully to team outcomes and decisions	Use structured facilitation, rotate roles, and highlight contributions from diverse perspectives to reinforce inclusion and shared ownership
Teams resolve disagreements constructively, maintaining trust and momentum during challenging or ambiguous moments	Train teams in conflict resolution, reinforce respectful dialogue, and celebrate successful collaboration during difficult periods
Employees willingly support colleagues during high-pressure periods, strengthening team cohesion and reliability	Encourage reciprocal support, recognize collaborative behavior, and maintain balanced workloads across teams and functions
Cross-functional groups achieve strong results because they value diverse expertise, perspectives, and problem-solving approaches	Create mixed-discipline teams, clarify shared goals, and reinforce the importance of varied viewpoints in decision-making

There is a unique joy that comes from a culture that supports and celebrates continuous learning. This bliss is felt when a "difficult" new skill becomes a standard capability across the team. It's the feeling of collective growth. In your daily life, this means you are surrounded by curious, engaged minds who are always looking for a better way. This blissful learning environment reduces the fear of the unknown, as everyone knows they have the support and the tools to master whatever comes next. It creates a youthful, optimistic energy in the workforce, ensuring that the organization remains intellectually vibrant and ready for whatever disruptions the future might hold. This bliss is the knowledge that your team's intelligence is constantly expanding, ensuring your company's long-term survival.

DO'S FOR THE LEADER	DON'TS FOR THE LEADER
Create a culture of continuous learning.	See learning as a one-time event.
Invest in training and education.	Cut training budgets to save money.
Reward learning and new skills.	Only reward the skills people already have.
Encourage sharing new knowledge.	Let people hoard their own knowledge.
Make it fun and easy to learn.	Make learning a chore or a punishment.
Use mistakes as learning opportunities.	Blame people for all mistakes.

SITUATION	SUGGESTED ACTION
Employees eagerly pursue new skills because learning is embedded naturally into daily work routines	Provide micro-learning resources, integrate learning prompts, and celebrate continuous improvement across teams and functions
Teams share insights regularly, strengthening collective intelligence and capability across the organization	Establish knowledge-sharing rituals, recognize contributors, and maintain accessible learning repositories for ongoing development and collaboration
Employees feel energized by opportunities to explore new tools, technologies, and emerging methods	Offer guided experimentation, provide coaching, and highlight successful adoption stories to reinforce confidence and curiosity
Managers encourage learning by modeling curiosity and participating actively in development activities	Recognize leaders who learn visibly, integrate learning goals into performance plans, and reinforce growth expectations consistently
Teams reflect on successes and failures to extract meaningful lessons for future work and improvement	Facilitate structured retrospectives, document insights, and integrate learnings into planning and execution processes
Employees feel supported in their learning journeys, regardless of experience level or organizational role	Provide equitable access to resources, tailor learning paths, and track progress consistently across individuals and teams

The feeling of collective achievement and pride in a team's diverse accomplishments is the pinnacle of workforce harmony. This bliss is experienced when a major transformation goal is reached, and everyone—from the front line to the C-suite—knows they played a vital role. It's a moment of shared "brilliance" where the diverse talents of the organization have converged to create something extraordinary. This bliss provides a lasting sense of pride and professional fulfillment. It validates all the hard work and difficult shifts, leaving the organization with a reinforced sense of its own capability and a deep, collective confidence. This is the ultimate "win" for a leader—seeing the organization shine with a brilliance that is powered by the unique, unified talents of its entire workforce.

DO'S FOR THE LEADER	DON'TS FOR THE LEADER
Create a feeling of collective pride.	Only praise a few star performers.
Highlight the team's accomplishments.	Let individual success be the only focus.
Celebrate diversity and new ideas.	Ignore the benefits of varied perspectives.
Share success stories with the company.	Keep a team's success a secret.
Reward both individual and team success.	Create a competition for success.
Let everyone share in the glory.	Take all the credit for yourself.

SITUATION	SUGGESTED ACTION
Teams celebrate major milestones together, reinforcing pride in shared accomplishments and collective success	Create recognition rituals, highlight contributions broadly, and connect achievements to organizational goals and long-term direction
Employees feel proud of their work because leaders acknowledge excellence consistently and meaningfully across teams	Provide specific recognition, share success stories, and reinforce the value of high-quality performance throughout the organization
Teams showcase diverse strengths, demonstrating how varied expertise drives exceptional outcomes and innovation	Highlight cross-disciplinary wins, rotate spotlight features, and reinforce appreciation for diverse contributions across functions
Employees take collective ownership of results, strengthening unity and shared purpose across the organization	Encourage team-based recognition, align incentives, and celebrate collaborative achievements that reflect shared commitment and capability
Managers highlight team brilliance during organizational forums, boosting visibility, morale, and long-term confidence	Share accomplishments publicly, provide platforms for teams to present, and reinforce pride in collective success and contribution
Teams reflect on achievements to build confidence and momentum for future challenges and opportunities	Facilitate celebratory retrospectives, document wins, and integrate insights into future planning and execution cycles

BONUS CHAPTER

Ai6x6™ Transformation Blueprint

Workforce Diversity

"We are building products that people with very diverse backgrounds use, and I think we all want our company makeup to reflect the makeup of the people who use our products."

— Sheryl Sandberg, Former COO of Meta (Facebook). For digital products to be successful globally, the teams creating them must possess the diverse perspectives necessary to understand and serve a diverse user base, avoiding blind spots and bias in technology.

Successful **Workforce Transformation** depends on diversity to maximize human capital and accelerate the acquisition of new digital skills. A workforce that is diverse in age, culture, technical expertise, and socioeconomic background brings a vast array of unique problem-solving techniques and market insights to the table. This is critical for moving toward agile, cross-functional ways of working, where learning from disparate disciplines is essential. When teams are inclusive, they are better at developing solutions that are relevant and accessible across various global markets and user types. Diversity ensures that the newly implemented digital tools and processes are embraced by all employees, not just a select few, because the teams building them understand varied learning styles and adoption barriers. By deliberately fostering an inclusive environment, organizations unlock the full potential of their talent, creating a culture of constant adaptation and innovation. Ultimately, prioritizing diversity guarantees that the transformed organization is not just technologically advanced, but rich in the human ingenuity.

In today's interconnected world, diversity is a fundamental pillar of strength and innovation. Often simplified to visible differences, true diversity is a rich tapestry of attributes shaping how individuals perceive and contribute. Understanding this multifaceted nature is crucial for fostering environments where every voice thrives. This essay explores diversity through six critical dimensions: **demographic, cognitive, experiential, geographic, functional, and perspective diversity**. By examining these distinct facets, we can better appreciate how a holistic approach to diversity enriches teams and communities, driving robust solutions.

First, **Demographic Diversity** encompasses varied inherent and acquired characteristics. This includes age, bringing diverse generational insights, and gender, race, ethnicity, sexual orientation, ensuring a wide range of lived experiences. Socioeconomic status, religion, disability, and nationality further contribute to a richer collective understanding. Valuing demographic diversity ensures equitable representation and leverages unique insights from varied personal journeys.

Secondly, **Cognitive Diversity** highlights how people think, process information, and approach problems. This includes varied thinking styles like analytical vs. intuitive approaches, and different problem-solving or learning methods. When a group embraces cognitive diversity, it tackles problems from multiple angles, leading to more innovative solutions and a thorough understanding of complex issues, free from a single dominant thought process.

Next, **Experiential Diversity** focuses on the breadth of life and work experiences individuals bring. This covers varied work histories across industries and roles, providing a rich pool of professional knowledge. Diverse educational backgrounds, from different disciplines or institutions, also offer distinct frameworks. Beyond formal credentials, personal life experiences like travel and cultural immersion contribute significantly, equipping teams with a wide array of skills and adaptable approaches.

Geographic Diversity emphasizes the impact of place on an individual's perspective. This includes differences from regional or local backgrounds, shaping values and priorities. National backgrounds introduce distinct cultural norms, while urban versus rural upbringings offer varying insights into community dynamics and societal needs. These varied geographical lenses provide a broader understanding of global and local markets and behaviors.

Furthermore, **Functional Diversity** pertains to the variety of professional roles, specialized areas, and departmental expertise within a team. This dimension recognizes contributions from different functions like engineering, marketing, or finance, each bringing specific domain knowledge. Variations in hierarchical roles and industry experience also inform cross-pollination of best practices. Leveraging functional diversity ensures all necessary technical skills are present, fostering comprehensive problem-solving.

Finally, **Perspective Diversity** encompasses the fundamental differences in viewpoints and mental models individuals hold, often influenced by other diversity forms. This includes varied ideological stances, cultural interpretations, and distinct value systems. Understanding diverse stakeholder perspectives—from customers to employees—is crucial for holistic decision-making. Ultimately, perspective diversity allows for problems to be framed in multiple ways, challenging assumptions and encouraging empathetic, effective outcomes.

By intentionally cultivating and integrating these six dimensions—demographic, cognitive, experiential, geographic, functional, and perspective—organizations and societies can harness the profound power of true diversity. This comprehensive approach not only fosters equity but also enhances creativity, resilience, and innovation, demonstrating that a multitude of viewpoints is the clearest path to collective success.

1. DEMOGRAPHIC DIVERSITY

Embracing **demographic diversity** across the entire workforce means intentionally cultivating a talent pool that mirrors the rich tapestry of society and the organization's customer base. This goes beyond mere representation; it's about leveraging the inherent value that comes from employees with varied ages, genders, racial and ethnic backgrounds, physical abilities, socio-economic upbringings, and sexual orientations. These visible and non-visible differences aren't just checkboxes; they are sources of unique experiences, perspectives, and problem-solving approaches. A demographically diverse workforce brings distinct insights into market needs, cultural sensitivities, and customer experiences, which are crucial for effective transformation. When a workforce lacks this breadth, it risks developing products, services, or strategies that are narrow in scope, fail to resonate with diverse markets, or overlook the needs of significant employee populations. True demographic diversity fosters a more empathetic, relatable, and ultimately, a more effective organization, better equipped to navigate complex global challenges and connect authentically with diverse stakeholders. It enhances the legitimacy and credibility of transformation initiatives by demonstrating a genuine commitment to inclusivity throughout all levels.

Topic	Tangible steps	Challenges you may face
AGE DIVERSITY IN WORKFORCE PLANNING		
Varied age groups impact skill transfer, innovation, and knowledge retention across the workforce. Diverse generational views enhance organizational resilience and adaptability.	Implement intergenerational mentorship programs. Offer flexible work arrangements for all ages. Provide upskilling/reskilling programs accessible to all.	Resistance to cross-generational learning. Overcoming stereotypes about younger/older workers. Managing different career expectations across generations.
GENDER DIVERSITY IN TEAMS		
Balanced gender representation influences team dynamics, problem-solving, and innovation. Diverse gender perspectives lead to comprehensive solutions and broader market appeal.	Set targets for gender balance in teams/departments. Promote equal pay and promotion opportunities. Implement inclusive language and communication training.	Unconscious biases in hiring/promotion. Lack of support for work-life balance hindering participation. Resistance to changing established team norms.
RACIAL & ETHNIC DIVERSITY IN CULTURE		
Diverse racial & ethnic backgrounds foster cultural intelligence and empathy within the organization. Improves global collaboration and customer understanding.	Develop cultural awareness training programs. Establish employee resource groups (ERGs). Prioritize diverse representation in company-wide initiatives.	Systemic bias in HR processes. Difficulty fostering psychological safety. Insufficient understanding of cultural differences leading to miscommunication.

Cognitive diversity across the workforce is paramount for navigating the inherent complexities of transformation. It acknowledges that true innovation and robust problem-solving stem not just from varied experiences or demographics, but fundamentally from differences in how people think, process information, and approach challenges. A team composed solely of analytical thinkers might excel at optimizing existing processes but could struggle with disruptive innovation. Conversely, a team too reliant on intuitive thinking might miss critical data points or overlook practical constraints. By fostering a mix of analytical, creative, systemic, and neurodiverse cognitive styles across all roles, organizations are better equipped to challenge assumptions, identify unseen opportunities, predict unforeseen risks, and design more comprehensive and resilient strategies. It's about building a collective intelligence where different cognitive strengths complement each other, leading to more thorough evaluations, more innovative solutions, and a greater capacity for adaptive learning in the face of rapid change. This richness in thought ensures that operational and strategic decisions are thoroughly vetted from multiple angles, preventing blind spots and fostering intellectual agility crucial for sustained transformation success.

Topic	Tangible steps	Challenges you may face
ANALYTICAL VS. INTUITIVE THINKING IN TEAMS		
Balancing data-driven & gut-feeling approaches within teams shapes project outcomes & problem-solving. Leads to nuanced decision-making, combining factual rigor with insightful foresight.	Design team structures requiring diverse cognitive input. Encourage employees to articulate their reasoning. Provide training on leveraging different thinking styles.	Friction between styles. Difficulty quantifying intuition. Culture valuing one style over another hinders collaboration.
SYSTEMIC THINKERS IN PROCESS IMPROVEMENT		
Employees seeing interconnectedness & long-term implications drive comprehensive process improvement. Ensures holistic, sustainable operational changes.	Provide systems thinking training for all levels. Assign systemic thinkers to cross-functional process redesign initiatives. Use "value stream mapping" to identify impacts.	Difficulty identifying/developing systemic thinking. Resistance to long-term views. Overwhelm from too much interconnected information.
CREATIVE PROBLEM-SOLVERS IN WORKFLOW DESIGN		
Employees with imaginative thinking inspire novel solutions to daily operational challenges & new product development. Crucial for breakthrough innovations.	Establish ideation labs or hackathons. Implement brainstorming sessions for unconventional ideas. Create dedicated time for "blue-sky" thinking.	Risk aversion stifling new ideas. Difficulty translating creative ideas into actionable plans. Perceived lack of "seriousness" for certain roles.

3. EXPERIENTIAL DIVERSITY

Experiential diversity across the workforce is a powerful catalyst for successful transformation, providing a rich tapestry of practical insights drawn from varied career paths, industries, educational backgrounds, and personal life stages. Employees who have navigated different organizational cultures, worked across diverse sectors (e.g., startup to corporate, public to private), or gained international exposure bring a broader repertoire of solutions and a more nuanced understanding of complex challenges. This breadth of experience prevents insular thinking, encourages the adoption of "best practices" from seemingly unrelated fields, and fosters resilience by drawing on lessons learned from past successes and failures in varied contexts. For instance, an employee with experience in crisis management can bring invaluable composure and clear-headedness during periods of intense organizational upheaval. By intentionally cultivating a workforce with diverse professional and personal journeys, an organization gains a formidable collective wisdom, enhancing its ability to foresee opportunities, mitigate risks, and execute transformation initiatives with greater adaptability and effectiveness. It promotes a learning culture where varied practical wisdom is valued and applied to new strategic challenges.

Topic	Tangible steps	Challenges you may face
CROSS-INDUSTRY INSIGHTS IN PROBLEM-SOLVING		
Employees from different industries bring fresh perspectives & transferable solutions to transformation hurdles. Broadens organizational problem-solving.	Recruit talent with diverse industry backgrounds. Organize peer-learning forums for cross-industry insights. Encourage participation in external industry associations.	Difficulty translating insights to new contexts. Resistance to "outsider" approaches. Lack of understanding of new industry nuances for integration.
STARTUP VS. CORPORATE BACKGROUNDS IN TEAMS		
Impact of employees from agile startup vs. structured corporate environments on team dynamics & innovation. Blending these fosters agility & stability.	Integrate startup talent into corporate innovation projects. Create joint ventures pairing corporate/startup teams. Foster internal "intrapreneurship" programs.	Culture clash between agile & traditional structures. Difficulty integrating fast decisions into slower processes. High expectations for immediate results.
GLOBAL VS. LOCAL EXPERIENCE IN OPERATIONS		
Employees with diverse international & local market exposure balance global standardization & regional adaptation. Optimizes operational efficiency & market relevance.	Implement global talent rotation programs. Establish regional knowledge hubs with local market insights. Encourage dual reporting lines for international projects.	Complex balance of global vs. local. Resistance from local teams to global mandates. Misinterpretation of cultural nuances in operations.

4. GEOGRAPHIC DIVERSITY

Geographic diversity within the workforce is indispensable for organizations operating in a globalized world, especially during periods of significant transformation. Employees deeply rooted in varied regions, countries, or even urban versus rural settings bring critical local market understanding, cultural nuances, and insights into unique regulatory environments and logistical challenges. Without this localized perspective, global strategies risk being homogenous, culturally insensitive, or simply impractical in execution. For example, an employee experienced in emerging markets might bring crucial insights into scalable, low-cost solutions, while someone from a highly regulated economy understands the complexities of compliance. This diversity ensures that transformation initiatives are not just conceptualized centrally but are designed with built-in adaptability for regional variations. It empowers the organization to navigate geopolitical risks, understand localized customer behaviors, and foster stronger relationships with diverse regional stakeholders. Employees with varied geographic exposure contribute to strategies that are globally coherent yet locally relevant, maximizing impact and minimizing missteps in a fragmented world.

Topic	Tangible steps	Challenges you may face
REGIONAL MARKET UNDERSTANDING IN SALES		
Employees from different regions offer unique insights into local market dynamics, consumer behaviors, & competition. Vital for effective sales and marketing.	Recruit talent with local market knowledge. Mandate global strategies be reviewed/adapted by regional teams. Support employee immersion in diverse regions.	Overlooking nuances for standardization. Difficulty adapting to rapid changes in local markets. Resistance from central teams to local adaptations.
NAVIGATING GLOBAL REGULATORY LANDSCAPES		
Employees familiar with diverse international legal/compliance frameworks are crucial for multi-national operations & transformations. Ensures adherence & mitigates risks.	Create global compliance training programs for relevant roles. Invest in continuous education on international regulations. Consult local legal experts for every project.	Keeping up with rapidly changing laws. Reconciling conflicting requirements across jurisdictions. Risk of non-compliance due to inadequate local understanding.
LOCAL INFRASTRUCTURE AND RESOURCE KNOWLEDGE		
Employees understanding local infrastructure (digital, logistics) & resources impact operational transformation. Optimizes local implementation & supply chains.	Empower local teams for infrastructure investments. Develop centralized knowledge of regional capabilities. Foster local service provider partnerships.	Underestimating varied local infrastructure impact. Inaccurate resource assumptions. Difficulty scaling global solutions to diverse local conditions.

Functional diversity across the workforce is critical for executing comprehensive and interconnected business transformations. It ensures that strategic and operational decisions are informed by the deep expertise and unique perspectives of various professional disciplines, rather than being confined to a single departmental silo. A workforce comprising experts from finance, technology, human resources, operations, marketing, and legal functions can collectively identify opportunities, anticipate challenges, and design solutions that are holistic, practical, and sustainable. For instance, a finance analyst ensures fiscal viability, while a tech specialist champions digital adoption, and an HR generalist focuses on people-centric change management. Without this blend, transformation efforts might optimize one area at the expense of another, leading to fragmented implementation or unintended negative consequences. By fostering cross-functional collaboration and valuing diverse functional expertise, an organization can break down traditional barriers, enhance teamwork, and create truly integrated strategies that leverage the strengths of every part of the business, ensuring that the transformation is not just conceptually sound but also executable across all organizational facets.

Topic	Tangible steps	Challenges you may face
FINANCE-LED TRANSFORMATION GOALS FOR TEAMS		
Employees with strong financial backgrounds prioritize cost optimization & ROI as primary transformation drivers. Ensures fiscal responsibility in projects.	Appoint finance professionals to cross-functional project teams. Establish clear financial metrics & ROI targets for initiatives. Implement rigorous cost-benefit analyses for all investments.	Over-emphasis on short-term cuts. Resistance if financial goals overshadow innovation. Difficulty projecting ROI for intangible benefits in specific functions.
TECHNOLOGY-DRIVEN INNOVATION ACROSS WORKFORCE		
Employees with deep tech expertise champion digital adoption & disruptive innovation. Essential for driving tech-enabled change throughout the organization.	Elevate technologists to strategic roles in projects. Foster "tech literacy" for all employees. Create cross-functional innovation teams combining tech/business experts.	Resistance to tech change. Rapid evolution challenges skills. Difficulty integrating new tech with legacy systems and traditional workflows.
HR'S ROLE IN PEOPLE-CENTRIC CHANGE MANAGEMENT		
HR professionals focus on cultural shifts, talent management, & employee well-being during transformation. Ensures human capital is central to success.	Position HR teams as strategic partners in all change initiatives. Develop robust change management/communication plans. Implement well-being/reskilling programs for all employees.	Business units resist prioritizing people over profit. Difficulty measuring HR ROI. Challenges addressing employee anxiety/resistance to change.

Perspective diversity within the workforce is arguably one of the most subtle yet profound drivers of successful transformation. It goes beyond demographics, experience, or function, delving into the varied worldviews, core values, ethical frameworks, and risk tolerances that shape how employees interpret information and prioritize objectives. A team with diverse perspectives challenges inherent biases, avoids groupthink, and considers a wider spectrum of possibilities and potential consequences. For example, balancing employees who prioritize short-term gains with those focused on long-term sustainability, or those who are risk-seeking with those who are risk-averse, leads to more robust and balanced strategic decisions. This diversity ensures that transformation is not just about efficiency or growth, but also about ethical responsibility, employee well-being, and genuine stakeholder value. It encourages employees to approach problems from fundamentally different angles, leading to more resilient strategies, better stakeholder engagement, and a more adaptive and morally grounded organization capable of navigating complex and ambiguous future landscapes. It cultivates a culture of constructive challenge and holistic decision-making.

Topic	Tangible steps	Challenges you may face
CUSTOMER-CENTRIC VS. PRODUCT-CENTRIC VIEWS		
Workforce balances internal product focus with external customer needs. Ensures products solve customer problems & resonate with market.	Implement "Voice of Customer" programs for all employees. Conduct customer immersion days. Establish KPIs balancing product features with customer satisfaction.	Internal resistance to shifting from product-centric. Difficulty capturing diverse customer needs. Employees struggle balancing desires vs. feasibility.
SHORT-TERM VS. LONG-TERM VISION IN PROJECTS		
Employees' differing time horizons influence project pace, scope, & success. Balancing these ensures both immediate gains & future viability.	Develop multi-year roadmaps with short-term milestones. Assign teams with complementary focuses. Use scenario planning to visualize impacts.	Pressure for immediate results. Employees may prioritize short-term gains. Difficulty articulating long-term value to all stakeholders.
RISK-SEEKING VS. RISK-AVERSE APPROACHES IN INNOVATION		
Varied risk tolerance impacts pursuit of disruptive vs. incremental improvements. A healthy balance drives calculated growth and controlled experimentation.	Foster culture where calculated risks are encouraged. Establish "fail fast, learn faster" practices. Create project portfolios balancing high-risk/reward with safer bets.	Decision paralysis when tolerances conflict. Dominance of extreme risk views. Difficulty setting clear, unified risk parameters across teams.

SECTION G

BLUEPRINT SERIES

Moral Compass

Ai6x6™ Transformation Blueprint

Story 1 — The Nuremberg Trials and Ethical Accountability (1945–1946): After World War II, the world confronted atrocities committed under the guise of legal authority. The Nuremberg Trials established that individuals and institutions are accountable for actions that violate fundamental human rights. Legal frameworks were strengthened by moral reasoning, not replaced by it. Judges balanced evidence with ethical principles to define new global norms. The trials demonstrated that morality must guide power. They became a foundation for modern ethical governance.

Story 2 — The Tuskegee Syphilis Study Exposure (1932–1972): For decades, researchers withheld treatment from Black men to observe disease progression, violating consent and dignity. Public exposure triggered national outrage and forced reforms in medical ethics. Oversight boards, informed consent, and patient rights became mandatory. Machines and data had played roles in tracking outcomes, but human ethics had been absent. The scandal showed how systems fail when morality is excluded. It reshaped global standards for ethical research.

Story 3 — The Environmental Movement and Earth Day (1970): Industrial growth created pollution that threatened ecosystems and public health. Scientists provided data, but moral arguments mobilized action. Citizens demanded accountability, and governments introduced regulations to protect air, water, and wildlife. The movement reframed environmental protection as a shared ethical duty. It demonstrated how moral clarity can transform policy, industry, and public behavior. The cultural shift established sustainability as a long-term societal obligation.

Story 4 — The Anti-Apartheid Global Solidarity Movement (1950s–1994: Apartheid enforced racial segregation and oppression in South Africa. Activists, governments, and institutions worldwide united to challenge the regime through sanctions, protests, and moral pressure. Human stories amplified the injustice, while global networks coordinated action. The movement showed how collective moral conviction can influence political systems. It proved that ethical solidarity can reshape international norms. The end of apartheid became a landmark in moral mobilization.

Story 5 — The Rise of Corporate Social Responsibility (1990s–2000s): Globalization exposed labor abuses, environmental harm, and opaque supply chains. Consumers demanded transparency, and companies adopted codes of conduct, sustainability reporting, and ethical sourcing. Machines tracked compliance, but humans defined the standards. CSR shifted from public relations to strategic responsibility. Organizations learned that long-term value depends on ethical behavior. It marked a turning point where morality became integral to corporate identity.

Story 6 — The AI Ethics Movement (2015–2025): As AI systems influenced decisions in hiring, policing, healthcare, and finance, concerns grew about bias, transparency, and accountability. Researchers, policymakers, and civil society pushed for ethical frameworks to guide development. Machines provided capability, but humans defined boundaries. The movement emphasized fairness, explainability, and human oversight. It demonstrated that emerging technologies require moral governance as much as technical innovation. AI ethics became a global priority shaping the future of intelligent systems.

LESSONS FROM A WORKFORCE TRANSFORMATION PERSPECTIVE

Workforce maturity in moral contexts develops when employees internalize ethical standards that guide behavior beyond compliance. Over time, organizations learned that ethical cultures depend on workers who speak up, challenge harmful practices, and act with integrity even when incentives misalign. Mature workforces understand the human impact of their decisions and treat ethics as part of professional identity. They balance efficiency with fairness, confidentiality with transparency, and loyalty with accountability. Maturity is reached when employees feel responsible not only for outcomes but for the means by which those outcomes are achieved. Ethics becomes a shared commitment, not a procedural requirement.

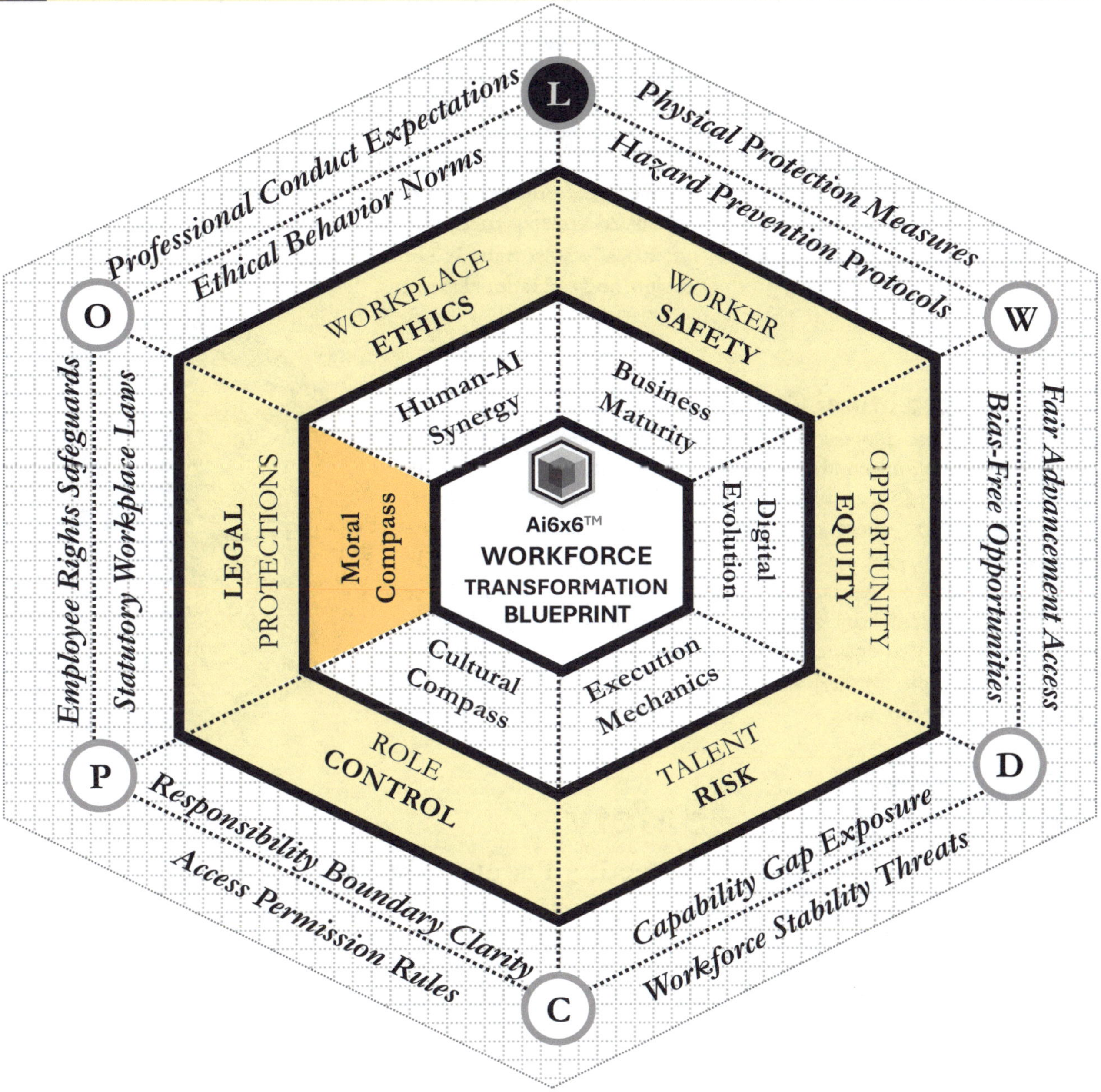

The **Ai6x6 Research Lab** identified the Moral Compass as the most vital lens because it serves as the final guardian of the worker's dignity and the organization's ethical honor. This is the ultimate filter for every decision, ensuring the drive for efficiency never outpaces the bedrock of labor protections and pay equity. We chose this because an organization's legacy is built on how it protected its people during the storm of change, not just the profits it generated. It is vital because it provides the moral authority required to manage the high responsibility of power, safety, and delegation in a data-driven world.

This lens is the heart's final check—a rigorous audit of OSHA compliance, talent risk, and the ethics of influence. It ensures the transformed enterprise is not only a titan of industry but a beacon of fairness and opportunity. The Lab believes that in the 2026 landscape, the greatest risk to a workforce is not attrition, but a loss of trust in the system's equity. By focusing on labor protections and RACI controls, we provide the tools to navigate the complex social contract between employer and employee. This lens is about the courage to be honorable and equitable when the pressure for output is highest. Ultimately, it ensures that every worker is treated with the respect and protection they deserve.

Morality has always shaped how organizations behave, but in the age of AI, it becomes a structural requirement rather than an aspirational ideal. The **Moral Compass** lens forces organizations to confront a truth that is often ignored: intelligence—whether human or machine—does not create ethical behavior. It amplifies whatever moral foundation already exists. When an organization is grounded in integrity, clarity, and principled action, AI strengthens it. When the foundation is inconsistent, opaque, or misaligned, AI exposes those weaknesses with unforgiving speed. Morality is no longer a philosophical preference; it is the architecture that determines whether accelerated decisions, automated systems, and scaled consequences lead to progress or harm. This lens reframes morality as an enterprise-wide discipline, not a leadership trait. It applies equally to workforce practices, data governance, customer interactions, product design, and operational systems.

Ethics

Establishing the Moral Center

Ethics forms the first dimension of the Moral Compass. Ethics is not simply about avoiding misconduct; it is about establishing a moral center that guides every decision, every interaction, and every use of technology. Across all six pillars, ethics determines how data is collected, how employees are treated, how customers are served, how products are built, and how operations are executed. Organizations that prioritize ethics create cultures where integrity is expected, not encouraged. Ethical clarity becomes the anchor that ensures AI is deployed with intention, restraint, and respect. In this dimension, principled behavior becomes the foundation of moral authority.

Safety

Protecting People, Systems, and Society

Safety represents the second dimension. As AI accelerates decision-making and expands operational reach, the potential for harm increases. Safety is not limited to physical environments; it includes psychological safety for the workforce, data safety for customers, model safety for AI systems, and operational safety for mission-critical processes. Safety ensures that no individual, team, or automated system can cause harm through unchecked influence or unintended consequences. It is not about limiting innovation; it is about ensuring that innovation does not outpace responsibility. In this dimension, responsible protection becomes essential to moral operation.

Equity

Ensuring Fairness Across Stakeholders

Equity forms the third dimension. Influence in modern organizations is no longer tied solely to hierarchy; it flows through data pipelines, algorithmic recommendations, and automated workflows. Equity ensures that these systems do not privilege certain groups, perspectives, or outcomes at the expense of others. It applies to workforce advancement, customer treatment, product accessibility, and data representation. Equity is not about equalizing results; it is about ensuring that the system itself is not biased. In this dimension, fairness becomes a structural requirement rather than a social aspiration.

Risk

Navigating Uncertainty With Discipline

Risk represents the fourth dimension. AI-enabled decisions carry greater speed, greater reach, and greater uncertainty. Risk is not about avoiding uncertainty; it is about navigating it with discipline. Organizations must anticipate consequences, stress-test assumptions, and understand the risks embedded in both human and machine judgment. This applies to operational risks, product risks, customer risks, workforce risks, and data risks. Organizations that embrace this dimension act decisively without acting blindly. In this dimension, risk-aware decision-making becomes a moral responsibility.

Control

Defining Governance and Accountability

Control forms the fifth dimension. As organizations adopt AI, authority becomes more complex: who decides, who approves, who monitors, and who intervenes when systems drift? Control ensures that governance structures are clear, decision rights are defined, and accountability is transparent. It applies to data governance, product governance, operational governance, and workforce governance. Control is not about bureaucracy; it is about ensuring that power—human or machine—is exercised with clarity and accountability. In this dimension, structured oversight becomes the backbone of moral operation.

Legal

Upholding Lawful and Responsible Conduct

Legal completes the Moral Compass. In a world where data, automation, and intelligence shape every action, legal obligations become inseparable from organizational obligations. Compliance is not a constraint; it is a safeguard. Legal integrity ensures that decisions align with regulatory expectations, societal norms, and ethical standards across all six pillars. It protects the organization from missteps and reinforces public trust. In this dimension, lawful stewardship becomes the final test of moral credibility.

The Moral Compass lens reminds organizations that moral authority is not a byproduct of good intentions; it is the result of deliberate, disciplined choices. Morality must be engineered into systems, not assumed in people. When organizations prioritize ethics, enforce safety, ensure equity, manage risk, establish control, and uphold legal accountability, they create environments capable of navigating complexity with confidence and integrity. They build systems where people feel protected, customers feel respected, and society feels considered. They create the moral conditions for transformation to take root and endure. In this maturity, morality becomes an enterprise capability—one that strengthens trust, accelerates adoption, and stabilizes transformation. It becomes the differentiator that determines whether AI amplifies value or amplifies harm. Organizations that master the Moral Compass do not merely deploy intelligent systems; they deploy them responsibly. They create futures where intelligence and integrity advance together, ensuring that progress is not only possible but principled.

MORAL COMPASS ANALOGY – NURTURING THE FAMILY HOME

Family Values ETHICS

The foundational moral code—religious, cultural, and social—that dictates the ground rules for "Right vs. Wrong." Establishing the core integrity of the home so that all who belong understand the non-negotiable standards of conduct.

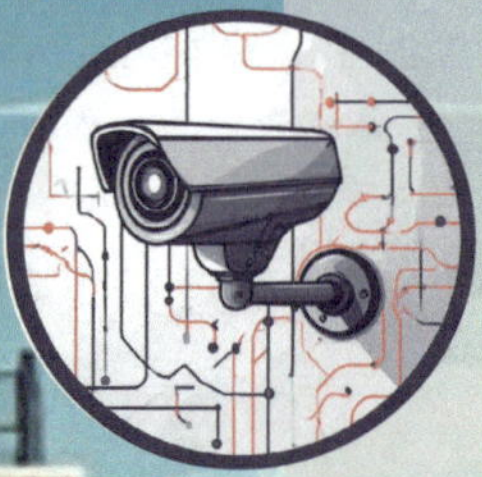

Home Security SAFETY

Integrated protection—cameras, sensors, and smoke detectors—that provide total awareness of the premises. Deploying proactive warning systems and delegation guardrails to protect the inhabitants from "hidden fires."

Balanced Contribution EQUITY

Ensuring everyone helps with the cooking and upkeep, with expenses shared equitably by each earning member. Maintaining participation and representation so that no member feels excluded from the collective success.

The Insurance Shield RISK

Comprehensive fire, flood, and auto coverage to provide long-term peace of mind for the entire household. Using stress testing and "what-if" scenarios to protect the family assets from unforeseen market storms.

Access & Keys CONTROL

Ensuring all adults have keys and codes, with clear emergency rules and a defined line of command for disasters. Establishing rigid RACI controls to ensure that a human always has the final word on the most critical locks.

The Estate Will LEGAL

Recording Wills, Trusts, and Medical Mandates with a legal custodian to safeguard the valuables for the future. Securing the organization's legacy through regulatory accountability, ensuring the estate is legally unassailable for generations.

Ai6x6™ Transformation Blueprint

Workplace Ethics

"A trustworthy algorithm begins with a trustworthy society."
Kofi Annan, former UN Secretary-General and Nobel Peace Prize laureate.

Workplace Ethics requires consistent behavioral standards, and Professional Conduct Expectations define how employees must act in interactions, decisions, and daily responsibilities. They establish the norms that govern respect, honesty, and accountability across the organization. This matters because conduct shapes culture more directly than policy. Professional Conduct Expectations prevent behavioral drift by clarifying what is acceptable and what is not. They also reinforce fairness by ensuring all employees operate under the same ethical rules. When ignored, conduct becomes inconsistent and unpredictable. Teams experience confusion about boundaries and expectations. Over time, Workplace Ethics deteriorates as norms weaken and behavior becomes misaligned. Ethical Behavior Norms ensure employees understand the principles that guide appropriate workplace actions.

The workplace is the pulse of the engine and the frontline of brand integrity. Leaders use Anonymous Reporting Channels and Code of Conduct Training Logs to ensure the human element remains healthy during rapid AI adoption. By sitting with "Ethics Investigation Summaries," a leader can detect the "rot" of toxic culture before it is amplified by technology. It is a commitment to a shared vernacular of integrity. This topic ensures that as execution is automated, the human workforce remains anchored in a code that machines cannot replicate, maintaining the organizational "Why" across all levels. In a print environment, you can move between conduct reports and disclosure forms, seeing the ethical health of your culture as a tangible asset. This deep engagement ensures that "Workplace Ethics" is not just a poster on a wall, but a clinical reality that protects the organization from the internal fragmentation that usually derails major transformations.

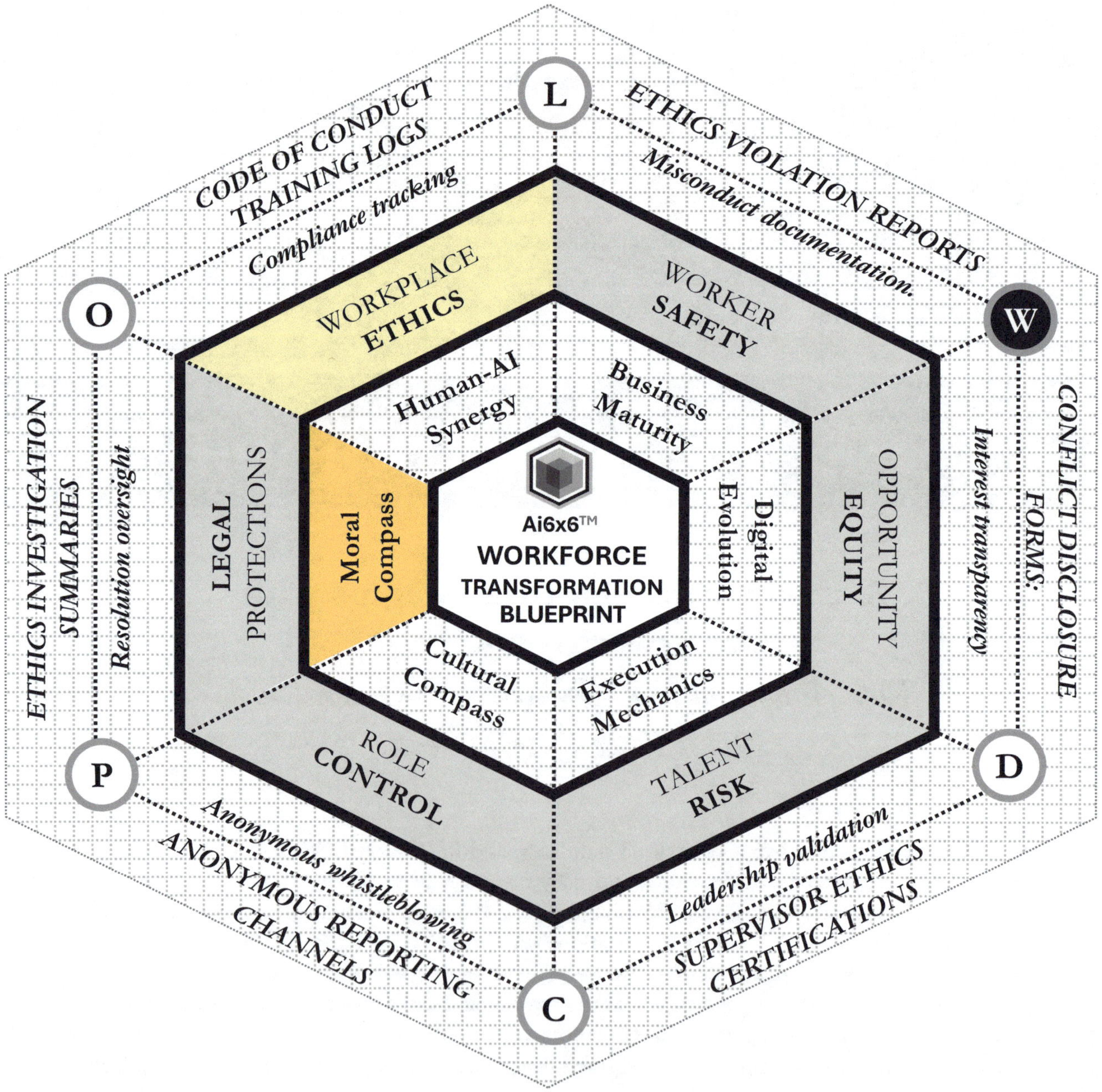

Ai6x6 WORKPLACE ETHICS FOR MORAL COMPASS | WORKFORCE TRANSFORMATION

1. CODE-OF-CONDUCT TRAINING LOGS

Code of Conduct Training Logs document employee participation in ethics training and confirm understanding of behavioral expectations. They ensure that every employee receives consistent guidance on acceptable conduct, reporting obligations, and organizational values. This structure ensures that Code of Conduct Training Logs serve as evidence of ethical readiness across the workforce. They matter because training without documentation creates gaps in accountability and compliance. Logs prevent ambiguity about who has been trained, who needs reinforcement, and where ethical risks may emerge. They also reinforce fairness by ensuring all employees receive the same ethical foundation. When ignored, training becomes inconsistent and ethical expectations vary widely. Employees may claim ignorance or misunderstanding when issues arise. Over time, the organization becomes vulnerable to preventable misconduct and regulatory exposure.

When done well, Code of Conduct Training Logs create transparency, consistency, and disciplined ethical onboarding. People trust that everyone is held to the same standards and receives the same preparation. This strengthens confidence in organizational fairness and ethical alignment. Training logs reinforce the principle that ethics must be taught, tracked, and reinforced. They remind leaders that ethical culture requires structured investment, not informal assumption. Ultimately, they build a workforce grounded in shared understanding and accountable behavior.

MORAL COMPASS SCENARIO	LEADERSHIP ACTION
Marisol notices several employees skipped mandatory conduct training but still marked themselves as complete. She worries the team is treating compliance as optional.	The leader audits the training logs and requires all incomplete employees to finish the course within 48 hours. They also implement system-verified attendance to prevent self-reporting.
Dante sees that new hires are going weeks without receiving conduct training. He realizes onboarding gaps are creating risk.	The leader updates the onboarding workflow to trigger automatic enrollment on day one. They also assign HR to monitor completion weekly.
Felicia discovers that remote employees are consistently behind on training. She suspects time-zone scheduling is the barrier.	The leader offers flexible, asynchronous training options and updates the log to track remote completion separately. They also brief managers on reinforcing expectations.
Rowan notices that managers are not reviewing their team's training status. He senses a lack of ownership.	The leader assigns each manager responsibility for verifying their team's logs. They also add training compliance to quarterly performance metrics.
Aiden realizes the training content hasn't been updated in years. Employees are completing outdated material.	The leader commissions a content refresh and logs the update cycle. They also schedule annual reviews to keep the material current.
Talia sees that contractors are not included in training logs. She worries they're operating without conduct expectations.	The leader expands the log to include contractors and mandates completion before system access. They also assign procurement to enforce compliance.

2. ETHICS VIOLATION REPORTS

Ethics Violation Reports document instances where employee behavior breaches organizational standards, policies, or moral expectations. They create a structured record of misconduct, patterns, and leadership responses. This documentation ensures that Ethics Violation Reports serve as a transparent mechanism for accountability. They matter because organizations cannot correct what they do not formally acknowledge. Reports reveal systemic issues, recurring risks, and areas requiring stronger controls. They also reinforce fairness by ensuring violations are addressed consistently rather than selectively. When ignored, misconduct becomes normalized and ethical standards lose credibility. Employees lose trust in leadership's willingness to enforce rules. Over time, the organization experiences cultural decay and increased legal exposure.

When done well, Ethics Violation Reports create accountability, visibility, and ethical discipline. People see that misconduct is taken seriously and addressed with consistency. This strengthens trust in the integrity of the system and encourages responsible behavior. Reports reinforce the principle that ethical breaches must be confronted, not minimized. They remind leaders that transparency is essential to moral authority. Ultimately, they help build a culture where ethical expectations are enforced and respected.

MORAL COMPASS SCENARIO	LEADERSHIP ACTION
Jairo receives a report that a supervisor is making inappropriate comments but hesitates to escalate it. He fears damaging the relationship.	The leader instructs Jairo to file the report immediately and assures him that neutrality is required. They also assign an independent investigator to avoid bias.
Linh notices repeated minor violations from the same employee. She worries the pattern signals deeper issues.	The leader directs Linh to consolidate the incidents into a formal report. They then initiate a behavioral review to address root causes.
Omar receives a report about misuse of company equipment but dismisses it as trivial. He later realizes the issue may be widespread.	The leader requires Omar to reopen the case and document all related incidents. They also expand the investigation to include other teams.
Serena learns that an employee is retaliating against a colleague who filed a complaint. She knows retaliation must be handled urgently.	The leader escalates the case to HR and enforces immediate separation between the parties. They also activate retaliation-prevention protocols.
Mateo receives a vague report about financial misconduct but lacks details. He's unsure whether to proceed.	The leader instructs Mateo to log the report and begin preliminary fact-finding. They also request additional information from the reporter through a protected channel.
Giselle discovers that a manager has been quietly dismissing ethics reports without documentation. She recognizes the governance risk.	The leader audits all dismissed reports and disciplines the manager for bypassing protocol. They also implement mandatory tracking for every ethics submission.

3. CONFLICT-DISCLOSURE FORMS

Conflict Disclosure Forms require employees to reveal personal, financial, or relational interests that could influence workplace decisions. They clarify potential conflicts, document risks, and guide mitigation strategies. This structure ensures that Conflict Disclosure Forms serve as a safeguard against hidden bias or undue influence. They matter because undisclosed conflicts distort fairness, decision quality, and trust. Forms prevent ethical ambiguity by making employees accountable for transparency. They also reinforce integrity by ensuring decisions are made in the organization's interest rather than personal gain. When ignored, conflicts remain hidden and compromise decision making. Teams question whether outcomes are fair or influenced by private interests. Over time, the organization becomes vulnerable to ethical breaches and reputational harm.

When done well, Conflict Disclosure Forms create transparency, impartiality, and disciplined decision governance. People trust that conflicts are identified and managed responsibly. This strengthens confidence in fairness and reduces the risk of compromised decisions. Disclosure reinforces the principle that integrity requires openness. It reminds employees that transparency is a moral obligation, not a procedural formality. Ultimately, it builds a workplace where trust is protected through honest self reporting.

MORAL COMPASS SCENARIO	LEADERSHIP ACTION
Priya forgets to disclose that her cousin works for a vendor bidding on a major contract. She realizes the omission only after shortlisting begins.	The leader pauses the evaluation and requires Priya to submit a formal disclosure. They then reassign the review to a conflict-free evaluator.
Malik hesitates to disclose his side business because he fears it might be seen as disloyal. He doesn't realize it overlaps with company services.	The leader guides Malik through the disclosure process and evaluates the overlap. They also set boundaries to prevent competitive conflict.
Sofia discloses that she is dating a colleague in her reporting chain. She worries the relationship will affect perceptions of fairness.	The leader restructures reporting lines to eliminate the conflict. They also document the disclosure to maintain transparency.
Ethan submits a disclosure about owning stock in a partner company. He's unsure whether the amount is material.	The leader reviews the financial threshold and confirms whether divestment or recusal is required. They also update the disclosure form with clearer guidance.
Lila discloses that she volunteers for an organization applying for a grant. She fears her involvement could bias the review.	The leader removes Lila from the evaluation process and logs the conflict. They also assign a neutral reviewer to ensure fairness.
Jordan forgets to update his disclosure after joining a nonprofit board. He realizes the oversight during an audit.	The leader requires immediate form resubmission and logs the late update. They also implement quarterly reminders for all employees.

4. SUPERVISOR ETHICS CERTIFICATIONS

Supervisor Ethics Certifications require managers to formally attest to their adherence to ethical standards, leadership responsibilities, and compliance obligations. They create a structured mechanism for reinforcing ethical leadership expectations. This certification ensures that Supervisor Ethics Certifications elevate ethical accountability from assumption to commitment. They matter because supervisors shape daily culture and influence how ethics is interpreted and enforced. Certifications prevent ethical drift by requiring leaders to reaffirm their responsibilities. They also reinforce fairness by ensuring supervisors are held to higher, explicit standards. When ignored, supervisors may apply ethics inconsistently or model behavior that contradicts organizational values. Teams lose trust in leadership integrity. Over time, the organization experiences uneven enforcement and cultural fragmentation.

When done well, Supervisor Ethics Certifications create clarity, accountability, and ethical leadership discipline. People see that supervisors are held to rigorous standards and must demonstrate integrity. This strengthens confidence in leadership and reinforces ethical expectations. Certifications reinforce the principle that leadership requires explicit moral commitment. They remind supervisors that ethical influence is a responsibility, not a privilege. Ultimately, they build a culture where ethical leadership is visible, consistent, and credible.

MORAL COMPASS SCENARIO	LEADERSHIP ACTION
Amara delays signing her certification because she's unsure whether a past decision violated policy. She fears signing inaccurately.	The leader arranges a confidential review of the decision and clarifies her obligations. They also add a pre-certification consultation step for all supervisors.
Declan signs his certification quickly but later realizes he overlooked a conflict in his team. He worries about amending it.	The leader instructs Declan to file a corrected certification immediately. They also reinforce that transparency strengthens—not weakens—credibility.
Nisha hesitates to certify because she has observed minor misconduct but isn't sure whether it qualifies. She fears over-reporting.	The leader walks her through the reporting thresholds and helps her document the concerns. They also update the certification guide with clearer examples.
Rowan signs his certification but later learns his team used outdated compliance procedures. He worries this invalidates his attestation.	The leader helps Rowan amend the certification and initiates a process update. They also schedule refresher training for his team.
Giselle avoids signing because she feels the certification implies personal blame for inherited issues. She wants clarity on scope.	The leader clarifies that certifications reflect current oversight, not historical fault. They also help her document inherited risks separately.
Mateo signs but admits he didn't read the full document. He didn't realize the legal implications.	The leader requires him to complete a guided review and re-sign. They also implement a comprehension check for future certifications.

5. ANONYMOUS REPORTING CHANNELS

Anonymous Reporting Channels provide confidential pathways for employees to raise concerns without fear of retaliation. They ensure that sensitive issues can be surfaced even when power dynamics or fear inhibit direct reporting. This structure ensures that Anonymous Reporting Channels serve as a critical safety valve for ethical transparency. They matter because employees often hesitate to report misconduct when identity exposure feels risky. Anonymous channels prevent silence, concealment, and unresolved ethical issues. They also reinforce trust by demonstrating leadership's commitment to protecting whistleblowers. When ignored, employees remain silent, and misconduct spreads unchecked. Leaders lose visibility into emerging risks. Over time, the organization becomes vulnerable to ethical collapse and reputational damage.

When done well, Anonymous Reporting Channels create psychological safety, transparency, and early intervention. People trust that concerns can be raised safely and taken seriously. This strengthens ethical vigilance and organizational integrity. Anonymous reporting reinforces the principle that every employee must have a protected voice. It reminds leaders that retaliation is both unethical and destructive. Ultimately, it builds a culture where speaking up is safe, supported, and valued.

MORAL COMPASS SCENARIO	LEADERSHIP ACTION
Lior notices that employees rarely use the anonymous hotline. He suspects they don't trust confidentiality.	The leader reinforces anonymity protections and shares examples of safely handled cases. They also run a communication campaign explaining how reports are protected.
Tessa discovers that hotline messages are being routed to a shared inbox. She worries identities could be inferred.	The leader moves the channel to a secure, restricted system. They also audit past access to ensure no breaches occurred.
Omar learns that employees think anonymous reports are ignored. He senses a credibility gap.	The leader publishes aggregated outcomes from past reports to demonstrate follow-through. They also commit to acknowledging every submission.
Priya sees that hotline instructions are buried deep in the intranet. Employees can't find them.	The leader relocates the instructions to a prominent location and adds QR codes in common areas. They also include hotline details in onboarding.
Jonas notices that reports spike after major announcements. He suspects employees are using the channel for general feedback.	The leader clarifies the purpose of the hotline and creates a separate channel for non-ethics feedback. They also update training materials to reinforce proper use.
Hana discovers that some employees fear digital tracking. She realizes the hotline platform looks too corporate.	The leader switches to a third-party provider with stronger anonymity assurances. They also communicate the change to rebuild trust.

6. ETHICS INVESTIGATION SUMMARIES

Ethics Investigation Summaries document the findings, conclusions, and corrective actions resulting from formal reviews of misconduct. They provide transparency into how ethical issues are evaluated and resolved. This structure ensures that Ethics Investigation Summaries demonstrate consistent, fair, and accountable decision making. They matter because opaque investigations erode trust and create perceptions of favoritism. Summaries prevent speculation by clarifying what occurred and how leadership responded. They also reinforce fairness by ensuring similar cases receive similar treatment. When ignored, investigations appear inconsistent or biased. Employees lose confidence in the integrity of the process. Over time, the organization experiences declining trust and increased ethical risk.

When done well, Ethics Investigation Summaries create clarity, accountability, and cultural stability. People see that ethical issues are handled with rigor and transparency. This strengthens trust in leadership and reinforces ethical expectations. Summaries reinforce the principle that integrity requires visible follow through. They remind leaders that accountability must be demonstrated, not implied. Ultimately, they build a culture where ethical enforcement is consistent, credible, and respected.

MORAL COMPASS SCENARIO	LEADERSHIP ACTION
Camila writes investigation summaries that are overly vague. Stakeholders can't understand what actions were taken.	The leader trains Camila on structured summary writing and requires clear findings, evidence, and outcomes. They also implement a review step before summaries are finalized.
Rowan notices that summaries vary widely in quality across investigators. He worries inconsistency undermines credibility.	The leader standardizes the summary template and trains all investigators on its use. They also conduct periodic audits for consistency.
Priya sees that some summaries omit key witness statements. She fears this creates an incomplete record.	The leader requires all summaries to include a full evidence inventory. They also add a checklist to ensure completeness.
Elias realizes that summaries are being stored in unsecured folders. He recognizes the confidentiality risk.	The leader moves all summaries to a secure, access-controlled repository. They also restrict permissions to essential personnel only.
Lila notices that investigation timelines are missing from summaries. She worries this hides delays.	The leader mandates timeline documentation in every summary. They also track investigation duration as a performance metric.
Mateo sees that summaries rarely include recommendations for prevention. He feels the organization is missing learning opportunities.	The leader updates the template to require preventive recommendations. They also review trends quarterly to identify systemic issues.

Ai6x6™ Transformation Blueprint

Worker Safety

"Sustainability is not a feature of AI — it is a requirement for civilization."

Greta Thunberg, climate activist known for global environmental advocacy.

Worker Safety requires proactive protection, and Physical Protection Measures define the safeguards that keep employees safe from harm in their daily work environment. They establish the standards for equipment, workspace design, and protective practices. This matters because unsafe conditions create immediate physical and operational risk. Physical Protection Measures prevent injury by ensuring hazards are identified and mitigated. They also reinforce fairness by ensuring all employees receive equal protection regardless of role or location. When ignored, safety incidents become more frequent and severe. Teams experience fear, distraction, and reduced productivity. Over time, Worker Safety deteriorates as risks accumulate unchecked. Hazard Prevention Protocols ensure safety threats are systematically identified and addressed before harm occurs.

Operational velocity is secondary to human safety. Reviewing Occupational Safety Checklists and Incident Logs in a focused, print-based environment reminds the leader that an antifragile workforce starts with physical and psychological security. This topic involves mapping "Hazard Identification" and "Safety Drill Schedules" to ensure that the drive for efficiency never compromises the well-being of the people. When a leader prioritizes safety, they build a culture of trust that allows employees to engage fully with the transformation. Safety is the prerequisite for the "Bliss Zone," ensuring the organization remains robust even under extreme market pressure. The print series allows you to physically audit your "armor" against "Process Safety Checklists," ensuring that your operations move fast without breaking the humans who run them. This clinical focus on safety transforms your workforce from a point of vulnerability into a bedrock of organizational resilience and high-performance execution.

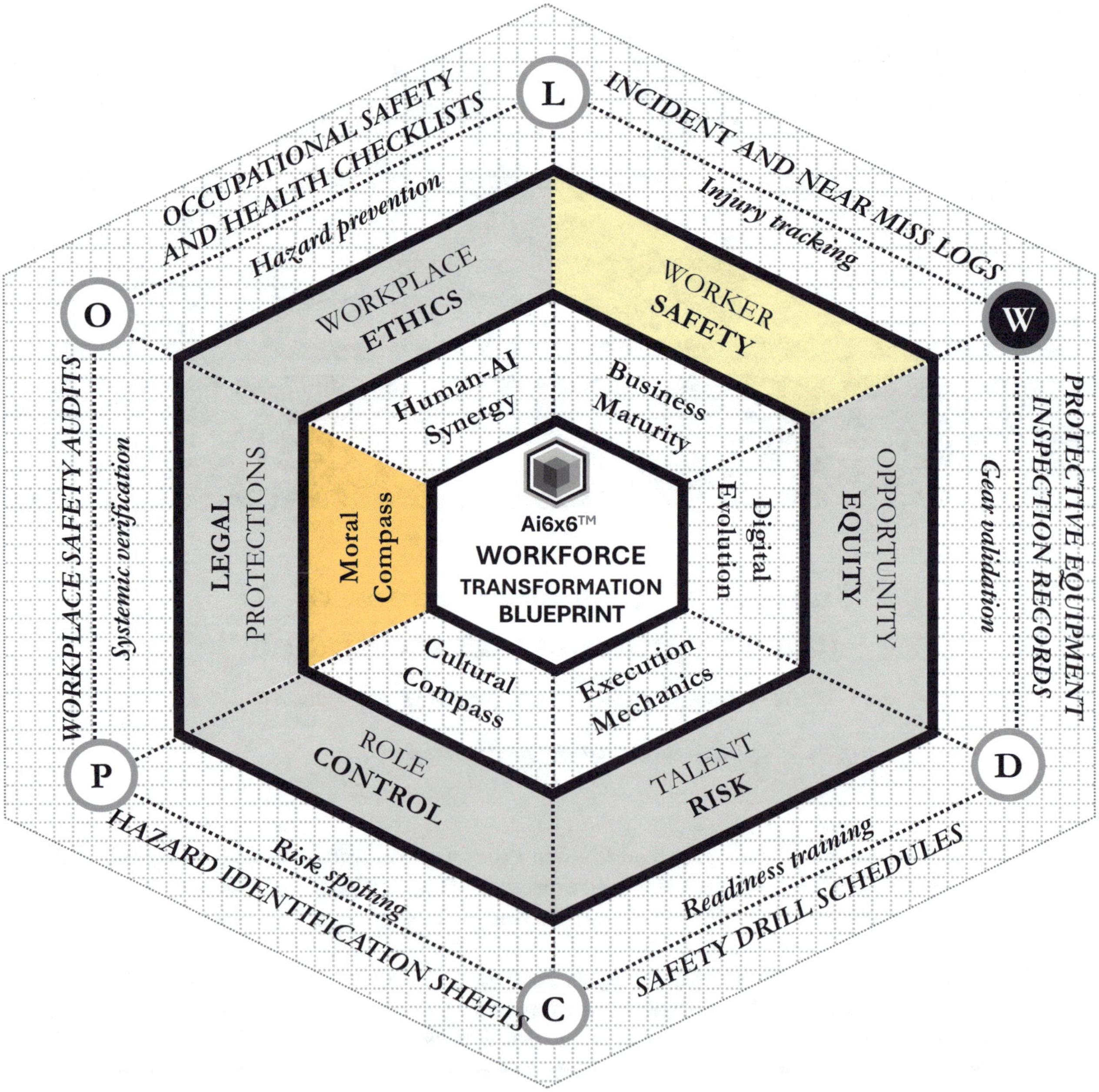

Ai6x6 WORKER SAFETY FOR MORAL COMPASS | WORKFORCE TRANSFORMATION

1. OCCUPATIONAL SAFETY AND HEALTH CHECKLISTS

Occupational Safety and Health Checklists define the required steps, safeguards, and verifications that ensure workplace environments remain safe and compliant. They translate regulatory expectations into daily operational behaviors that protect employees from harm. This structure ensures that Occupational Safety and Health Checklists function as a frontline defense against preventable incidents. They matter because safety failures often stem from overlooked details, skipped steps, or inconsistent practices. Checklists prevent complacency by forcing systematic attention to hazards. They also reinforce fairness by ensuring all employees receive the same level of protection. When ignored, safety risks accumulate silently and expose workers to injury. Teams lose confidence in leadership's commitment to their well being. Over time, the organization becomes vulnerable to accidents, legal exposure, and cultural deterioration.

When done well, Occupational Safety and Health Checklists create consistency, vigilance, and a disciplined safety culture. People trust that safety is not optional but embedded into every task. This strengthens morale and reduces operational disruptions caused by incidents. Checklists reinforce the principle that worker protection is a moral and organizational obligation. They remind leaders that safety requires structure, not assumption. Ultimately, they build a workplace where risk is minimized through disciplined, repeatable practice.

MORAL COMPASS SCENARIO	LEADERSHIP ACTION
Rina notices her team rushing through daily safety checklists without actually verifying equipment. She worries they're treating the checklist as a formality.	The leader requires a supervised walkthrough to ensure each item is physically checked. They also implement random spot-checks to reinforce accountability.
Jamal sees that new employees don't understand several checklist items. He realizes the terminology is too technical.	The leader rewrites the checklist in plain language and adds visual examples. They also train new hires on how to complete it correctly.
Felicity discovers that the checklist hasn't been updated since new machinery was installed. She fears critical steps are missing.	The leader updates the checklist to reflect the new equipment and validates it with safety experts. They also schedule quarterly reviews to keep it current.
Omar notices that night-shift workers skip checklist items due to time pressure. He senses a growing safety risk.	The leader adjusts shift staffing to allow proper completion and reinforces non-negotiable safety expectations. They also track completion rates by shift.
Priya finds that contractors are using outdated versions of the checklist. She worries they're missing key hazards.	The leader distributes the updated checklist to all contractors and requires acknowledgment. They also assign a coordinator to monitor contractor compliance.
Mateo sees that checklist results are being stored inconsistently across teams. He fears gaps in documentation.	The leader centralizes checklist storage in a shared system. They also train teams on proper logging procedures.

2. INCIDENT AND NEAR-MISS LOGS

Incident and Near Miss Logs document events where harm occurred or was narrowly avoided, capturing critical insights into workplace vulnerabilities. They create a structured record that reveals patterns, root causes, and systemic weaknesses. This documentation ensures that Incident and Near Miss Logs serve as a learning engine for safety improvement. They matter because unreported near misses often precede serious incidents. Logs prevent complacency by making invisible risks visible. They also reinforce accountability by ensuring leadership responds to early warning signs. When ignored, hazards remain unaddressed and escalate into preventable injuries. Employees lose trust in the organization's commitment to safety. Over time, the workplace becomes reactive rather than preventive.

When done well, Incident and Near Miss Logs create transparency, learning, and proactive risk reduction. People see that every safety signal is taken seriously and acted upon. This strengthens trust and encourages consistent reporting. Logs reinforce the principle that safety improvement requires honest visibility into failures and close calls. They remind leaders that prevention depends on learning from what almost happened. Ultimately, they build a culture where vigilance is continuous, and safety evolves through insight.

MORAL COMPASS SCENARIO	LEADERSHIP ACTION
Lila witnesses a near-miss involving a forklift but sees no one log it. She realizes the team doesn't understand the importance of near-miss reporting.	The leader conducts a safety briefing explaining why near-misses must be logged. They also require immediate documentation and follow-up analysis.
Andre notices repeated minor slips in the warehouse but no formal reports. He suspects employees fear being blamed.	The leader reinforces a no-fault reporting culture and anonymizes near-miss submissions. They also review the logs weekly to identify patterns.
Serena sees that incident logs are missing key details like time and location. She worries investigations will be incomplete.	The leader updates the log template to require essential fields. They also train supervisors on proper documentation.
Rowan discovers that near-miss logs are only reviewed quarterly. He feels this delays corrective action.	The leader moves to weekly reviews and assigns owners for each identified risk. They also track closure rates for accountability.
Priya notices that contractors rarely report incidents. She suspects they don't know the process.	The leader trains contractors on reporting expectations and integrates them into the same log system. They also assign a liaison to monitor compliance.
Hugo finds that several incidents were logged but never investigated. He worries risks are compounding.	The leader assigns investigators to each open case and sets deadlines for completion. They also implement a tracking dashboard for visibility.

3. PROTECTIVE EQUIPMENT INSPECTION RECORDS

Protective Equipment Inspection Records track the condition, functionality, and compliance of gear designed to safeguard employees from harm. They document inspections, defects, replacements, and maintenance actions. This structure ensures that Protective Equipment Inspection Records verify that safety equipment performs as intended. They matter because faulty or neglected equipment creates a false sense of security and increases injury risk. Records prevent oversight by ensuring inspections occur regularly and thoroughly. They also reinforce accountability by documenting who inspected what and when. When ignored, equipment failures go unnoticed until an incident occurs. Employees lose confidence in the reliability of their protective gear. Over time, the organization faces heightened risk and declining safety culture.

When done well, Protective Equipment Inspection Records create reliability, transparency, and disciplined safety management. People trust that their equipment is functional and properly maintained. This strengthens confidence in workplace protection and reduces preventable injuries. Inspection records reinforce the principle that safety requires verification, not assumption. They remind leaders that protective gear is only effective when consistently monitored. Ultimately, they build a workplace where equipment integrity is treated as a non negotiable priority.

MORAL COMPASS SCENARIO	LEADERSHIP ACTION
Amina discovers that several helmets are past their expiration date. She realizes inspection records haven't been updated in months.	The leader orders an immediate full-equipment inspection and replaces expired items. They also assign a dedicated owner for monthly record updates.
Victor notices that employees are sharing damaged gloves because replacements weren't logged. He fears injuries are imminent.	The leader updates the inspection records and orders new equipment. They also implement a "no-use" tagging system for damaged items.
Hana sees that inspection logs are handwritten and often illegible. She worries critical details are being lost.	The leader digitizes the inspection process and standardizes the format. They also train the team on using the new system.
Elias finds that respiratory equipment hasn't been tested for proper fit. He knows this violates safety standards.	The leader schedules mandatory fit-testing and updates inspection records accordingly. They also add annual fit-test reminders.
Mireya notices that night-shift workers don't have access to inspection tools. She fears their equipment is going unchecked.	The leader provides duplicate inspection kits for all shifts. They also require supervisors to verify completion.
Jonah discovers that equipment failures are logged but not linked to maintenance requests. He sees a dangerous disconnect.	The leader integrates inspection records with the maintenance system. They also assign a coordinator to track follow-through.

Safety Drill Schedules define the cadence, scope, and execution of emergency preparedness exercises. They ensure that employees practice responses to fires, evacuations, medical events, and other critical scenarios. This structure ensures that Safety Drill Schedules transform preparedness from theory into muscle memory. They matter because emergencies demand immediate, coordinated action that cannot be improvised. Schedules prevent complacency by ensuring drills occur regularly and realistically. They also reinforce accountability by requiring leaders to maintain readiness across all teams. When ignored, employees freeze, panic, or act inconsistently during real emergencies. Response times slow and risks escalate. Over time, the organization becomes dangerously unprepared for high stakes events.

When done well, Safety Drill Schedules create confidence, coordination, and rapid response capability. People know exactly what to do when emergencies arise. This strengthens resilience and reduces harm during critical moments. Drill schedules reinforce the principle that preparedness is a moral obligation to protect life. They remind leaders that readiness requires repetition, not intention. Ultimately, they build a workplace where emergency response is practiced, reliable, and effective.

MORAL COMPASS SCENARIO	LEADERSHIP ACTION
Priya realizes her team hasn't practiced an evacuation drill in over a year. She worries they won't know what to do in an emergency.	The leader reinstates quarterly drills and publishes the schedule. They also assign floor wardens to guide each drill.
Rowan sees that drills are always held during day shifts. Night-shift employees have never participated.	The leader schedules drills across all shifts and documents participation. They also brief night-shift supervisors on their responsibilities.
Serena notices that employees treat drills casually and don't follow instructions. She fears this will translate to real emergencies.	The leader reinforces drill seriousness and assigns observers to evaluate compliance. They also provide feedback after each drill.
Hugo discovers that new hires are not included in scheduled drills. He worries they're unprepared.	The leader adds new-hire drill participation to onboarding. They also track attendance to ensure coverage.
Amina sees that drill routes are outdated due to recent renovations. She fears employees will follow incorrect paths.	The leader updates evacuation maps and revises drill procedures. They also conduct a walkthrough to familiarize teams with new routes.
Victor notices that drill results are not being analyzed. He feels the organization is missing improvement opportunities.	The leader implements a post-drill review process and assigns action items. They also track improvements over time.

5. HAZARD IDENTIFICATION SHEETS

Hazard Identification Sheets document potential risks, unsafe conditions, and environmental threats present in the workplace. They provide structured visibility into hazards that require mitigation, monitoring, or redesign. This structure ensures that Hazard Identification Sheets function as a proactive safety diagnostic. They matter because many injuries stem from hazards that were known but not formally addressed. Sheets prevent oversight by forcing systematic identification and documentation. They also reinforce fairness by ensuring all employees benefit from the same level of hazard awareness. When ignored, risks remain unmitigated and accidents become more likely. Employees feel unsafe and unsupported. Over time, the organization develops a culture of tolerance toward unsafe conditions.

When done well, Hazard Identification Sheets create awareness, prevention, and disciplined risk management. People understand where hazards exist and how they are being addressed. This strengthens trust in leadership's commitment to safety. Hazard documentation reinforces the principle that risk must be identified before it can be controlled. It reminds leaders that visibility is the first step toward protection. Ultimately, it builds a workplace where hazards are recognized early and resolved systematically.

MORAL COMPASS SCENARIO	LEADERSHIP ACTION
Lila sees that employees are skipping hazard sheets because they think hazards are "obvious." She worries critical risks are being ignored.	The leader trains the team on proper hazard identification and reinforces mandatory completion. They also add visual examples to the sheets.
Andre notices that hazard sheets are filled out identically every day. He suspects employees are copying old entries.	The leader audits the sheets and requires real-time observations. They also rotate observers to ensure fresh perspectives.
Serena discovers that chemical hazards are not being documented properly. She fears regulatory violations.	The leader updates the hazard sheet to include chemical-specific fields. They also train employees on proper handling and documentation.
Rowan finds that ergonomic hazards are never listed, even though employees complain about discomfort. He senses underreporting.	The leader adds ergonomic prompts to the sheet and conducts workstation assessments. They also assign follow-up actions for identified risks.
Priya notices that contractors are not completing hazard sheets. She worries they're unaware of the requirement.	The leader mandates contractor participation and integrates them into the same hazard-tracking system. They also brief contractor supervisors.
Hugo sees that hazard sheets are submitted but never reviewed. He fears risks are going unaddressed.	The leader assigns a reviewer and sets weekly review cycles. They also track closure of identified hazards.

6. WORKPLACE SAFETY AUDITS

Workplace Safety Audits evaluate the effectiveness of safety practices, controls, and compliance across the organization. They assess whether policies are followed, risks are mitigated, and safety culture is functioning as intended. This structure ensures that Workplace Safety Audits provide objective insight into real world safety performance. They matter because organizations often assume safety is strong until an incident proves otherwise. Audits reveal gaps, inconsistencies, and vulnerabilities that require correction. They also reinforce accountability by ensuring leaders cannot overlook safety obligations. When ignored, risks accumulate unnoticed and safety standards deteriorate. Employees lose trust in leadership's commitment to their well being. Over time, the organization becomes exposed to preventable harm and regulatory consequences.

When done well, Workplace Safety Audits create rigor, transparency, and continuous improvement. People see that safety is evaluated with seriousness and follow through. This strengthens confidence in the organization's protective systems. Audits reinforce the principle that safety requires verification, not assumption. They remind leaders that oversight is essential to sustaining a safe environment. Ultimately, they build a workplace where safety excellence is measured, maintained, and continuously elevated.

MORAL COMPASS SCENARIO	LEADERSHIP ACTION
Amina discovers that safety audits are being rushed to meet deadlines. She worries critical issues are being missed.	The leader extends audit timelines and assigns additional auditors. They also require evidence-based verification for each audit item.
Victor notices that audit findings repeat every quarter. He suspects corrective actions are not being implemented.	The leader assigns owners for each finding and tracks progress until closure. They also escalate unresolved items to senior leadership.
Hana sees that some departments resist audits and hide issues. She recognizes a cultural barrier.	The leader reinforces audit expectations and conducts unannounced spot audits. They also meet with resistant managers to address concerns.
Elias discovers that audit reports are stored in different locations. He fears important findings are being lost.	The leader centralizes audit documentation in a secure repository. They also standardize the reporting format.
Mireya notices that safety audits don't include mental-health risks. She feels the scope is too narrow.	The leader expands the audit criteria to include psychological safety indicators. They also train auditors on recognizing these risks.
Jonah finds that audit recommendations are vague and hard to act on. He sees teams struggling to interpret them.	The leader revises the recommendation format to include specific actions and timelines. They also review recommendations with teams to ensure clarity.

Ai6x6™ Transformation Blueprint

Opportunity Equity

"The moral test of AI is whether it protects the most vulnerable, not the most powerful."

Bryan Stevenson, civil rights lawyer and founder of the Equal Justice Initiative.

Opportunity Equity requires equal access to growth, and Fair Advancement Access ensures employees have transparent pathways to development, promotion, and recognition. It establishes the conditions that prevent favoritism, exclusion, or hidden criteria. This matters because opportunity distribution shapes long term career outcomes. Fair Advancement Access prevents inequity by ensuring advancement is based on merit rather than proximity or bias. It also reinforces trust by demonstrating that opportunities are available to all employees. When ignored, advancement becomes inconsistent and unfair. Teams experience disengagement as growth becomes unpredictable. Over time, Opportunity Equity deteriorates as disparities widen. Bias Free Opportunities ensure decisions about advancement are grounded in objective, inclusive criteria.

Equity is the antidote to the skills gap. By engrossing themselves in Pay Equity Audits and Bias-Free Job Descriptions, leaders ensure that talent is not being throttled by invisible barriers or silos. This topic focuses on "Promotion Transparency" and "Inclusive Hiring Scorecards" to maximize the ROI of human capital. It is a strategic realization that bias is a form of operational friction. By removing that friction, the leader unlocks the full potential of the workforce. Equity ensures that the organization's "engine" is fueled by the best available talent, regardless of origin or background. Flipping through these reports in the print series helps you see the "Kill Zones" in your talent pipeline—where potential is being lost to systemic flaws. By clinicalizing equity, you create a meritocracy that scales innovation faster than your competitors, ensuring that your organization is populated by the most capable strategists, executors, and creators available in the market.

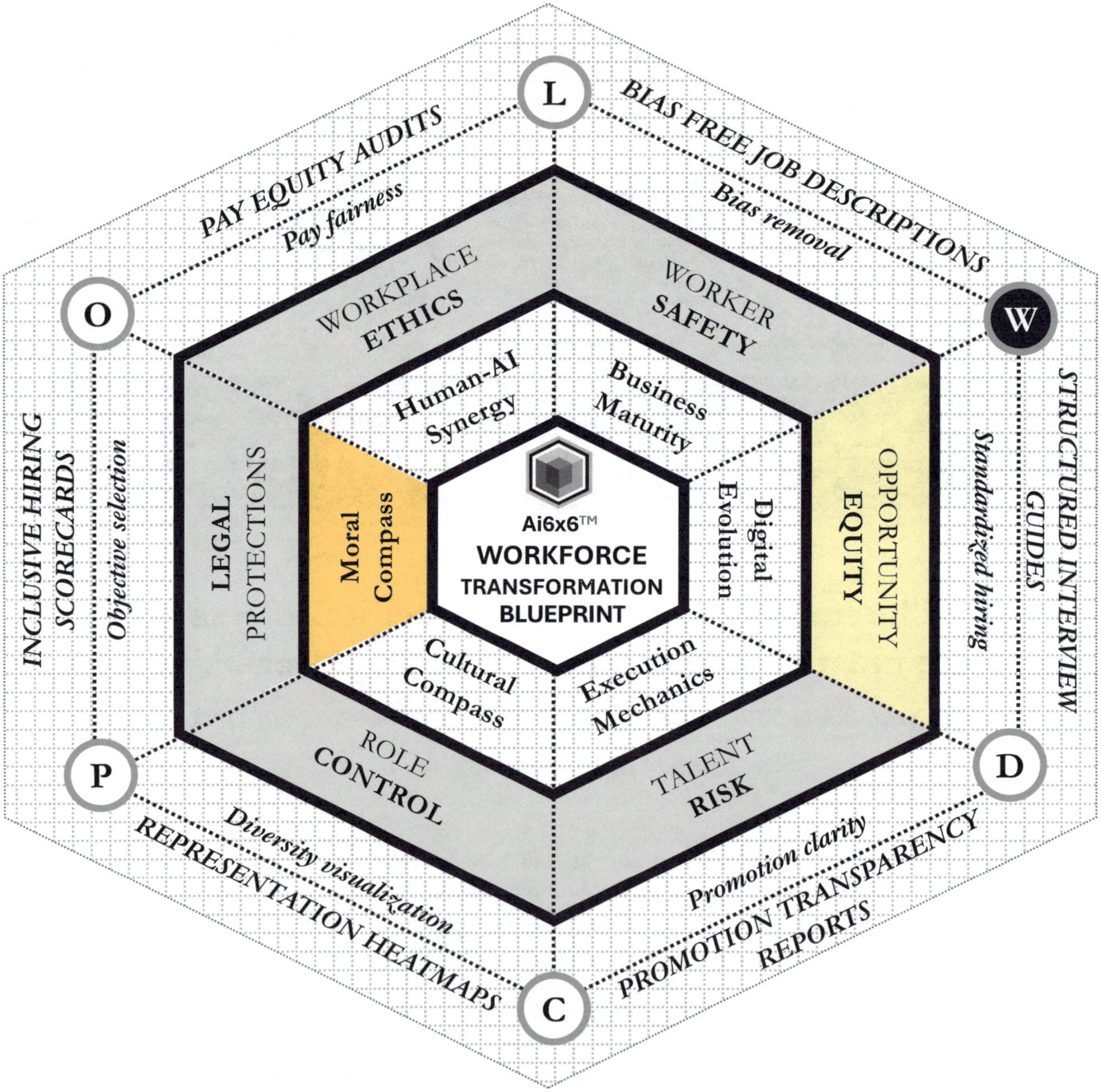

Ai6x6 OPPORTUNITY EQUITY FOR MORAL COMPASS | WORKFORCE TRANSFORMATION

1. PAY EQUITY AUDITS

Opportunity Equity requires structural fairness in compensation, and Pay Equity Audits evaluate whether employees are paid consistently for comparable work. They analyze patterns across roles, demographics, and performance to identify disparities. This structure ensures that Opportunity Equity is grounded in measurable, data driven insight. It matters because compensation inequity erodes trust, damages morale, and signals deeper systemic bias. Audits prevent hidden disparities from persisting unchallenged. They also reinforce fairness by requiring leadership to confront uncomfortable truths about pay practices. When ignored, inequities widen and become embedded in organizational norms. Employees question whether advancement is truly merit based. Over time, Opportunity Equity collapses under the weight of unaddressed disparities.

When done well, Pay Equity Audits strengthen Opportunity Equity by creating transparency and corrective action. People trust that compensation decisions are monitored and aligned with principles of fairness. This improves retention and reinforces cultural integrity. Audits reinforce the principle that equity must be verified, not assumed. They remind leaders that compensation fairness is a moral and strategic obligation. Ultimately, they build a workplace where pay reflects value, not bias.

MORAL COMPASS SCENARIO	LEADERSHIP ACTION
Nadia discovers two analysts with identical roles and tenure are paid differently. She feels uneasy knowing the disparity has gone unnoticed.	The leader initiates an immediate pay-equity correction and documents the rationale. They also schedule quarterly audits to prevent future gaps.
Trevor notices that pay increases are consistently higher for employees who negotiate aggressively. He realizes quieter employees are being disadvantaged.	The leader standardizes raise criteria and removes negotiation-based variability. They also train managers on equitable compensation practices.
Aisha sees that returning parents are coming back at lower pay bands than peers. She suspects systemic bias.	The leader reviews all return-from-leave compensation and corrects discrepancies. They also add parental-leave equity checks to the audit.
Marco finds that pay bands differ across regions without clear justification. He worries the inconsistencies are arbitrary.	The leader harmonizes pay bands and documents region-specific adjustments transparently. They also publish the updated structure for all employees.
Evelyn notices that long-tenured employees are paid below market due to outdated salary structures. She fears retention risk.	The leader adjusts salaries to market benchmarks and logs the changes in the audit. They also implement annual market-rate reviews.
Jonah discovers that equity adjustments are being made informally by managers. He sees a governance breakdown.	The leader centralizes all pay adjustments through HR and documents every change. They also retrain managers on compensation governance.

2. BIAS-FREE JOB DESCRIPTIONS

Opportunity Equity begins at the point of entry, and Bias Free Job Descriptions ensure that roles are defined without language that discourages or excludes qualified candidates. They clarify responsibilities, expectations, and qualifications in neutral, inclusive terms. This structure ensures that Opportunity Equity is embedded before hiring even begins. It matters because biased language shapes who applies, who feels welcome, and who sees themselves represented. Bias free descriptions prevent unintentional barriers that limit diversity. They also reinforce fairness by ensuring all candidates receive an equal invitation to compete. When ignored, job descriptions subtly filter out talent and perpetuate homogeneity. Candidates self select out due to perceived misalignment. Over time, Opportunity Equity erodes at the earliest stage of the talent pipeline.

When done well, Bias Free Job Descriptions strengthen Opportunity Equity by expanding access to qualified applicants. People feel welcomed and represented in the hiring process. This improves diversity, innovation, and long-term talent strength. Descriptions reinforce the principle that equity must begin with language. They remind leaders that inclusion is shaped by the words they choose. Ultimately, they build a hiring system where opportunity is open, accessible, and unbiased.

MORAL COMPASS SCENARIO	LEADERSHIP ACTION
Rhea reviews a job description filled with aggressive language like "dominant," "fearless," and "commanding." She realizes it may deter qualified candidates.	The leader rewrites the description using neutral, inclusive language. They also implement a bias-screening step for all future postings.
Omar notices that job descriptions consistently require degrees even when not necessary. He suspects this excludes non-traditional talent.	The leader removes unnecessary degree requirements and replaces them with skill-based criteria. They also train hiring managers on competency-based evaluation.
Lila sees that a posting lists "native English speaker" as a requirement. She recognizes this as discriminatory.	The leader removes the requirement and replaces it with "proficient communication skills." They also review all active postings for similar issues.
Andre finds that job descriptions emphasize long hours and constant availability. He worries this discourages caregivers.	The leader revises the descriptions to reflect realistic expectations and flexible options. They also add a work-life balance statement to all postings.
Serena notices that leadership roles use masculine-coded terms like "assertive" and "decisive." She fears this narrows the applicant pool.	The leader replaces coded language with behavior-based competencies. They also run all descriptions through a bias-detection tool.
Hugo discovers that job descriptions for technical roles assume prior experience with niche tools. He realizes this excludes capable learners.	The leader updates the descriptions to distinguish required skills from trainable ones. They also add a "willingness to learn" competency.

3. STRUCTURED INTERVIEW GUIDES

Opportunity Equity depends on consistent evaluation, and Structured Interview Guides standardize how candidates are assessed. They define questions, scoring criteria, and behavioral anchors that reduce subjective interpretation. This structure ensures that Opportunity Equity is protected from interviewer bias. It matters because unstructured interviews amplify personal preference, affinity bias, and inconsistent judgment. Guides prevent evaluators from improvising criteria or shifting expectations. They also reinforce fairness by ensuring all candidates face the same evaluation process. When ignored, interviews become unpredictable and inequitable. Candidates receive different questions, different standards, and different opportunities to demonstrate capability. Over time, Opportunity Equity deteriorates into inconsistency and favoritism.

When done well, Structured Interview Guides strengthen Opportunity Equity by creating transparency and uniform evaluation. People trust that hiring decisions reflect merit rather than personal bias. This improves confidence in the fairness of the selection process. Guides reinforce the principle that equity requires structure and discipline. They remind leaders that consistency is essential to fair assessment. Ultimately, they build a hiring environment where every candidate competes on equal footing.

MORAL COMPASS SCENARIO	LEADERSHIP ACTION
Priya notices interviewers are asking different questions to each candidate. She worries this creates inconsistent evaluations.	The leader implements a structured interview guide with standardized questions. They also train interviewers on consistent scoring.
Rowan sees that interviewers rely heavily on "gut feeling." He recognizes this as a bias risk.	The leader replaces subjective assessments with behavior-based questions. They also require written justification for each score.
Serena discovers that interviewers are asking legally risky questions about family status. She knows this exposes the company to liability.	The leader updates the guide to include only compliant questions. They also retrain interviewers on prohibited topics.
Omar notices that technical interviews vary wildly in difficulty. He fears candidates are being evaluated unfairly.	The leader standardizes technical questions and calibrates difficulty levels. They also require interviewers to follow the guide exactly.
Lila sees that interviewers skip follow-up questions for certain candidates. She suspects unconscious bias.	The leader adds mandatory follow-up prompts to the guide. They also monitor interviews for adherence.
Andre realizes interviewers are not documenting answers thoroughly. He worries decisions lack evidence.	The leader requires detailed note-taking and integrates the guide into the applicant-tracking system. They also audit interview records regularly.

4. PROMOTION TRANSPARENCY REPORTS

Opportunity Equity requires visibility into advancement pathways, and Promotion Transparency Reports reveal how promotions are awarded, who is advancing, and why. They document criteria, patterns, and outcomes across teams and levels. This structure ensures that Opportunity Equity is not left to informal networks or subjective impressions. It matters because opaque promotion processes breed suspicion and disengagement. Reports prevent favoritism by making advancement decisions reviewable and accountable. They also reinforce fairness by clarifying what is required to progress. When ignored, promotions appear arbitrary or politically driven. Employees lose trust in leadership and disengage from development efforts. Over time, Opportunity Equity collapses as advancement becomes uneven and exclusionary.

When done well, Promotion Transparency Reports strengthen Opportunity Equity by illuminating pathways to growth. People understand what drives advancement and how to prepare. This improves motivation, retention, and leadership pipeline diversity. Reports reinforce the principle that opportunity must be visible to be equitable. They remind leaders that transparency is a cornerstone of trust. Ultimately, they build a workplace where advancement is earned through clarity and merit.

MORAL COMPASS SCENARIO	LEADERSHIP ACTION
Amina notices employees are confused about how promotions are decided. She senses frustration building.	The leader publishes a clear promotion-criteria report and holds a Q&A session. They also require managers to review criteria with their teams.
Victor sees that promotion decisions are announced without explanation. Employees feel blindsided.	The leader adds rationale summaries to promotion reports. They also require managers to provide individual feedback to non-selected candidates.
Hana discovers that promotion cycles vary by department. She worries this creates inequity.	The leader standardizes promotion cycles and documents them in the transparency report. They also align all departments to the same timeline.
Elias notices that high-potential employees are not aware of development expectations. He fears they're missing opportunities.	The leader adds development pathways to the transparency report. They also require managers to discuss readiness criteria during check-ins.
Mireya sees that promotion data lacks demographic breakdowns. She suspects hidden disparities.	The leader adds demographic analysis to the report and reviews trends quarterly. They also assign accountability for closing gaps.
Jonah realizes that employees don't know who sits on promotion committees. He senses mistrust.	The leader publishes committee membership and selection criteria. They also rotate committee members annually to ensure fairness.

Opportunity Equity requires awareness of who is present, who is missing, and where disparities exist, and Representation Heatmaps visualize these patterns across roles and levels. They highlight demographic distribution, advancement gaps, and areas requiring intervention. This structure ensures that Opportunity Equity is monitored with precision. It matters because representation disparities often persist unnoticed without visual analysis. Heatmaps prevent blind spots by making inequities impossible to ignore. They also reinforce accountability by showing where leadership must act. When ignored, disparities widen and become normalized. Employees question whether the organization values diversity. Over time, Opportunity Equity erodes as representation stagnates.

When done well, Representation Heatmaps strengthen Opportunity Equity by creating clarity and urgency. People see that leadership tracks representation with seriousness and transparency. This builds trust and motivates corrective action. Heatmaps reinforce the principle that equity requires measurement. They remind leaders that representation is a strategic indicator, not a symbolic metric. Ultimately, they build a workforce that reflects the organization's values and commitments.

MORAL COMPASS SCENARIO	LEADERSHIP ACTION
Nadia notices the heatmap shows almost no representation of women in senior engineering roles. She realizes the issue has been ignored.	The leader launches a targeted development program and sets representation goals. They also track progress quarterly through the heatmap.
Trevor sees that early-career roles are diverse but mid-career roles are not. He suspects a promotion bottleneck.	The leader investigates the bottleneck and adjusts promotion pathways. They also add mid-career representation metrics to the heatmap.
Aisha discovers that one region has significantly lower representation than others. She worries local hiring practices are biased.	The leader audits regional hiring processes and retrains managers. They also set region-specific representation targets.
Marco notices that representation drops sharply in customer-facing roles. He suspects biased assumptions about "fit."	The leader reviews role requirements and removes subjective criteria. They also train hiring managers on inclusive evaluation.
Evelyn sees that employees with disabilities are nearly absent from leadership roles. She suspects accessibility barriers.	The leader conducts an accessibility audit of leadership pathways. They also add disability representation metrics to the heatmap.
Jonah finds that representation varies widely between teams with similar functions. He suspects inconsistent hiring practices.	The leader standardizes hiring processes and monitors team-level representation. They also hold managers accountable for disparities.

that hiring decisions reflect consistent, equitable criteria. They define competencies, scoring anchors, and behavioral indicators that reduce bias. This structure ensures that Opportunity Equity is protected during candidate selection. It matters because unstructured evaluation amplifies subjective judgment and inconsistent standards. Scorecards prevent evaluators from relying on intuition or personal preference. They also reinforce fairness by ensuring all candidates are assessed using the same rubric. When ignored, hiring decisions become unpredictable and inequitable. Candidates with similar qualifications receive different evaluations. Over time, Opportunity Equity deteriorates and talent pipelines narrow.

When done well, Inclusive Hiring Scorecards strengthen Opportunity Equity by creating disciplined, transparent evaluation. People trust that hiring decisions reflect merit and structured assessment. This improves diversity, fairness, and long term talent quality. Scorecards reinforce the principle that equity requires intentional design. They remind leaders that fairness must be engineered into hiring systems. Ultimately, they build a selection process where opportunity is consistent, unbiased, and principled.

MORAL COMPASS SCENARIO	LEADERSHIP ACTION
Rhea notices interviewers are scoring candidates inconsistently. She fears bias is influencing decisions.	The leader implements a standardized scorecard with clear criteria. They also train interviewers on objective scoring.
Omar sees that scorecards emphasize culture “fit” over skills. He recognizes this as a bias risk.	The leader replaces “fit” with behavior-based competencies. They also require justification for each score.
Lila discovers that interviewers skip sections of the scorecard when they like a candidate. She worries this creates unfair advantages.	The leader enforces full completion of the scorecard and audits submissions. They also coach interviewers on discipline and consistency.
Andre notices that scorecards don’t include accessibility considerations. He feels candidates with disabilities are being evaluated unfairly.	The leader updates the scorecard to include accessibility-neutral criteria. They also train interviewers on inclusive evaluation.
Serena sees that scorecards lack weighting for critical skills. She fears strong candidates are being overlooked.	The leader adds weighted scoring for essential competencies. They also recalibrate the scoring system with hiring managers.
Hugo realizes that scorecards are stored inconsistently across teams. He worries data is being lost.	The leader centralizes scorecard storage in the applicant-tracking system. They also require digital submission for every interview.

BLUEPRINT SERIES

Ai6x6™ Transformation Blueprint

Talent Risk

"AI will reflect our values, so we must choose them with intention, not convenience."

Sheryl Sandberg, business leader and author known for work on leadership and equity.

Talent Risk requires awareness of workforce vulnerabilities, and Capability Gap Exposure identifies where skills, experience, or capacity are insufficient to meet organizational demands. It establishes visibility into the weaknesses that threaten performance. This matters because capability gaps create operational delays and quality issues. Capability Gap Exposure prevents disruption by revealing where training, hiring, or restructuring is needed. It also reinforces fairness by ensuring workload expectations match actual capacity. When ignored, teams become overwhelmed and underprepared. Employees experience burnout as gaps widen. Over time, Talent Risk increases as capability issues compound. Workforce Stability Threats highlight where turnover, disengagement, or shortages jeopardize continuity.

This topic identifies where the "Armor" of the workforce is thin. Using Succession Bench Strength Maps and Attrition Risk Dashboards helps leaders prioritize human capital investments where they matter most. It is a clinical diagnostic of the talent pipeline. By studying "Skill Gap Assessments" on a printed page, the leader can see the long-term fragility of the organization's human resources. Mapping "Workforce Scenario Plans" allows the leader to build resilience, ensuring that when key talent shifts, the organization's "orchestration" continues without interruption. Talent risk management is the insurance policy for cultural continuity. In the print series, you can look at the whole talent ecosystem simultaneously, spotting "Kill Zones" where attrition could lead to systemic failure. This engrossed focus allows you to move capital and talent to the most critical growth areas, ensuring that your workforce moves as fast as your thoughts, protected from the shock of sudden talent loss.

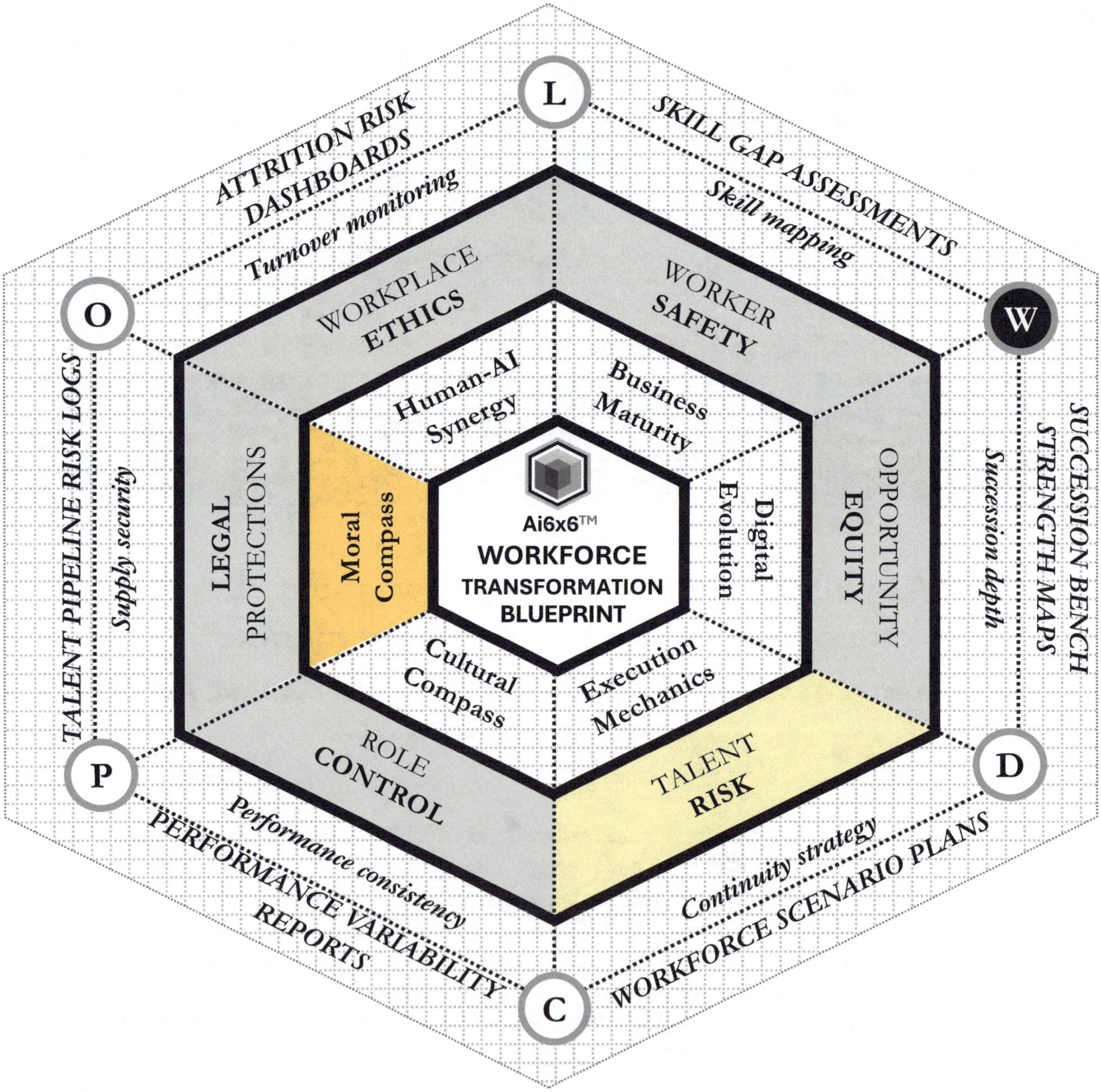

Ai6x6 TALENT RISK FOR MORAL COMPASS | WORKFORCE TRANSFORMATION

1. ATTRITION RISK DASHBOARDS

Talent Risk requires visibility into workforce stability, and Attrition Risk Dashboards reveal patterns that signal potential departures. They consolidate data on turnover trends, engagement indicators, and risk factors across teams. This structure ensures that Talent Risk is monitored with real time insight rather than reactive guesswork. It matters because unexpected attrition disrupts continuity, increases costs, and weakens institutional knowledge. Dashboards prevent blind spots by highlighting where retention vulnerabilities are emerging. They also reinforce fairness by ensuring leaders address systemic issues rather than blaming individuals. When ignored, attrition accelerates and becomes unpredictable. Teams lose confidence in leadership's ability to maintain stability. Over time, Talent Risk compounds into operational fragility and cultural decline.

When done well, Attrition Risk Dashboards strengthen Talent Risk management by enabling early intervention and targeted support. People see that leadership monitors workforce health with seriousness and transparency. This improves trust, retention, and long term workforce planning. Dashboards reinforce the principle that Talent Risk requires proactive visibility. They remind leaders that stability must be measured, not assumed. Ultimately, they build a workforce where retention is protected through insight and action.

MORAL COMPASS SCENARIO	LEADERSHIP ACTION
Keisha notices the dashboard shows a sudden spike in attrition risk for her high-performing analysts. She realizes workload pressure has quietly escalated.	The leader meets with the team to identify root causes and adjusts workload distribution. They also set up weekly pulse checks to monitor risk movement.
Damon sees that one department's attrition risk is consistently higher than others. He suspects a leadership issue.	The leader conducts a targeted climate assessment and coaches the department head on corrective behaviors. They also track improvements through the dashboard over the next quarter.
Liora notices that early-career employees show rising attrition risk after six months. She worries onboarding isn't sticking.	The leader redesigns the onboarding program and adds a 90-day reinforcement cycle. They also assign mentors to new hires to stabilize engagement.
Mateo sees that remote employees have higher attrition risk scores. He suspects they feel disconnected.	The leader implements structured touchpoints and virtual engagement routines. They also monitor risk scores monthly to measure improvement.
Serena notices that attrition risk spikes after performance review season. She fears the feedback process is demotivating.	The leader audits the review process and trains managers on constructive feedback. They also add a post-review support plan for at-risk employees.
Omar sees that attrition risk is high for employees with unclear career paths. He realizes development planning is inconsistent.	The leader requires managers to create individualized growth plans. They also track completion and impact through the dashboard.

2. SKILL GAP ASSESSMENTS

Talent Risk depends on capability readiness, and Skill Gap Assessments identify where current competencies fall short of future needs. They analyze roles, performance, and strategic requirements to reveal development priorities. This structure ensures that Talent Risk is addressed through targeted skill building rather than reactive hiring. It matters because unaddressed skill gaps slow execution and weaken competitiveness. Assessments prevent misalignment by clarifying which capabilities require investment. They also reinforce fairness by ensuring development opportunities are based on objective needs. When ignored, skill deficiencies accumulate and hinder organizational performance. Employees feel unprepared and unsupported. Over time, Talent Risk escalates into strategic vulnerability.

When done well, Skill Gap Assessments strengthen Talent Risk management by guiding development, training, and workforce planning. People understand what skills matter and how to grow. This improves confidence, capability, and long term adaptability. Assessments reinforce the principle that Talent Risk requires foresight. They remind leaders that capability gaps must be identified before they become crises. Ultimately, they build a workforce equipped for evolving demands.

MORAL COMPASS SCENARIO	LEADERSHIP ACTION
Talia discovers her team lacks critical data-analysis skills needed for upcoming projects. She worries they'll fall behind.	The leader conducts a full skill gap assessment and prioritizes training for high-impact skills. They also assign stretch assignments to accelerate learning.
Rowan notices that senior employees are struggling with new digital tools. He fears resistance is masking capability gaps.	The leader runs a targeted skills assessment and creates a tailored upskilling plan. They also provide hands-on coaching to build confidence.
Priya sees that junior employees excel technically but lack communication skills. She worries this limits their growth.	The leader adds communication competencies to the assessment and launches a development workshop. They also pair juniors with mentors for practice.
Andre realizes his team's skill gaps don't match the organization's future strategy. He senses misalignment.	The leader updates the assessment to reflect strategic priorities. They also reassign training budgets to future-critical skills.
Serena finds that skill gaps vary widely across regions. She suspects inconsistent training access.	The leader standardizes the assessment across all locations and equalizes training resources. They also track progress region by region.
Hugo notices that employees self-report skills inaccurately. He worries decisions are based on flawed data.	The leader introduces objective skill assessments and calibrates self-ratings. They also train managers on validating skill levels.

3. SUCCESSION BENCH STRENGTH MAPS

Talent Risk requires leadership continuity, and Succession Bench Strength Maps visualize the readiness of future leaders across critical roles. They assess potential, preparedness, and development needs for key positions. This structure ensures that Talent Risk is mitigated through deliberate succession planning. It matters because leadership gaps disrupt strategy, culture, and execution. Maps prevent overreliance on a few individuals by revealing where depth is lacking. They also reinforce fairness by ensuring succession decisions are based on capability rather than proximity or favoritism. When ignored, leadership transitions become chaotic and destabilizing. Teams lose confidence in organizational continuity. Over time, Talent Risk becomes a structural weakness that limits growth.

When done well, Succession Bench Strength Maps strengthen Talent Risk management by creating clarity and readiness. People see that leadership development is intentional and equitable. This improves morale, retention, and long-term stability. Maps reinforce the principle that Talent Risk requires planned continuity. They remind leaders that future capability must be cultivated, not assumed. Ultimately, they build an organization prepared for seamless leadership transitions.

MORAL COMPASS SCENARIO	LEADERSHIP ACTION
Lila realizes her leadership pipeline has only one ready successor for three critical roles. She fears a continuity gap.	The leader expands the bench strength map and identifies high-potential employees for accelerated development. They also assign targeted stretch roles to build readiness.
Marcus sees that successors are concentrated in one department. He suspects cross-functional mobility is lacking.	The leader broadens the map to include talent from other departments. They also create rotational assignments to diversify the pipeline.
Amina notices that several successors are nearing retirement. She realizes the map is outdated.	The leader refreshes the map and identifies next-generation talent. They also schedule annual updates to prevent future gaps.
Victor finds that diverse employees are underrepresented in the succession pool. He worries about long-term equity.	The leader reviews selection criteria and removes biased filters. They also add diverse high-potential employees to the development pipeline.
Hana sees that successors lack experience in crisis management. She fears they're unprepared for real-world challenges.	The leader assigns crisis-simulation training and scenario-based development. They also update the map to reflect readiness improvements.
Elias notices that managers are nominating the same people repeatedly. He suspects favoritism.	The leader requires evidence-based nominations and cross-validates with performance data. They also rotate nomination responsibilities to reduce bias.

4. WORKFORCE SCENARIO PLANS

Talent Risk requires adaptability, and Workforce Scenario Plans prepare the organization for multiple future labor conditions. They outline responses to shifts in demand, technology, skills, and workforce availability. This structure ensures that Talent Risk is managed through strategic foresight rather than reactive improvisation. It matters because workforce disruptions can emerge suddenly and unpredictably. Scenario plans prevent unpreparedness by exploring alternative futures and required actions. They also reinforce fairness by ensuring employees are not blindsided by abrupt changes. When ignored, organizations struggle to respond to talent shortages, surpluses, or capability shifts. Teams experience instability and uncertainty. Over time, Talent Risk becomes a barrier to strategic execution.

When done well, Workforce Scenario Plans strengthen Talent Risk management by enabling agility and informed decision making. People trust that leadership anticipates change and prepares responsibly. This improves resilience and reduces disruption during transitions. Scenario planning reinforces the principle that Talent Risk requires strategic imagination. It reminds leaders that uncertainty must be planned for, not feared. Ultimately, it builds a workforce capable of navigating evolving conditions.

MORAL COMPASS SCENARIO	LEADERSHIP ACTION
Priya realizes her team has no plan for sudden workforce shortages. She fears operations would stall during disruptions.	The leader builds a scenario plan outlining backup staffing, cross-training, and emergency protocols. They also run a simulation to test readiness.
Rowan sees that rapid growth scenarios haven't been modeled. He worries the team won't scale effectively.	The leader creates a growth-scenario plan with hiring triggers and onboarding capacity. They also align HR and finance to support rapid expansion.
Serena notices that remote-work disruptions aren't included in current plans. She senses a blind spot.	The leader adds remote-failure scenarios and defines contingency workflows. They also test remote-resilience plans quarterly.
Omar finds that succession scenarios don't account for sudden leadership exits. He fears a leadership vacuum.	The leader updates the scenario plan with emergency leadership assignments. They also brief interim leaders on expectations.
Lila sees that economic downturn scenarios are outdated. She worries the team is unprepared for budget cuts.	The leader refreshes the plan with updated financial assumptions. They also create a tiered reduction strategy to minimize disruption.
Andre realizes that cross-department dependencies aren't mapped. He fears cascading failures.	The leader integrates inter-team dependencies into the scenario plan. They also run a cross-functional stress test to validate assumptions.

5. PERFORMANCE VARIABILITY REPORTS

Talent Risk requires understanding how performance fluctuates across individuals, teams, and time, and Performance Variability Reports reveal these patterns. They analyze consistency, volatility, and underlying drivers of performance differences. This structure ensures that Talent Risk is addressed through insight rather than assumption. It matters because unpredictable performance undermines reliability and execution. Reports prevent misinterpretation by distinguishing between systemic issues and individual challenges. They also reinforce fairness by ensuring interventions are based on data, not bias. When ignored, performance issues persist unaddressed and create uneven workloads. Teams experience frustration and declining morale. Over time, Talent Risk escalates into operational instability.

When done well, Performance Variability Reports strengthen Talent Risk management by enabling targeted support and balanced expectations. People see that performance is evaluated with rigor and fairness. This improves trust, development, and accountability. Reports reinforce the principle that Talent Risk requires understanding patterns, not reacting to symptoms. They remind leaders that consistency must be cultivated. Ultimately, they build a workforce where performance reliability is strengthened through insight and action.

MORAL COMPASS SCENARIO	LEADERSHIP ACTION
Nadia notices that performance varies wildly between two teams with similar roles. She suspects inconsistent management practices.	The leader reviews both teams' workflows and coaching approaches. They also standardize best practices across teams.
Trevor sees that one employee's performance fluctuates month to month. He worries burnout may be the cause.	The leader meets with the employee to understand root causes and adjusts workload. They also schedule regular check-ins to stabilize performance.
Aisha discovers that variability spikes during peak seasons. She realizes the team lacks surge-capacity planning.	The leader adds seasonal staffing and cross-training to reduce variability. They also monitor performance trends to refine planning.
Marco notices that variability correlates with unclear goals. He suspects misalignment.	The leader clarifies expectations and sets measurable targets. They also train managers on goal-setting discipline.
Evelyn sees that variability is highest among new hires. She worries onboarding isn't effective.	The leader strengthens onboarding with structured milestones. They also assign mentors to support early performance stability.
Jonah finds that variability increases after organizational changes. He suspects communication gaps.	The leader improves change-management communication and provides transition support. They also track variability post-change to measure impact.

6. TALENT PIPELINE RISK LOGS

Talent Risk requires visibility into future workforce supply, and Talent Pipeline Risk Logs track vulnerabilities in hiring, development, and succession. They document shortages, bottlenecks, and emerging gaps across critical roles. This structure ensures that Talent Risk is managed through continuous monitoring. It matters because pipeline weaknesses often go unnoticed until they disrupt operations. Logs prevent surprises by highlighting where future capability is at risk. They also reinforce fairness by ensuring development and hiring decisions address systemic needs. When ignored, talent shortages become crises that slow execution. Teams struggle to fill roles and maintain continuity. Over time, Talent Risk becomes a persistent constraint on organizational growth.

When done well, Talent Pipeline Risk Logs strengthen Talent Risk management by enabling proactive planning and targeted investment. People trust that leadership monitors future capability with seriousness. This improves confidence in long term workforce stability. Logs reinforce the principle that Talent Risk requires continuous visibility. They remind leaders that future talent must be cultivated deliberately. Ultimately, they build a workforce prepared for sustained performance and growth.

MORAL COMPASS SCENARIO	LEADERSHIP ACTION
Rhea notices the pipeline log shows declining interest from university recruits. She fears the brand is losing appeal.	The leader refreshes the campus strategy and increases engagement efforts. They also track pipeline recovery through the log.
Omar sees that mid-career hiring pipelines are thin. He suspects sourcing channels are outdated.	The leader diversifies sourcing channels and updates the log with new partners. They also measure conversion rates to refine the strategy.
Lila discovers that diverse candidates drop off at the final interview stage. She suspects bias in evaluation.	The leader audits interview panels and retrains evaluators. They also track drop-off patterns to ensure improvement.
Andre notices that critical roles have no backup candidates. He worries about continuity risk.	The leader prioritizes those roles in the pipeline and accelerates sourcing. They also add risk flags to the log for visibility.
Serena sees that internal mobility is low despite strong internal talent. She suspects managers are hoarding talent.	The leader enforces mobility expectations and updates the log with internal candidates. They also hold managers accountable for development.
Hugo finds that pipeline data is inconsistent across regions. He fears decisions are based on unreliable information.	The leader standardizes data collection and trains regional teams. They also audit the log quarterly for accuracy.

Ai6x6™ Transformation Blueprint

Role Control

"Ethics is the compass that keeps innovation from becoming destruction."

Ban Ki-moon, former UN Secretary-General known for global diplomacy.

Role Control requires structural precision, and Responsibility Boundary Clarity defines what each role is accountable for and where its authority ends. It establishes the limits that prevent overlap, confusion, or conflict. This matters because unclear roles create operational inefficiency and ethical risk. Responsibility Boundary Clarity prevents misalignment by ensuring employees understand their scope. It also reinforces fairness by ensuring responsibilities are assigned intentionally rather than informally. When ignored, boundaries blur and decisions become inconsistent. Teams experience frustration as work is duplicated or neglected. Over time, Role Control deteriorates as structure weakens. Access Permission Rules ensure employees have the appropriate level of system and information access for their responsibilities.

Clarity is the ultimate cure for "Execution Drift." This topic maps Responsibility Boundary Maps and Access Control Lists to ensure every employee knows exactly how their work fuels the broader orchestration. By defining "Task Ownership" and "Workload Allocation" through a physical blueprint, the leader eliminates the confusion that slows down transformation. It's about creating a "Permission Matrix" that empowers employees to act within their domain without constant oversight. Role control ensures that the organization moves as a synchronized unit, where the gap between leadership intent and workforce execution is finally and permanently closed. Using the print framework, you can physically flip between "Task Ownership Logs" and "Escalation Path Diagrams," seeing the whole map of human intent. This prevents the "Kill Zone" of operational silos, ensuring that the "How" and the "Why" of every job description are physically and mentally linked, fostering an environment of radical, human-driven accountability.

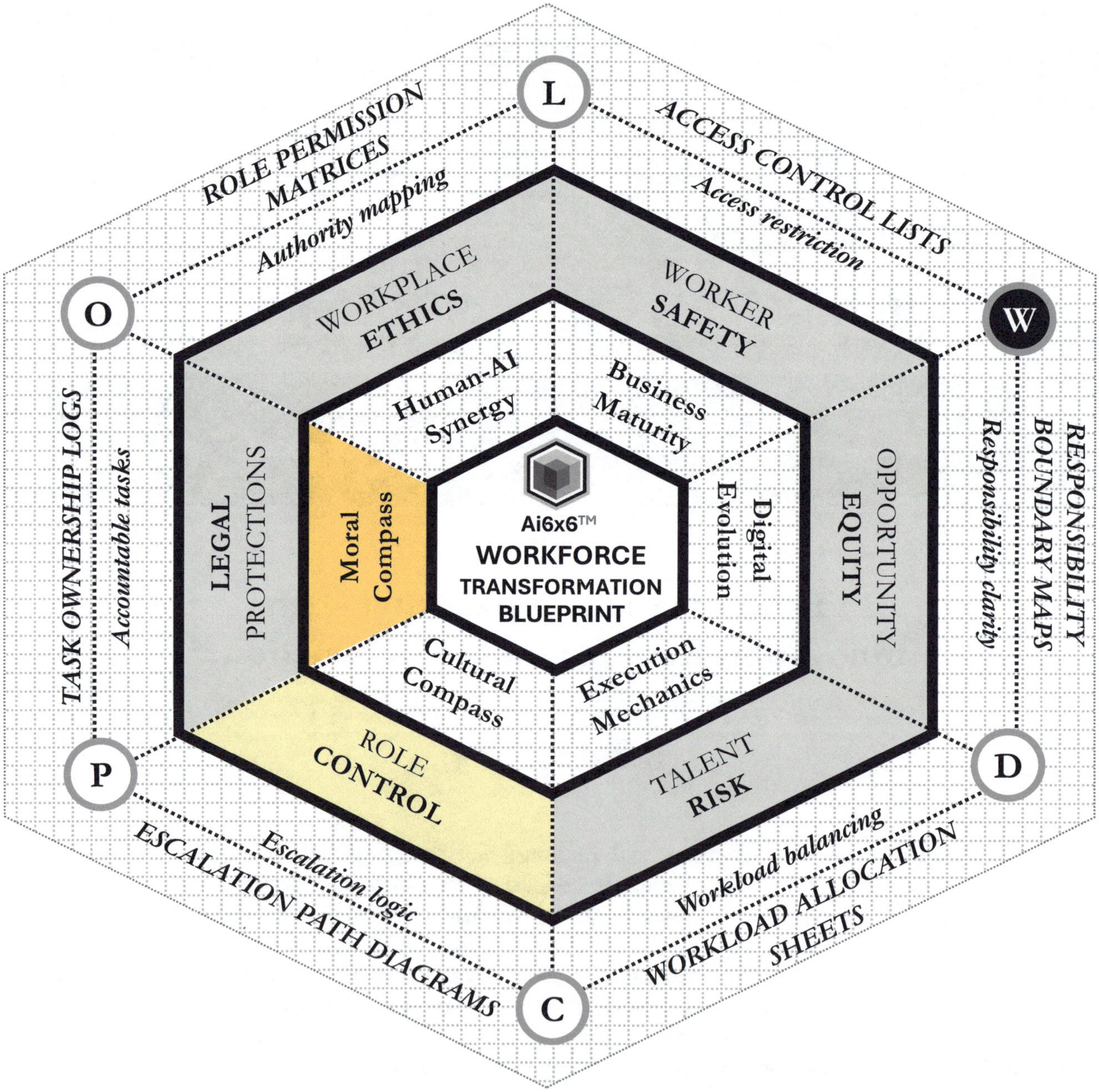

Ai6x6 ROLE CONTROL FOR MORAL COMPASS | WORKFORCE TRANSFORMATION

1. ROLE PERMISSION MATRICES

Role Control requires clarity in authority boundaries, and Role Permission Matrices define what each role is allowed to access, decide, and execute. They map permissions to responsibilities, so employees understand the limits and expectations of their position. This structure ensures that Role Control is enforced through explicit, documented authority. It matters because unclear permissions create operational confusion, ethical risk, and inconsistent decision making. Matrices prevent overreach by ensuring employees do not act outside their mandate. They also reinforce fairness by ensuring permissions are assigned based on role, not personal influence. When ignored, authority becomes informal and unevenly distributed. Teams experience conflict, duplication, and gaps in accountability. Over time, Role Control deteriorates into ambiguity and operational instability.

When done well, Role Permission Matrices strengthen Role Control by creating transparency and disciplined authority. People understand their scope and act confidently within defined boundaries. This improves coordination, reduces conflict, and accelerates execution. Matrices reinforce the principle that authority must be structured, not improvised. They remind leaders that clarity is essential to ethical and operational integrity. Ultimately, they build a workplace where roles are respected, and authority is exercised responsibly.

MORAL COMPASS SCENARIO	LEADERSHIP ACTION
Kendra realizes several junior employees have permissions far beyond their roles. She worries this exposes the team to operational and security risks.	The leader immediately resets permissions to match the matrix and documents the correction. They also schedule a quarterly permission audit to prevent drift.
Malik discovers that a new hire cannot access essential tools because their permissions were never assigned. He sees productivity dropping.	The leader updates the permission matrix and assigns the correct access within the hour. They also integrate permission setup into onboarding.
Sienna notices that a manager granted themselves elevated permissions without approval. She recognizes the governance breach.	The leader revokes the unauthorized access and logs the incident. They also reinforce approval rules and require manager retraining.
Omar finds that contractors have lingering permissions long after their engagement ended. He fears data exposure.	The leader removes all outdated contractor access and updates the matrix with expiration rules. They also assign IT to run monthly access sweeps.
Priya sees that cross-functional teams are confused about who can approve what. She senses decision bottlenecks forming.	The leader clarifies permission tiers in the matrix and distributes a simplified reference guide. They also hold a walkthrough session for all teams.
Hugo notices that temporary project permissions were never revoked. He worries this creates long-term risk.	The leader resets all temporary permissions and adds automatic expiration dates to the matrix. They also require project leads to verify access at project close.

Role Control depends on secure and appropriate access, and Access Control Lists specify which individuals can view, modify, or use specific systems and information. They align access privileges with job responsibilities to prevent misuse or accidental exposure. This structure ensures that Role Control is enforced through precise access governance. It matters because excessive or misaligned access creates security vulnerabilities and ethical risks. Lists prevent unauthorized actions by restricting access to those who genuinely need it. They also reinforce fairness by ensuring access is granted based on role, not hierarchy or personal relationships. When ignored, sensitive information becomes exposed and operational integrity weakens. Teams lose trust in the organization's ability to protect data. Over time, Role Control collapses into uncontrolled access and elevated risk.

When done well, Access Control Lists strengthen Role Control by creating disciplined, role aligned access. People trust that systems are protected, and privileges are assigned responsibly. This improves security, compliance, and operational reliability. Access governance reinforces the principle that authority must be limited to necessity. It reminds leaders that protection requires intentional restriction. Ultimately, it builds a workplace where access is controlled, accountable, and aligned with responsibility.

MORAL COMPASS SCENARIO	LEADERSHIP ACTION
Lila discovers that sensitive documents are accessible to employees outside the department. She fears accidental leaks.	The leader restricts access immediately and updates the access control list. They also require document owners to review permissions monthly.
Rowan notices that terminated employees still appear on access lists. He recognizes the compliance risk.	The leader removes all inactive users and integrates HR offboarding with access removal. They also implement automated deactivation triggers.
Serena finds that team members are sharing login credentials to bypass access restrictions. She sees a major security gap.	The leader enforces unique credential use and resets all shared passwords. They also retrain the team on access-control policies.
Omar realizes that access lists haven't been updated since the last reorg. He worries people have access to systems they no longer need.	The leader rebuilds the access list based on current roles. They also assign system owners to maintain ongoing accuracy.
Priya sees that access approvals are being granted informally through chat messages. She knows this violates governance rules.	The leader mandates formal approval workflows and logs all access changes. They also disable informal access-granting channels.
Hugo discovers that privileged accounts are not being monitored. He fears misuse could go undetected.	The leader enables monitoring for all privileged accounts and updates the access list accordingly. They also schedule weekly reviews of activity logs.

3. RESPONSIBILITY BOUNDARY MAPS

Role Control requires clear ownership, and Responsibility Boundary Maps define where one role's duties end and another's begin. They visualize the distribution of tasks, decisions, and obligations across teams. This structure ensures that Role Control is grounded in explicit, shared understanding. It matters because unclear boundaries lead to duplication, conflict, and accountability gaps. Maps prevent confusion by clarifying who is responsible for what. They also reinforce fairness by ensuring responsibilities are not unevenly assigned or informally shifted. When ignored, teams operate with conflicting assumptions and unresolved tensions. Work slows as employees negotiate ownership instead of executing. Over time, Role Control erodes into inefficiency and frustration.

When done well, Responsibility Boundary Maps strengthen Role Control by creating alignment and operational clarity. People understand their obligations and how their work connects to others. This improves coordination, reduces conflict, and accelerates execution. Boundary mapping reinforces the principle that clarity is essential to accountability. It reminds leaders that responsibility must be defined, not assumed. Ultimately, it builds a workplace where roles operate smoothly and collaboratively.

MORAL COMPASS SCENARIO	LEADERSHIP ACTION
Keisha notices her team keeps stepping into tasks owned by another department. She sees tension rising.	The leader updates the boundary map and clarifies ownership in a joint meeting. They also establish cross-team escalation rules.
Damon realizes that two managers believe they own the same workflow. Conflicting instructions are confusing employees.	The leader redraws the boundary map and assigns a single accountable owner. They also communicate the updated boundaries to all stakeholders.
Liora sees that new hires are unsure which responsibilities fall under their role. She senses onboarding gaps.	The leader integrates the boundary map into onboarding and reviews it with each new hire. They also add role-specific examples for clarity.
Mateo discovers that responsibilities shift informally during busy periods. He worries this creates hidden accountability gaps.	The leader formalizes temporary boundary adjustments and documents them in the map. They also require managers to review boundaries after peak periods.
Serena notices that cross-functional projects lack clear responsibility lines. She sees delays and duplicated work.	The leader creates a project-specific boundary map and assigns owners for each workstream. They also review boundaries at each project milestone.
Omar finds that employees escalate issues to the wrong teams because boundaries are unclear. He sees frustration building.	The leader updates the boundary map with clear escalation paths. They also hold a walkthrough session to reinforce understanding.

4. WORKLOAD ALLOCATION SHEETS

Role Control depends on balanced distribution of work, and Workload Allocation Sheets document how tasks are assigned across individuals and teams. They reveal whether responsibilities align with capacity, skill, and role expectations. This structure ensures that Role Control is supported by equitable workload management. It matters because uneven workloads create burnout, disengagement, and performance variability. Sheets prevent overload by making imbalances visible and correctable. They also reinforce fairness by ensuring work is distributed based on role requirements rather than favoritism. When ignored, some employees become overwhelmed while others remain underutilized. Teams experience resentment and declining morale. Over time, Role Control weakens as workload inequity becomes normalized.

When done well, Workload Allocation Sheets strengthen Role Control by creating transparency and balance. People trust that work is assigned thoughtfully and equitably. This improves performance, well being, and team cohesion. Allocation sheets reinforce the principle that fairness must be operationalized. They remind leaders that workload distribution is a moral and managerial responsibility. Ultimately, they build a workplace where capacity is respected and productivity is sustainable.

MORAL COMPASS SCENARIO	LEADERSHIP ACTION
Nadia notices that one employee is overloaded while others have capacity. She worries burnout is imminent.	The leader redistributes tasks using the allocation sheet and sets workload limits. They also review the sheet weekly to maintain balance.
Trevor sees that high-visibility tasks are always assigned to the same people. He suspects favoritism.	The leader rotates task assignments and documents the rationale in the sheet. They also train managers on equitable workload distribution.
Aisha discovers that urgent tasks are not being prioritized correctly. She sees deadlines slipping.	The leader updates the allocation sheet with priority levels. They also require daily standups to review shifting priorities.
Marco realizes that remote employees are receiving fewer assignments. He worries they're being unintentionally sidelined.	The leader adjusts the sheet to ensure equal distribution across locations. They also monitor assignment patterns for bias.
Evelyn notices that tasks requiring specialized skills are being assigned to untrained employees. She fears quality issues.	The leader updates the sheet to match tasks with skill profiles. They also schedule training to expand capability.
Jonah sees that managers are not updating the sheet regularly. He recognizes the data is becoming unreliable.	The leader enforces weekly updates and ties compliance to performance expectations. They also assign a coordinator to validate accuracy.

5. ESCALATION PATH DIAGRAMS

Role Control requires structured decision flow, and Escalation Path Diagrams define how issues move upward when they exceed an employee's authority. They clarify thresholds, triggers, and the sequence of escalation steps. This structure ensures that Role Control is maintained through disciplined upward communication. It matters because unclear escalation leads to delays, misjudgments, and unresolved risks. Diagrams prevent confusion by showing exactly when and how to elevate issues. They also reinforce fairness by ensuring escalation is based on criteria, not personal comfort or hierarchy. When ignored, problems linger at the wrong level or escalate inconsistently. Teams experience frustration and uncertainty about decision ownership. Over time, Role Control deteriorates into reactive and unpredictable escalation behavior.

When done well, Escalation Path Diagrams strengthen Role Control by creating clarity and timely intervention. People understand when to escalate and whom to involve. This improves responsiveness, reduces risk, and enhances decision quality. Diagrams reinforce the principle that authority must be supported by structured escalation. They remind leaders that timely transparency is essential to responsible governance. Ultimately, they build a workplace where issues move efficiently to the right level.

MORAL COMPASS SCENARIO	LEADERSHIP ACTION
Rhea notices employees escalate issues directly to senior leadership, bypassing intermediate steps. She sees chaos forming.	The leader updates the escalation diagram and reinforces proper routing. They also train employees on when and how to escalate.
Omar discovers that the escalation path for technical issues is outdated. Teams are contacting the wrong support groups.	The leader refreshes the diagram and aligns it with current team structures. They also publish it in all relevant channels.
Lila sees that urgent issues are being escalated too slowly because employees don't know the correct sequence. She fears operational delays.	The leader simplifies the diagram and highlights time-critical paths. They also run a simulation to reinforce understanding.
Andre notices that cross-department escalations stall because no one knows who owns the next step. He sees bottlenecks everywhere.	The leader clarifies cross-team escalation owners and updates the diagram. They also assign escalation coordinators for complex cases.
Serena finds that employees escalate interpersonal conflicts directly to HR. She suspects managers are avoiding their responsibilities.	The leader updates the diagram to route interpersonal issues through managers first. They also train managers on conflict resolution.
Hugo realizes that after-hours escalations are unclear. Employees don't know who is on call.	The leader adds an after-hours escalation path and publishes the on-call schedule. They also require teams to review it monthly.

6. TASK OWNERSHIP LOGS

Role Control requires traceability, and Task Ownership Logs document who is responsible for each task, milestone, and deliverable. They create visibility into accountability and progress across teams. This structure ensures that Role Control is reinforced through explicit ownership. It matters because unclear task ownership leads to delays, duplication, and blame shifting. Logs prevent ambiguity by assigning responsibility to specific individuals. They also reinforce fairness by ensuring accountability aligns with role expectations. When ignored, tasks fall through the cracks and deadlines slip. Teams struggle to coordinate and resolve issues. Over time, Role Control collapses into confusion and inconsistent execution.

When done well, Task Ownership Logs strengthen Role Control by creating transparency and disciplined follow through. People understand who owns what and how progress is tracked. This improves coordination, accountability, and performance reliability. Logs reinforce the principle that responsibility must be explicit and visible. They remind leaders that execution depends on clear ownership. Ultimately, they build a workplace where tasks are completed with clarity, consistency, and accountability.

MORAL COMPASS SCENARIO	LEADERSHIP ACTION
Keisha notices tasks are being completed but no one knows who actually owned them. She sees accountability fading.	The leader updates the ownership log with named owners for every task. They also require owners to confirm completion in writing.
Damon discovers that tasks are being reassigned informally without documentation. He worries this creates confusion.	The leader mandates that all reassignments be logged and approved. They also audit the log weekly for accuracy.
Liora sees that high-priority tasks are assigned to employees who lack capacity. She fears deadlines will be missed.	The leader updates the log to include workload indicators. They also redistribute tasks based on capacity.
Mateo notices that recurring tasks are not documented. He realizes knowledge is trapped with individuals.	The leader adds recurring tasks to the log and assigns permanent owners. They also create backup ownership for continuity.
Serena finds that task owners are unclear during cross-functional projects. She sees delays and finger-pointing.	The leader assigns explicit owners for each project task and updates the log. They also review ownership at each project checkpoint.
Omar discovers that completed tasks are not being marked as done. He worries this distorts progress tracking.	The leader enforces real-time updates and ties log accuracy to performance expectations. They also assign a coordinator to verify status weekly.

Ai6x6™ Transformation Blueprint

Legal Protections

"Trust is built in transparency, and AI is no exception."
Margrethe Vestager, EU Commissioner known for tech regulation leadership.

Legal Protections require adherence to employee rights, and Employee Rights Safeguards define the legal standards that protect workers from discrimination, retaliation, and unsafe conditions. They establish the baseline obligations employers must meet to maintain lawful operations. This matters because violations expose employees to harm and organizations to legal risk. Employee Rights Safeguards prevent misconduct by ensuring rights are clearly defined and enforced. They also reinforce fairness by ensuring all employees receive equal legal protection. When ignored, rights violations become more frequent and severe. Teams experience fear and distrust as protections weaken. Over time, Legal Protections deteriorate as compliance becomes inconsistent. Statutory Workplace Requirements ensure organizational practices align with labor laws and regulatory mandates.

Anchoring the workforce in rights to build a foundation for growth. By mastering FLSA, ADA, and FMLA compliance, the leader builds a culture that grows specifically because its people feel legally and morally protected. This topic ensures that "Wages," "Leave," and "Representation" are managed with clinical precision. It prevents the organization from being blindsided by labor disputes that could derail the 90-day roadmap. Legal protection is not about avoiding lawsuits; it is about creating a "Safe Zone" for the workforce that fosters the loyalty and stability required for high-velocity, AI-integrated business operations. In the print series, you can sit with these regulations away from digital distractions, seeing how they form the "Moral Compass" of your human capital strategy. This clinical mastery ensures that your "Workforce Transformation" is not just efficient, but legally unassailable and morally sound, providing the stability required for your people to embrace the self-disruption of the AI revolution.

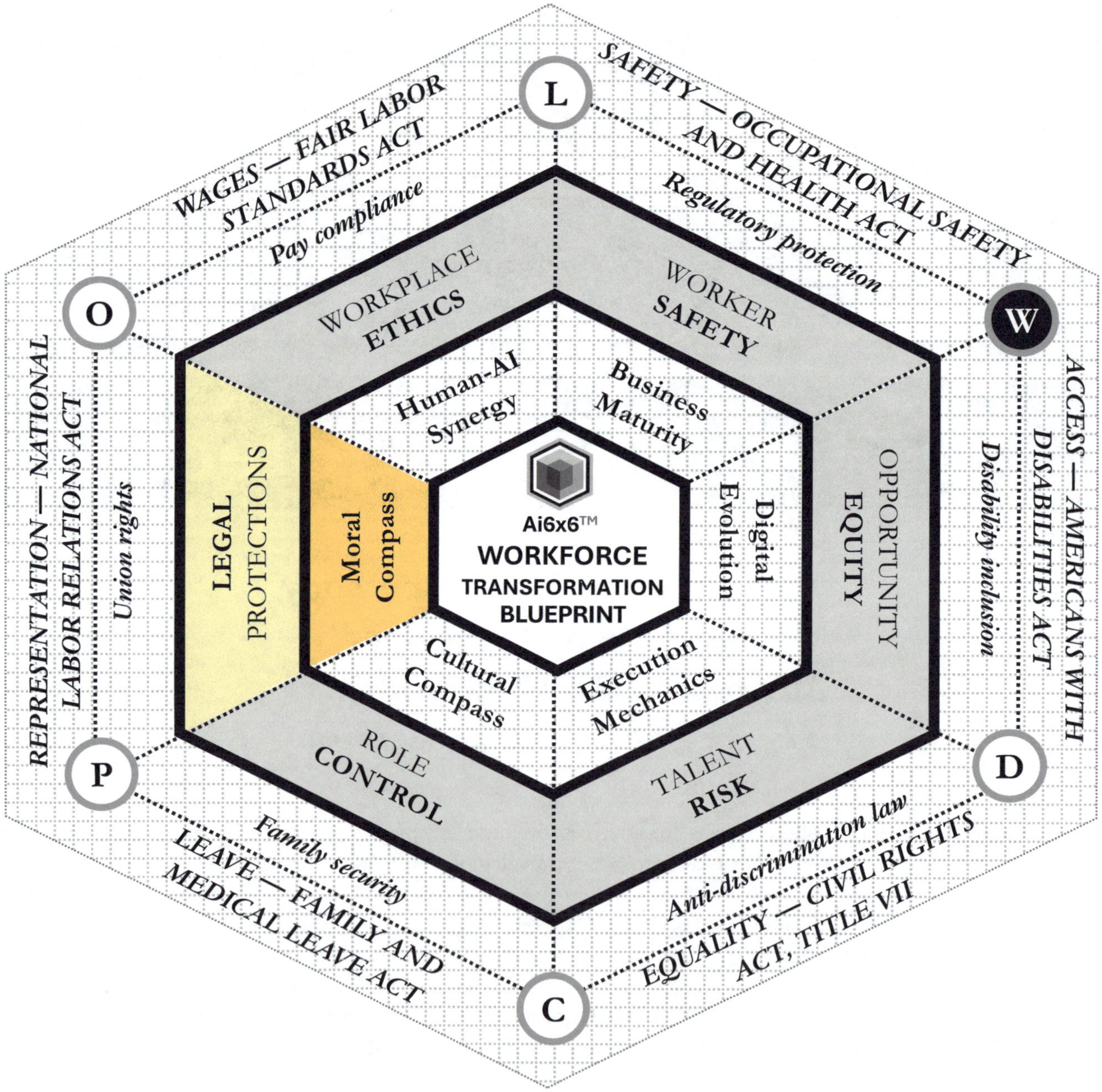

Ai6x6 LEGAL PROTECTIONS FOR MORAL COMPASS | WORKFORCE TRANSFORMATION

1. WAGES — FAIR LABOR STANDARDS ACT

Legal Protections ensure that workers are treated fairly under the law, and the Fair Labor Standards Act establishes the wage, overtime, and compensation standards that safeguard employee rights. It defines minimum pay, regulates overtime eligibility, and sets expectations for accurate timekeeping. This structure ensures that Legal Protections are grounded in enforceable wage fairness. It matters because wage violations undermine trust, create inequity, and expose organizations to legal and ethical risk. FLSA compliance prevents exploitation by ensuring employees are compensated for the work they perform. It also reinforces fairness by standardizing pay practices across roles and industries. When ignored, wage theft, misclassification, and underpayment become systemic. Employees feel undervalued and mistreated. Over time, Legal Protections erode and organizational credibility collapses.

When done well, FLSA alignment strengthens Legal Protections by ensuring compensation is lawful, transparent, and equitable. People trust that their pay reflects both legal standards and organizational integrity. This improves morale, retention, and cultural stability. Wage compliance reinforces the principle that fairness must be codified, not assumed. It reminds leaders that compensation is both a legal obligation and a moral commitment. Ultimately, it builds a workplace where pay practices uphold dignity and justice.

MORAL COMPASS SCENARIO	LEADERSHIP ACTION
Taryn discovers that hourly employees are working through lunch without pay because they feel pressured to "keep things moving." She realizes this violates wage laws and risks burnout.	The leader restores all missed breaks as paid time and reinforces mandatory rest periods. They also retrain supervisors on FLSA requirements and monitor compliance weekly.
Malik notices that a manager is classifying employees as exempt to avoid paying overtime. He senses the misclassification is intentional.	The leader audits all classifications and corrects the improper exemptions immediately. They also escalate the manager's behavior to HR for disciplinary review.
Serena learns that employees are asked to clock out before finishing closing tasks. She recognizes this as off-the-clock work.	The leader compensates employees for all unpaid time and updates scheduling to prevent after-hours tasks. They also reinforce zero-tolerance for off-the-clock labor.
Omar sees that travel time between job sites isn't being paid. He knows this violates FLSA travel rules.	The leader updates payroll practices to include travel time and issues back pay. They also train managers on compensable travel requirements.
Lila discovers that a team is averaging 60-hour weeks without overtime approval. She fears systemic overwork.	The leader enforces overtime authorization rules and compensates all owed overtime. They also rebalance staffing to reduce excessive hours.
Andre notices that tipped employees are not receiving proper tip-credit documentation. He worries the company is out of compliance.	The leader corrects the documentation and ensures all tip-credit rules are followed. They also implement monthly audits for tipped roles.

2. SAFETY — OCCUPATIONAL SAFETY AND HEALTH ACT

Legal Protections require safe working conditions, and the Occupational Safety and Health Act establishes the standards that protect employees from workplace hazards. It mandates safety controls, training, and reporting mechanisms that reduce risk. This structure ensures that Legal Protections are enforced through regulated safety practices. It matters because unsafe environments lead to injuries, fatalities, and long term harm. OSHA compliance prevents negligence by requiring organizations to identify, mitigate, and monitor hazards. It also reinforces fairness by ensuring all workers receive the same level of protection. When ignored, safety risks escalate and employees lose trust in leadership's commitment to their well being. Teams become fearful, disengaged, and vulnerable. Over time, Legal Protections collapse as unsafe conditions become normalized.

When done well, OSHA alignment strengthens Legal Protections by creating a disciplined, proactive safety culture. People trust that their health is prioritized and protected. This improves morale, productivity, and organizational resilience. Safety compliance reinforces the principle that worker protection is a legal and ethical mandate. It reminds leaders that safety must be engineered, monitored, and continuously improved. Ultimately, it builds a workplace where every employee is safeguarded by law and leadership.

MORAL COMPASS SCENARIO	LEADERSHIP ACTION
Priya discovers that machine guards were removed to speed up production. She knows this creates a serious OSHA violation.	The leader halts operations immediately and reinstalls all guards. They also retrain the team on mandatory safety protocols.
Rowan notices that employees are lifting heavy materials without proper equipment. He fears injuries are imminent.	The leader provides lifting aids and updates training on ergonomic safety. They also assign supervisors to monitor compliance.
Serena sees that chemical labels are missing from several containers. She recognizes the hazard communication breach.	The leader relabels all chemicals and updates the hazard communication program. They also conduct refresher training on handling procedures.
Omar learns that emergency exits are blocked by inventory. He knows this is a major OSHA violation.	The leader clears all exits immediately and enforces a zero-blockage policy. They also add daily exit-path inspections.
Lila notices that noise levels exceed safe limits in the production area. She worries employees are at risk of hearing damage.	The leader provides hearing protection and schedules a noise assessment. They also implement engineering controls to reduce exposure.
Andre discovers that incident logs are incomplete. He fears OSHA reporting requirements are being ignored.	The leader updates all logs and trains supervisors on reporting rules. They also assign a compliance owner to maintain accuracy.

Legal Protections ensure equal access to opportunity, and the Americans with Disabilities Act mandates accommodations that enable individuals with disabilities to participate fully in the workplace. It defines accessibility requirements, prohibits discrimination, and ensures reasonable adjustments. This structure ensures that Legal Protections uphold dignity, inclusion, and fairness. It matters because inaccessible environments exclude talent and violate fundamental rights. ADA compliance prevents barriers that limit participation or advancement. It also reinforces fairness by ensuring all employees can contribute without disadvantage. When ignored, individuals with disabilities face exclusion, stigma, and unequal treatment. Teams lose valuable perspectives and capabilities. Over time, Legal Protections erode as inequity becomes embedded in daily operations.

When done well, ADA alignment strengthens Legal Protections by creating an inclusive, accessible workplace. People trust that their needs will be respected and accommodated. This improves engagement, retention, and cultural integrity. Accessibility reinforces the principle that opportunity must be available to everyone. It reminds leaders that inclusion requires structural support, not sentiment. Ultimately, it builds a workplace where all employees can thrive without barriers.

MORAL COMPASS SCENARIO	LEADERSHIP ACTION
Keisha notices that a new hire with mobility challenges struggles to navigate the office layout. She realizes the workspace isn't ADA-compliant.	The leader reconfigures the workspace and installs accessible pathways. They also conduct an accessibility audit for the entire floor.
Damon sees that an employee with a hearing impairment isn't receiving meeting transcripts. He worries they're being excluded.	The leader provides real-time captioning and ensures transcripts are delivered after every meeting. They also update accessibility protocols for all teams.
Liora learns that a manager denied a request for a modified schedule. She recognizes this as a potential ADA violation.	The leader reviews the request, approves the accommodation, and retrains the manager. They also implement a centralized accommodation review process.
Mateo notices that software used by the team is incompatible with screen readers. He fears visually impaired employees are being disadvantaged.	The leader procures accessible tools and updates procurement criteria. They also provide training on inclusive technology use.
Serena discovers that interview rooms are not wheelchair accessible. She worries candidates are being excluded before they even start.	The leader relocates interviews to accessible rooms and updates scheduling procedures. They also audit all interview spaces for compliance.
Omar sees that an employee with chronic pain is struggling with their workstation. He realizes ergonomic accommodations are overdue.	The leader arranges an ergonomic assessment and provides adaptive equipment. They also track accommodation effectiveness over time.

Legal Protections require nondiscrimination, and Title VII of the Civil Rights Act prohibits employment decisions based on race, color, religion, sex, or national origin. It establishes the legal foundation for equal treatment in hiring, promotion, and workplace conduct. This structure ensures that Legal Protections defend employees from bias and prejudice. It matters because discrimination undermines fairness, damages culture, and violates fundamental rights. Title VII compliance prevents inequitable practices that marginalize individuals or groups. It also reinforces accountability by requiring organizations to investigate and address discriminatory behavior. When ignored, bias becomes normalized and opportunity becomes uneven. Employees lose trust in leadership and disengage. Over time, Legal Protections collapse as discrimination shapes organizational outcomes.

When done well, Title VII alignment strengthens Legal Protections by ensuring equity is enforced, not optional. People trust that decisions are based on merit rather than identity. This improves morale, diversity, and organizational credibility. Civil rights compliance reinforces the principle that fairness must be protected by law. It reminds leaders that equality is a legal requirement and a moral imperative. Ultimately, it builds a workplace where dignity and justice guide every decision.

MORAL COMPASS SCENARIO	LEADERSHIP ACTION
Nadia hears that a manager consistently assigns less desirable shifts to employees of a specific ethnicity. She recognizes a discrimination pattern.	The leader reassigns shifts equitably and launches a formal investigation. They also retrain the manager on anti-discrimination laws.
Trevor discovers that a candidate was rejected because the hiring manager "didn't like their accent." He knows this is discriminatory.	The leader overturns the decision and retrains the hiring manager. They also add bias-screening steps to the hiring process.
Aisha notices that women are routinely excluded from high-visibility projects. She fears systemic bias.	The leader audits project assignments and redistributes opportunities. They also implement oversight to ensure equitable access.
Marco learns that an employee reported religious harassment but was told to "ignore it." He sees a serious compliance failure.	The leader escalates the case to HR and enforces anti-harassment protocols. They also discipline the supervisor who dismissed the complaint.
Evelyn sees that LGBTQ+ employees are being mocked in group chats. She recognizes a hostile environment.	The leader investigates the behavior and enforces disciplinary action. They also reinforce inclusive conduct expectations across the team.
Jonah discovers that promotion decisions rely heavily on subjective "likability." He suspects bias is influencing outcomes.	The leader replaces subjective criteria with structured evaluation. They also train managers on equitable promotion practices.

5. LEAVE — FAMILY AND MEDICAL LEAVE ACT

Legal Protections safeguard employee well being, and the Family and Medical Leave Act guarantees job protected leave for health, caregiving, and family needs. It defines eligibility, duration, and employer obligations. This structure ensures that Legal Protections support employees during critical life events. It matters because workers must balance personal responsibilities with professional demands. FMLA compliance prevents retaliation or job loss during periods of legitimate need. It also reinforces fairness by ensuring all eligible employees receive consistent leave rights. When ignored, employees face impossible choices between health, family, and employment. Stress, burnout, and turnover increase. Over time, Legal Protections erode as caregiving becomes a source of vulnerability.

When done well, FMLA alignment strengthens Legal Protections by supporting humane, responsible workforce practices. People trust that their jobs are secure during personal hardship. This improves loyalty, well being, and organizational stability. Leave protections reinforce the principle that work must coexist with life. They remind leaders that compassion is both a legal requirement and a cultural strength. Ultimately, they build a workplace where employees are supported through life's most challenging moments.

MORAL COMPASS SCENARIO	LEADERSHIP ACTION
Rhea learns that an employee was denied FMLA leave because the manager didn't understand eligibility rules. She sees a compliance gap.	The leader approves the leave retroactively and retrains the manager. They also centralize FMLA decisions through HR.
Omar notices that employees fear requesting leave because they think it will hurt their careers. He senses a culture problem.	The leader reinforces non-retaliation policies and communicates FMLA rights clearly. They also monitor for subtle retaliation.
Lila discovers that intermittent leave requests are being ignored. She worries employees with chronic conditions are being disadvantaged.	The leader processes all pending requests and updates the tracking system. They also train managers on intermittent leave rules.
Andre sees that medical documentation is being mishandled. He recognizes the privacy risk.	The leader secures all documentation and limits access to authorized personnel. They also retrain staff on confidentiality requirements.
Serena learns that an employee returning from leave was demoted. She suspects retaliation.	The leader restores the employee's role and investigates the demotion. They also reinforce return-to-work protections.
Hugo notices that FMLA tracking is inconsistent across departments. He fears errors in eligibility calculations.	The leader centralizes tracking and standardizes documentation. They also audit past cases for accuracy.

6. REPRESENTATION — NATIONAL LABOR RELATIONS ACT

Legal Protections ensure employee voice, and the National Labor Relations Act protects the right to organize, bargain collectively, and engage in concerted activity. It establishes the legal framework for representation and worker advocacy. This structure ensures that Legal Protections empower employees to influence workplace conditions. It matters because power imbalances can silence concerns and limit fairness. NLRA compliance prevents retaliation against employees who seek representation or collective action. It also reinforces accountability by ensuring leadership respects lawful organizing efforts. When ignored, workers lose their ability to negotiate for fair treatment. Tension, mistrust, and conflict escalate. Over time, Legal Protections collapse as employee voice is suppressed.

When done well, NLRA alignment strengthens Legal Protections by ensuring employees can advocate for their rights. People trust that their perspectives will be heard and respected. This improves engagement, fairness, and workplace stability. Representation rights reinforce the principle that employee voice is essential to ethical governance. They remind leaders that collaboration strengthens organizational health. Ultimately, they build a workplace where workers participate meaningfully in shaping their conditions.

MORAL COMPASS SCENARIO	LEADERSHIP ACTION
Keisha hears that a supervisor told employees they "aren't allowed" to discuss wages. She knows this violates NLRA rights.	The leader corrects the misinformation and retrains the supervisor. They also communicate employees' rights to engage in protected discussions.
Damon discovers that management removed union flyers from a break room. He recognizes this as unlawful interference.	The leader restores the materials and briefs managers on NLRA protections. They also review facility policies for compliance.
Liora learns that employees were disciplined for discussing working conditions in a group chat. She sees a clear violation.	The leader reverses the discipline and documents the correction. They also train managers on protected concerted activity.
Mateo notices that a manager is monitoring employee discussions about organizing. He fears surveillance concerns.	The leader stops the monitoring and documents the breach. They also reinforce strict prohibitions on surveillance of protected activity.
Serena hears that employees were told they must notify management before speaking to a union rep. She knows this is unlawful.	The leader corrects the policy and communicates employees' rights. They also retrain supervisors on NLRA compliance.
Omar discovers that employees fear retaliation for raising workplace concerns collectively. He senses a chilling effect.	The leader reaffirms NLRA protections and creates a safe reporting channel. They also monitor for retaliation patterns.

BONUS CHAPTER

Ai6x6™ Transformation Blueprint

The Humanity Paradox

"There is no such thing as 'away'. When we throw anything away, it must go somewhere." -- **Annie Leonard**, proponent of sustainability and creator of *The Story of Stuff*, exposed the hidden social and environmental costs of our industrial systems. Her insight is the heartbeat of the **Humanity Paradox**, guiding leaders to teach the workforce that every operational action has a physical consequence "somewhere," shifting the culture from mindless consumption to shared, systemic accountability.

The traditional HR approach to sustainability has relied on surface-level culture shifts and individual recycling habits, which fail to address the systemic anxiety of the AI era. This linear thinking ignores the growing friction between machine efficiency and human dignity. Leaders who treat the workforce as a mere tool for optimization will face catastrophic burnout and talent loss. Achieving Net Zero requires a multi-dimensional strategy that navigates the **Humanity Paradox**. It involves balancing the speed of AI-driven automation with a "Just Transition" for workers and addressing the emotional toll of eco-anxiety while demanding high-speed innovation. Only by valuing human fulfillment as a primary metric can leaders mobilize the collective intelligence needed to re-engineer a sustainable future.

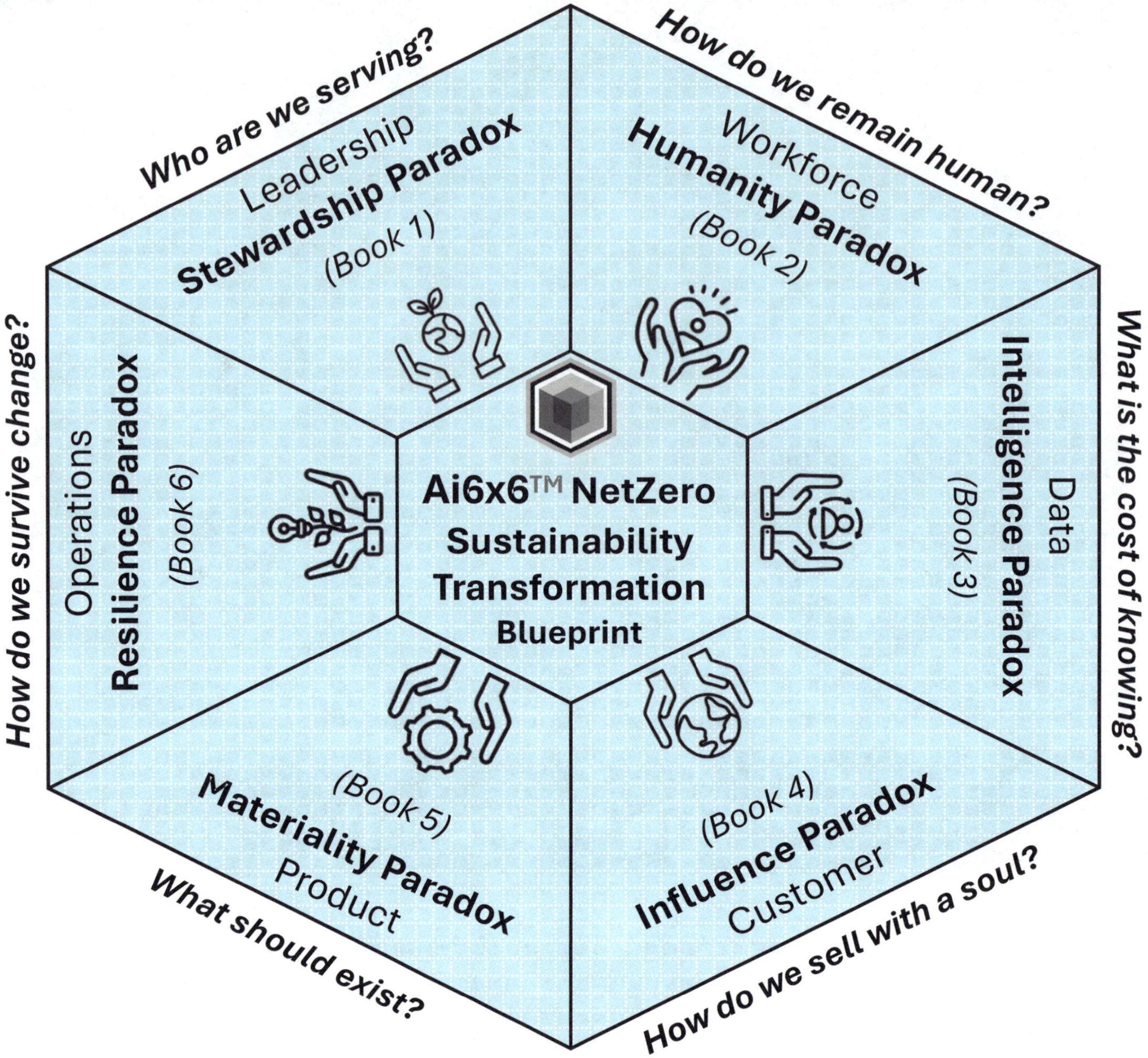

The **Ai6x6 NetZero Sustainability Transformation Blueprint** redefines the path to Net Zero by replacing the traditional "check-the-box" compliance mindset with a rigorous framework of **paradoxical leadership**. In a world accelerated by AI, sustainability is rarely a choice between a clear right and wrong; instead, it is a constant navigation of thirty-six competing tensions across six organizational lenses. By framing the journey through these paradoxes—such as the conflict between **Digital Intelligence and Energy Consumption** or **Global Targets and Local Infrastructure**—the framework prepares leaders for the messy, non-linear realities of systemic change.

Rather than viewing these tensions as obstacles to be avoided, the **Ai6x6 methodology** utilizes them as engines for innovation. It guides executives to move past the paralysis of "either/or" thinking, teaching them to embrace a "both/and" strategy that maintains fiduciary health while securing long-term planetary survival. This approach ensures that a company's moral compass remains calibrated even as technology and climate volatility shift the ground beneath them. Ultimately, the series provides a diagnostic roadmap that turns systemic friction into strategic advantage, empowering leaders to achieve radical Net Zero goals without sacrificing the operational resilience or human dignity required to thrive in the post-AI era.

THE HUMANITY PARADOX is the core struggle of the green transition: how to balance the cold efficiency of machines with the warm needs of people. As AI takes over the heavy lifting of carbon tracking and resource optimization, the workforce faces a crisis of value. Organizations must choose between using technology to simply cut costs or using it to empower humans for a higher purpose. This tension creates a workplace where people feel both essential to the mission and expendable to the algorithm. Navigating this requires a new social contract that protects human dignity while embracing the radical speed of a digital, Net Zero future.

1. THE AUTOMATION-EQUITY PARADOX (Fair Jobs vs. Fast Computers)

Using AI to radically slash waste and carbon is a planetary win, but it creates a moral crisis when those gains come at the expense of vulnerable workers. The paradox lies in the fact that the most "efficient" path to Net Zero often involves replacing human roles with low-carbon, automated systems. Leaders must decide if their environmental goals are worth the social cost of displacement. A truly sustainable business cannot just be green; it must be just. Balancing the speed of machine-led optimization with the responsibility to protect human livelihoods is the ultimate test of a company's social conscience.

2. THE INDIVIDUAL-SYSTEMIC PARADOX (Small Actions vs. Big Systems)

Employees are often told that their personal habits—like recycling or reducing travel—are the key to a sustainable culture. However, a demoralizing paradox emerges when workers realize that one hour of systemic industrial waste can outweigh a lifetime of their individual efforts. This gap between personal accountability and corporate reality can lead to "green fatigue." To solve this, organizations must bridge the divide by showing how individual contributions directly influence large-scale systemic changes. It is about moving from "do your part" to "let's redesign the whole," ensuring that no effort feels like a drop in a polluted ocean.

3. THE TALENT-ASPIRATION PARADOX (Green Promises vs. Real Life)

Companies win the war for talent by marketing a bold, sustainable purpose to the next generation. Yet, a painful paradox occurs when new hires enter the office and find legacy systems, plastic-heavy cultures, and slow-moving energy transitions. This "authenticity gap" destroys trust and leads to high turnover. Gen Z and Alpha are not looking for a mission statement; they are looking for evidence of action. For the workforce, the struggle is between the aspirational brand they joined and the messy reality of a firm that is still trying to figure out how to be good.

4. THE ECO-ANXIETY PARADOX (Working Hard vs. Feeling Worried)

The AI era demands that workers be more productive and cognitively agile than ever before. At the same time, the rising mental burden of "climate doom" is shrinking the very mental bandwidth needed to perform. This paradox creates a workforce that is expected to innovate at high speeds while carrying the heavy emotional weight of a planet in crisis. Organizations that ignore this "eco-anxiety" risk total burnout. Sustainability must extend inward to include the mental health of the people expected to build the future, recognizing that a stressed mind cannot solve a planetary emergency.

5. THE DISTRIBUTED-INTELLIGENCE PARADOX (Local Freedom vs. Global Rules)

Empowering workers to make local, agile decisions is the fastest way to spark green innovation on the front lines. However, this creates a paradox of control: if everyone follows their own "green" path, the company risks failing its unified, audit-grade global targets. Leadership must find a way to give employees the freedom to experiment without losing the cohesive moral compass that keeps the entire organization on track. It is a delicate balance of trusting the intelligence of the crowd while maintaining a strict, centralized accountability for the final results of the transformation.

6. THE SMART-FACILITY PRIVACY PARADOX (Smart Offices vs. Personal Privacy)

To reach Net Zero, modern offices use AI sensors to track every person's movement, micro-managing lighting and temperature for maximum efficiency. The paradox is that the more "helpful" the building becomes for the planet, the more "invasive" it becomes for the person. Workers can feel like they are being surveilled rather than supported. This tension requires a transparent approach to data where employees understand that their privacy is respected, even as their presence is used to save energy. The goal is a smart workspace that serves both the environment and the individual's sense of autonomy.

1. THE AUTOMATION-INCLUSION PARADOX (Efficiency vs. Equity)

- **Automotive Manufacturing:** A legacy plant implements a "Just Transition" program to retrain 5,000 internal combustion engine workers for high-tech EV battery assembly, ensuring no one is left behind by the shift to electric.
- **Warehousing:** A logistics giant uses "Exoskeletons" to reduce the physical strain on older employees, allowing them to remain productive and included in the workforce despite the high-speed demands of automated sorting.
- **Energy:** A North Sea oil firm transitions its offshore technicians to maintain floating wind farm arrays, leveraging their specialized maritime skills for the renewable era.
- **Agriculture:** A commercial farming operation trains seasonal laborers to operate AI-driven weeding drones, shifting their roles from manual toil to high-skill technical supervision.
- **Mining:** A mining group moves workers from dangerous underground pits to surface-level "Digital Cockpits," maintaining employment while drastically improving safety and digital inclusion.
- **Call Centers:** A customer service firm transitions entry-level agents into "AI Empathy Supervisors," where they audit bot interactions to ensure cultural sensitivity and complex problem-solving.

2. THE COGNITIVE-EFFICIENCY PARADOX (AI Speed vs. Human Depth)

- **Software Development:** A firm automates "grunt" code generation with AI, mandating that the time saved be used for "Deep Thinking" sessions focused on designing more energy-efficient algorithms.
- **Architecture:** A design studio uses AI to handle structural load calculations, freeing architects to spend their cognitive energy researching biomimetic materials that sequester carbon.
- **Law:** A global law firm uses AI for document discovery, allowing junior associates to focus on the high-level strategic nuances of emerging ESG regulations and climate litigation.
- **Education:** A university uses AI for grading and administrative tasks, allowing professors to dedicate their saved time to 1-on-1 mentorship of students on sustainability projects.
- **Graphic Design:** A creative agency automates repetitive layout tasks, prioritizing human talent for the complex work of evaluating the life-cycle impact of physical packaging designs.
- **Accounting:** An audit firm uses AI for basic transaction verification, shifting human auditors to the complex, non-linear work of "Natural Capital" accounting and biodiversity risk.

3. THE CULTURE-COMPLIANCE PARADOX (Mandates vs. True Belief)

- **Pharmaceuticals:** A lab moves away from "mandatory green training" toward employee-led "Green Labs" where researchers compete to identify and eliminate chemical waste in drug trials.
- **Retail:** A global supermarket chain launches a "Sustainability Shark Tank" where store-level employees pitch energy-saving ideas directly to the C-suite for funding.
- **Hospitality:** A hotel group empowers its housekeeping teams to redesign laundry and cleaning protocols, resulting in higher water savings than any top-down corporate mandate achieved.
- **Airlines:** A carrier incentivizes pilots with bonuses for finding "Green Glide" flight paths that save fuel, making sustainability a personal competitive challenge rather than a rule.
- **Consumer Tech:** A hardware firm creates "Green Guilds" where developers peer-review the carbon impact of each other's code, building a community of practice around efficiency.
- **Media:** A film studio empowers production crews to self-audit plastic waste on sets, rewarding teams with the highest reduction rates rather than enforcing a standard penalty.

4. THE ECO-ANXIETY PARADOX (Urgency vs. Mental Health)

- **Creative Agencies:** A firm provides "Climate Sabbaticals" and specialized mental health support for staff working on high-stress briefs for environmental NGOs and disaster relief.
- **Consulting:** A "Big Four" firm introduces "Purpose Days" where employees can take paid time off to work on local ecological restoration, reducing the feeling of helplessness in the face of the climate crisis.
- **Public Sector:** A city government implements "Eco-Resilience Coaching" for urban planners to help them process the emotional toll of designing for extreme weather and sea-level rise.
- **Insurance:** A firm offers group counseling for its climate-risk adjusters to mitigate the vicarious trauma of documenting recurring environmental destruction.
- **Education:** A secondary school system provides "Climate Cafe" sessions for teachers to discuss their fears about the future, ensuring they are emotionally grounded for their students.
- **NGOs:** A global conservation group implements mandatory "Digital Detox" periods for field workers to prevent burnout from the constant stream of negative environmental data.

5. THE DISTRIBUTED-INTELLIGENCE PARADOX (Global Rules vs. Local Genius)

- **Retail Banking:** A bank empowers branch-level tellers to approve "Green Micro-loans" for local solar installs without waiting for the slow, centralized head-office approval process.
- **Food & Beverage:** A global soda brand allows local plant managers to deviate from global water standards if they find a more efficient, indigenous way to recycle local groundwater.
- **Manufacturing:** A conglomerate creates a decentralized "Innovation Fund" where local factory floors can test their own green hardware modifications without corporate oversight.
- **Hotel Groups:** A chain allows individual island resorts to develop their own "Circular Breakfast" menus based on local seasonal availability rather than a global menu mandate.
- **Logistics:** A courier firm lets local delivery hubs design their own "last-mile" routes to account for specific city congestion and bike-lane accessibility.
- **Apparel:** A brand gives its regional sourcing agents the power to override global contracts if they find a local supplier with a significantly lower carbon footprint.

6. THE SMART-FACILITY PRIVACY PARADOX (Efficiency vs. Surveillance)

- **Real Estate:** A commercial landlord uses smart sensors to optimize HVAC based on occupancy but implements "Edge Privacy" where all individual tracking data is deleted every 24 hours.
- **Tech Campus:** A software giant uses gait-recognition for office security and energy lighting but siloes the data so it cannot be used by HR for performance monitoring.
- **Warehousing:** A fulfillment center uses heat-mapping to improve floor safety and energy use, providing workers with an "Opt-Out" zone where no digital tracking occurs.
- **Healthcare:** A smart hospital uses location sensors to ensure equipment is sterilized and available, but guarantees that the data is never used to track nurse "idling" time.
- **Co-working:** A space provider shares aggregated "Energy Impact" data back with its members, showing the collective win of the smart office without exposing individual habits.
- **Public Transport:** A city transit agency uses smart-camera analytics to manage station crowding and airflow, but uses "Privacy Masks" to blur all faces in the live data stream.

Ai6x6 COMPREHENSIVE TRANSFORMATION BOOK SERIES

Most comprehensive business transformation book series worldwide

Ai6x6™ Transformation Engine

The AI6x6 Transformation Engine™

Your Operating System for the AI Era

In an age of autonomous systems, fragmented strategy is a liability. The **AI6x6 Transformation Engine™** is a precision-engineered, 18-volume masterwork designed to move your enterprise from organizational inertia to scalable momentum. Built on a proprietary 6×6 geometry, this ecosystem unifies architecture, execution, and calibration into one high-performance unit.

The journey begins with the **Transformation Blueprint Series**, your **Intelligence Layer**. It codifies a "Navigation Deck" of Six Pillars and Six Lenses to anchor your strategic North Star. Next, the **Consulting Playbook Series** acts as the **Transmission Layer**, utilizing the IGNITE™ methodology to translate executive intent into high-torque, repeatable action. Finally, the **Maturity Diagnostic Series** serves as the **Sensor Layer**, providing a data-driven intake to quantify readiness and calibrate performance.

This isn't just a book collection; it is a rigid **Chassis** for institutional excellence. Secure the definitive architecture for sustained, high-stakes digital evolution. Drive the future with the Engine.

Transformation Series

To ensure this book series benefits your company holistically, we strongly recommend you buy the series and then distribute it to the appropriate transformation sponsors in each of these departments. Don't forget to browse through consulting offers at the end of the book.

WE NEED YOU!

WELCOMING OUR NEXT SET OF THOUGHT LEADERS

The remarkable success of the Ai6x6 Organization Transformation Series is built upon the **hard-won wisdom** of executive leaders who have successfully navigated the AI transformation within their industries, turning theory into a deeply practical, actionable guide. As we expand the series, we are extending a highly selective invitation to a new cohort of pioneers—individuals whose professional journey is at the **leading edge of AI strategy and execution**. We are looking for executives who possess a proven track record of rewriting business models, mastering unforeseen ethical and competitive challenges, and offering a **clear blueprint** for the next 3-5 years of the AI-Era.

Contributing to the Ai6x6 Series is a unique opportunity to **cement your professional legacy**, joining an exclusive, curated network of global authorities whose insights are distilled into executive-level knowledge cards, directly shaping the governance and innovation decisions of the world's most influential companies. Our selection process is based strictly on **pure expertise and demonstrated leadership** in the AI-Era, reinforcing our commitment to editorial integrity and authentic thought leadership—there is no monetary exchange involved. Select contributors will also receive an invitation to blog on our expert network microsite, further amplifying their platform. To confidentially inquire about contributing your expertise, please contact us via our dedicated portal at **www.Ai6x6.com/contact** or by scanning the adjacent QR code.

Contact

SUBSCRIBE TO OUR NEWSLETTERS: To stay ahead of the curve and receive continuous, executive-level intelligence on the rapid evolution of the AI-Era, we invite you to register for our **Ai6x6 Weekly Newsletter**. Each edition delivers curated insights directly from our thought leaders and highlights critical shifts in AI strategy, governance, and provides relevant excepts from our website and books. Keep your finger on the pulse of the future and ensure your company remains competitive by signing up on the same page

PLEASE REVIEW US ON AMAZON!

Workforce Blueprint

We know your time is valuable, but if the Ai6x6 Organization Transformation Series delivered the real-world, actionable strategies you needed, please consider leaving a brief review on Amazon. If we missed the mark, we would also like to know, and that is the only we we can improve. Your honest feedback is crucial; it helps fellow executives discover this unique framework and benefit from the insights of our thought leaders. Thank you for supporting our mission!

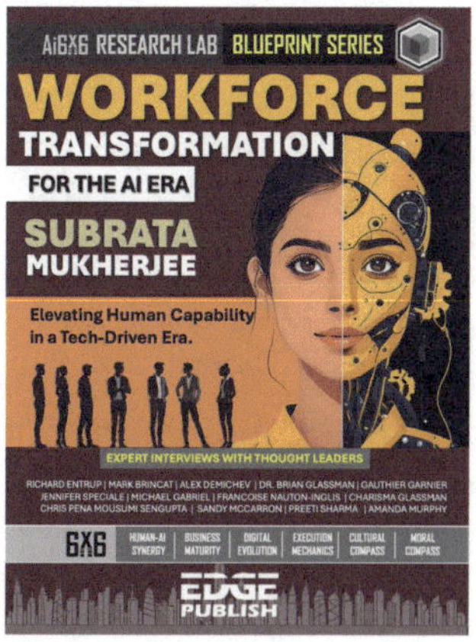

THE Ai6x6™ PRACTITIONER PROGRAM

Your Authority. Our Infrastructure. A Shared Legacy.

The **Ai6x6™ Practitioner Program** is designed for the high-level strategist—with proven industry leadership experience of over thirty years—who seeks a modern, standardized engine to deploy their operational wisdom. We provide the institutional infrastructure of a global consulting firm, delivered as a white-label solution that honors your professional identity and independence.

From $2000 per month membership

The "Un-Franchise" Model

We have re-engineered the consulting partnership for the modern era. Think of this as a **franchise without the friction or the high fees to get started**:

- **No Long-Term Commitments:** Maintain your agility with a model built for the independent architect.
- **No Loyalty Fees or Royalties:** You keep 100% of your engagement fees. We don't tax your success.
- **A Partnership of Mutual Trust:** Our relationship is powered by a transparent monthly membership, ensuring our incentives are perfectly aligned with your practice's growth.

The Foundations of Practice

All Practitioners gain immediate access to the **Ai6x6™ "Firm-in-a-Box"**:

- **White-Label Ecosystem:** Seamlessly integrate your own brand into our proprietary presentations, client pitches, and diagnostic audit templates.
- **The Clinical Engine:** Augment your strategic intuition with fact-based, quantitative scoring systems that replace subjective brainstorming with measurable transformation.

The White Glove Partnership

For those who wish to move from independent advisor to institutional authority, our **White Glove** tier offers a deep-tissue integration into the Ai6x6™ ecosystem:

- **Monthly 1-on-1 Strategic Check-ins:** Direct access to the **Ai6x6™ leadership team** to stress-test your pipeline, refine high-stakes offerings, or audit active client delivery.
- **Institutional Visibility:** Gain a global platform as a **Co-Author or Featured Expert** in upcoming volumes. We showcase your success stories on **Ai6x6.com**, cementing your legacy as a master of the 6 Lens Sequence.
- **The Bulk Purchase Program:** Exclusive access to deep discounts on the **Ai6x6™ Book Series**, enabling you to provide every client stakeholder with the "source code" for orchestration.

APPLY TO OUR PRACTITIONER PROGRAM AT:

Ai6x6.com/practitioner

Consulting Practitioner

THE Ai6x6™ TRANSFORMATION RETREAT

One Playbook. Your Leadership. Total Orchestration.

The **Ai6x6™ Business Transformation Series** is designed to eliminate the need for traditional, slow-moving consulting teams. Using our "Clinical Blueprints," your organization can self-organize, self-diagnose, and self-disrupt. Everything you need to facilitate a high-stakes, 2.5-day executive offsite is contained within the books.

The Self-Guided Mastery (100% Internal & Free)

You have the "Source Code." Your internal Offsite Coordinator or Chief of Staff can lead the entire 2.5-day agenda end-to-end using the **Consultant Playbook (Part 2)** and our **Supplemental Workbooks**.

- **No External Fees:** Run the entire diagnostic without "billable hours."
- **Full Control:** Use the 6-book series to "mix and match" exercises that fit your specific departmental frictions.
- **Tactile Execution:** Move from whiteboards to prioritized MoSCoW cards and RACI matrices using the physical templates provided in the series. Our Advisory section has all the inspiration your team needs.

Packages from $4999

The Orchestrated Upgrade (White-Glove Support)

For leadership teams seeking institutional weight and an objective "Cold Eye" review without the overhead of travel, we offer the **Orchestrated Retreat** package. This provides three critical layers of external authority:

- **Phase 1: The 30-Minute Architect's Briefing (Pre-Retreat):** A high-velocity strategic sprint with your internal Facilitator. we help you select the exact modules from across the 6-book series to target your unique friction points and set the tactical "Rules of Engagement."
- **Phase 2: The 30-Minute Executive Keynote (Live Remote):** A virtual address delivered by the **Ai6x6™ Leadership Team** to kick off Day 1. This keynote provides the external authority required to break through internal political silos and align the room's energy.
- **Phase 3: The 14-Day Clinical Roadmap Audit (Post-Retreat) :** Submit your digital outputs—prioritized lists, RACI matrices, and visualized Roadmaps. Our team performs a rigorous review and delivers a **Comprehensive Strategic Report** within **two weeks**, stress-testing your plan for diagnostic integrity and self-disruption readiness.

UPGRADE TO AN ORCHESTRATED RETREAT:

Ai6x6.com/retreat

Consulting Retreat

THE Ai6x6™ C-SUITE GHOST PAIRING

For Senior Executives

Intelligence Off the Record and Under the Radar

The **Ai6x6™** series gave you the diagnostic frameworks. The Roadmap Service gave you the blueprint. But the reality of the C-Suite is a high-friction environment where internal politics often suppress the "Clinical Truth." If you are ready to operate with a dedicated, silent partner who provides the radical candor your inner circle cannot, we invite you to engage the **Ai6x6™ Strategic Ghost**.

The Shadow Cabinet

We do not believe in public accolades or "boardroom visibility." We believe in being the invisible intellectual infrastructure for the modern leader. This is a high-fidelity, secure partnership designed to be the objective "second brain" that your legacy organization is politically incapable of providing.

The Protocol: Subscribed Strategy

The Ghost Partnership is a precision-engineered monthly engagement. We don't appear on your org chart, your cap table, or your public calendar.

- **The Weekly Calibration:** 60 minutes of raw, unfiltered strategic synthesis—a "Cold Eye" review of your current friction points and tactical moves.
- **The Monthly Artifact:** One bespoke, high-stakes strategic asset—a Red Team analysis of your board deck, a Vulnerability Audit of a new initiative, or a competitor "Disruption Map."
- **The Direct Feed:** 24/7 access via a secure, encrypted WhatsApp channel for real-time decision support and "sanity checks."
- **The Annual Pivot:** A comprehensive year-end roadmap review to ensure your personal trajectory remains aligned with the 6 Lens Sequence.

The Shield: Total Secrecy

The Ghost operates under a dual-layer legal framework. A rigorous NDA and a specialized **Secrecy Agreement** ensure that our involvement remains entirely invisible. In a world of corporate noise, we are your only pure signal.

The Commitment

- **Access:** 4 hours of dedicated sync-up time per month.
- **Output:** 1 high-fidelity strategic artifact per month.
- **Security:** Full encryption and total anonymity.
- **Focus:** Your personal asymmetric advantage.

Subscriptions from $2500 per month

Consulting Ghost

SECURE YOUR SHADOW PARTNER :

Ai6x6.com/ghost

"Peace is its own reward."

Mahatma Gandhi

Ai6x6 Research Lab
Publishing Team

NEW YORK OFFICE
(Editorial Hub)

Editorial Lead: Marisa Thompson
Research Lead: Talia Rosen
Line Editor: Jordan Alvarez

LONDON OFFICE
(Design Hub)

Design Lead: Keiko Tanaka
Layout Specialist: Oliver Whitmore
Illustrator: Aisha Rahman
Design Team from Sumu Arts

KOLKATA OFFICE
(Support Hub)

Project Manager: Sohini Chatterjee
Production : Arindam Ghosh
Quality Lead: Priya Deshmukh
Fact Checker: Ravi Patel
Research Assistant: Ananya Sengupta
Research Assistant: Rohan Mukherjee
Project Assistant: Madhumita Roy
Production Assistant: Sagnik Dutta
Quality Assistant: Ishita Banerjee
Series Lead (Blueprint): Amit Verma
Series Lead (Playbook): Neha Reddy
Series Lead (Diagnostic): Karan Mehta

The End